Rice Fortification

Rice fortification is a proven technique to address the prevalence of micronutrient malnutrition, i.e. 'hidden hunger'. This book, *Rice Fortification: Technology, Machinery and Quality Assurance*, aims to provide comprehensive technical literature on the production of fortified rice (FR) with emphasis on the raw ingredients, technology, equipment and machinery, micronutrient analysis, quality assurance and control (QA/QC), and government programs and policies for successful implementation.

Key features of the book include:

(i) Discussion on the importance of raw ingredients, i.e. broken rice, vitamins and mineral premix (VMP) for the manufacturing of fortified rice kernels (FRK), including the detailed description of each process operation, such as grinding, mixing, extrusion, drying, and polishing.
(ii) Production of FR; packaging, storage, and shelf-life evaluation of FRK and FR, along with the required plant and machinery to produce FRK and FR; QA/QC insights and the micronutrient analysis, consumer acceptance, GMP/GHP and HACCP protocols, as well as Indian standards.
(iii) Efficacy and safety of FR.

This book, written and edited by a team of experts who have worked on rice fortification technology, will be of immense help to industry professionals and academicians as well as undergraduate students and postgraduate researchers. This book will also guide through troubleshooting tips to support continued learning, professional development, and production of FR.

Rice Fortification

Technology, Machinery and Quality Assurance

Edited by
Hari Niwas Mishra, Chandrakant Genu Dalbhagat,
and Shubham Mandliya

CRC Press
Taylor & Francis Group
Boca Raton London New York

CRC Press is an imprint of the
Taylor & Francis Group, an **informa** business

Designed cover image: Hari Niwas Mishra, Chandrakant Genu Dalbhagat, and Shubham Mandliya

First edition published 2026
by CRC Press
2385 NW Executive Center Drive, Suite 320, Boca Raton FL 33431

and by CRC Press
4 Park Square, Milton Park, Abingdon, Oxon, OX14 4RN

CRC Press is an imprint of Taylor & Francis Group, LLC

ISBN: 9781041117612 (hbk)
ISBN: 9781041117315 (pbk)
ISBN: 9781003661245 (ebk)

DOI: 10.1201/9781003661245

Typeset in Times
by codeMantra

Contents

Preface

Malnutrition stands as a formidable global threat to public health and holds immense socioeconomic significance. Micronutrient malnutrition (MNM), often called hidden hunger, poses a significant global health challenge, affecting billions due to inadequate intake of essential vitamins and minerals. As per World Health Organisation (WHO), over 2 billion people globally are estimated to be deficient in key nutrients such as vitamin A, iodine, iron, and zinc. Among the many forms of MNM, iron deficiency stands out as a notable concern, leading to anaemia, a condition characterized by reduced haemoglobin levels and diminished oxygen-carrying capacity in the blood. Even the WHO paints a concerning picture, estimating that 37% of pregnant women and 40% of children under five are anaemic on a global scale. The situation is particularly severe in India, where about 58% of children, 57% of women, and 22% men are reported to be anaemic. Addressing this widespread health issue requires intervention and improved nutritional access.

Diverse strategies worldwide are employed to combat nutritional deficiencies, encompassing dietary diversification, supplementation, biofortification, and food fortification. Among these approaches, food fortification emerges as a proven, sustainable, cost-effective, and impactful solution. Staple foods like wheat flour, maize flour, rice, oil, pulses, and salt, consumed consistently by the global population, can be fortified with micronutrients – iron, folic acid, vitamin A, iodine, and others – without compromising taste, texture, or colour and incurring minimal costs for consumers. This inclusive solution extends its reach to nearly the entire population.

The success of global initiatives, such as universal salt iodization, showcases the efficacy of fortification in reducing iodine deficiency. Research data support fortification with folic acid, iron, vitamin A, and zinc, demonstrating reductions in birth defects, anaemia, child blindness, and associated risks. Over the past 80 years, staple food fortification has made substantial strides, with over 140 countries mandating fortification of at least one food vehicle – maize flour, rice, wheat flour, or salt. Rice, a staple food for billions, offers a powerful platform to address these challenges. Through innovative fortification technologies, rice can be enriched with essential vitamins and minerals, significantly enhancing its nutritional value. This has the potential to play a transformative role in combating malnutrition and promoting public health, particularly among vulnerable communities.

This book, *Rice Fortification: Technology, Machinery and Quality Assurance*, is specially designed to cover all aspects of rice fortification technology – from raw ingredients to finished products, packaging and storage to the efficacy of fortified rice. It provides detailed information on different steps of rice fortification techniques, starting from raw materials to the end products, i.e. fortified rice kernels (FRK) and fortified rice (FR), to packaging and storage protocols, quality controls and analysis, clinical studies, and government policies and regulations. Besides, this book will also guide through troubleshooting tips to support continued learning and professional development. This book provides a comprehensive guide and complete technical resource for those involved in rice fortification, such as researchers, FRK and FR manufacturers, machinery fabricators, QA/QC personnel, government agencies, policymakers, and public health practitioners.

Hari Niwas Mishra
Chandrakant Genu Dalbhagat
Shubham Mandliya

About the Editors

Hari Niwas Mishra, Emeritus Professor of Food Technology at the Indian Institute of Technology, Kharagpur (India), has over 42 years of professional experience in teaching, research, and administration, as well as many laurels and awards to his credit. He has published 658 research papers including 281 in peer-reviewed international journals and 375 in conference proceedings. He has written 5 books, 4 e-books, 38 book chapters, 10 lecture compendia and laboratory manuals, 4 technology manuals, and several popular articles in newspapers and magazines. He has 19 Indian patents to his credit. Besides, he has been on the editorial boards of several reputed international journals. He has supervised more than 302 student research projects, including 12 postdoctoral fellow (PDF) and 50 PhD research scholars, including 2 PMRF and 7 GYTI awardees. He has handled several international and national sponsored research and industrial consultancy projects. He has created a state-of-the art research laboratory and pilot plants for RTE/RTC food processing and shelf-life extension. He has received several awards and honours such as AIFPA Presidents' Award, GYTI Award, NRDC Award, Best Teacher Award, and Dr JS Pruthi Award for his outstanding contribution in the field of food processing in the country. He has worked in different capacities on various academic and administrative committees of IIT Kharagpur and many other institutions in the country. Professor Mishra has visited several countries abroad, including Taiwan, Singapore, Thailand, Malaysia, Australia, Turkey, UAE, France, Switzerland, Denmark, Germany, Sweden, Norway, Ireland, UK, Belgium, Hungary, Trinidad and Tobago, USA, and Canada.

Chandrakant Genu Dalbhagat is working as an Assistant Professor in the Department of Food Process Engineering, National Institute of Technology Rourkela, Odisha, India. His core research area includes rice fortification, food extrusion, protein texturizations, drying, artificial intelligence-machine learning (AI-ML) applications, and computer-aided food engineering. With over 10 years' experience in teaching and research, he has published more than 30 papers in peer-reviewed national and international journals. He has actively presented his research at various national and international conferences, workshops, and symposia. He also serves as a reviewer for several reputed journals. He significantly contributed to developing a process for manufacturing fortified rice kernels (FRK) and fortified rice (FR). He received the Gandhian Young Technological Innovation (GYTI) Award in 2019 for research on manufacturing FRK using extrusion technology. Furthermore, he served as a Panel Member on a committee of the Bureau of Indian Standards, New Delhi, for developing Indian Standards for manufacturing FRK, FR, vitamin-mineral premix, and equipment for manufacturing FRK and FR.

Shubham Mandliya, a Postdoctoral Researcher at the Department of Food and Nutritional Sciences, University of Reading, UK is a former Training Officer in the Rice Fortification Project at the Agricultural and Food Engineering Department, Indian Institute of Technology Kharagpur, India. He has over eight years of professional experience in the broad area of food engineering and technology including rice fortification, plant protein texturization, plant-based products, and macromolecular interactions. A B.Tech degree holder in Agricultural Engineering from JNKVV, Jabalpur, and Joint M.Tech-PhD degree in Food Process Engineering from IIT Kharagpur. During his PhD, he received the prestigious Prime Minister Research Fellowship (PMRF) from Ministry of Education, Government of India for his work on plant-protein texturization. During his PhD, he undertook a research visit to the University of Debrecen, Hungary. He has been recognized with several accolades, including the Young Inventor Award, the Prof. H. Das Memorial Award, the DST Young Scientist Travel Grant, and multiple best oral and poster presentation awards at international conferences. His research portfolio includes over 20 peer-reviewed research and review publications, five Indian patents, nine book chapters, two training manuals, and two workshop proceedings. He serves as a reviewer for esteemed journals such as *Trends in Food Science and Technology*, *International Journal of Biological Macromolecules*, *Journal of Food Engineering*, and *Food Bioscience*. He has visited several countries abroad, including United Kingdom, Hungary, Austria, Czechia, Germany, the Netherlands, France, Switzerland, and Turkey. He has made significant contributions by training industry professionals, academicians, and researchers in the manufacturing and quality evaluation of fortified rice kernels and fortified rice.

Contributors

Nithya A.
Agricultural and Food Engineering Department
Indian Institute of Technology Kharagpur
West Bengal, India

Bindu Bala
Department of Zoology
Sri Guru Teg Bahadur Khalsa College
Sri Anandpur Sahib, Punjab, India

Yuvraj Bhosale
Agricultural and Food Engineering Department
Indian Institute of Technology Kharagpur
West Bengal, India

Ayushi Chatterjee
Agricultural and Food Engineering Department
Indian Institute of Technology Kharagpur
West Bengal, India

Chandrakant Genu Dalbhagat
Department of Food Process Engineering
National Institute of Technology Rourkela
Odisha, India

Indrajit Dalai
Agricultural and Food Engineering Department
Indian Institute of Technology Kharagpur
West Bengal, India

Arpan Dubey
Agricultural and Food Engineering Department
Indian Institute of Technology Kharagpur
West Bengal, India

Siddhesh Kumar Dubey
Agricultural and Food Engineering Department
Indian Institute of Technology Kharagpur
West Bengal, India

Madhuresh Dwivedi
Department of Food Process Engineering
National Institute of Technology Rourkela
Odisha, India

Sandip Sanjay Gite
Department of Food Process Engineering
National Institute of Technology Rourkela
Odisha, India

K. Kamaleeswari
Department of Food Process Engineering
National Institute of Technology Rourkela
Odisha, India

Balasubramanian K.
Quality Assurance Division
Food Safety and Standards Authority of India
New Delhi, India

Payal Karmakar
Agricultural and Food Engineering Department
Indian Institute of Technology Kharagpur
West Bengal, India

Devesh Kumar
Bureau of Indian Standards, Central Laboratory, Maharajpur
Sahibabad Industrial Area Site 4, Sahibabad, Ghaziabad, Uttar Pradesh, India

Pradyuman Kumar
Food Engineering and Technology
Sant Longowal Institute of Engineering and Technology
Longowal, Punjab, India

Sachin Kumar
Food Storage Depot
Food Corporation of India
Ludhiana, Punjab, India

Saurabh Kumar
Agricultural and Food Engineering Department
Indian Institute of Technology Kharagpur
West Bengal, India

Shubham Mandliya
Agricultural and Food Engineering Department
Indian Institute of Technology Kharagpur
West Bengal, India

Gayatri Mishra
Agricultural and Food Engineering Department
Indian Institute of Technology Kharagpur
West Bengal, India

Hari Niwas Mishra
Agricultural and Food Engineering Department
Indian Institute of Technology Kharagpur
West Bengal, India

Sourav Misra
Mechanical Processing Division
ICAR-National Institute of Natural Fibre Engineering and Technology
Kolkata, West Bengal, India

Sujosh Nandi
Agricultural and Food Engineering Department
Indian Institute of Technology Kharagpur
West Bengal, India

Brajesh Kumar Panda
Agricultural and Food Engineering Department
Indian Institute of Technology Kharagpur
West Bengal, India

Ghumika Pandita
Institute of Chemical Technology
Biotechnology Park, Additional MIDC Area, Aurangabad Road, Jalna, Maharashtra, India

Rahul Patel
Agricultural and Food Engineering Department
Indian Institute of Technology Kharagpur
West Bengal, India

Rakesh Kumar Raigar
College of Agricultural Engineering and Post Harvest Technology
Central Agricultural University Imphal
Ranipool, Gangtok, Sikkim, India

Ganga Kishore S.
Department of Food Process Engineering
National Institute of Technology
Rourkela
Odisha, India

Raju Kumar Saha
Agricultural and Food Engineering Department
Indian Institute of Technology
Kharagpur
West Bengal, India

Pradeep Sharma
Food and Agriculture Department
Bureau of Indian Standards
New Delhi, India

Vasudha Sharma
Department of Food Technology
Jamia Hamdard
New Delhi, India

Lavika Singh
Food and Agriculture Department
Bureau of Indian Standards
New Delhi, India

V. R. Sinija
Department of Food Process Engineering
National Institute of Food Technology, Entrepreneurship and Management – Thanjavur (NIFTEM-T),
Thanjavur, Tamil Nadu, India

Tanya Sonker
Food Engineering and Technology
Sant Longowal Institute of Engineering and Technology
Longowal, Punjab, India

Koppada Sunanda
Agricultural and Food Engineering Department
Indian Institute of Technology
Kharagpur
West Bengal, India

Prashanth Thankachan
Division of Nutrition
St Johns Research Institute
Bangalore, Karnataka, India

Suneeti Toteja
Food and Agriculture Department
Bureau of Indian Standards
New Delhi, India

Malla Vandana
Agricultural and Food Engineering Department
Indian Institute of Technology
Kharagpur
West Bengal, India

Siddharth Vishwakarma
Department of Food Science and Technology
The Ohio State University
Columbus, OH, USA

Gorenand Prasad Yadav
Bureau of Indian Standards
Parwanoo Branch Office (Himachal Pradesh)
Madhya Marg, Sector 27B, Chandigarh, India

1 Micronutrient Malnutrition
Current Status and Interventions

Tanya Sonker, Vasudha Sharma, and Pradyuman Kumar

1.1 INTRODUCTION TO MICRONUTRIENT MALNUTRITION

Micronutrients are essential nutrients, which are required by the body in very minute amounts for proper physiological functioning, including growth, development, and disease prevention. According to the World Health Organization (WHO), "Micronutrients are vitamins and minerals needed by the body in very small amounts. However, their impact on a body's health is critical, and deficiency in any of them can cause severe and even life-threatening conditions." (WHO 2019). Micronutrients, including vitamins and trace elements (Figure 1.1), are crucial for numerous physiological and biochemical processes in the human body and are essential for maintaining overall health. They support tissue repair, contribute to the formation and maintenance of bones and teeth, function as cofactors and coenzymes in enzymatic reactions, and help regulate a wide range of bodily functions (Awuchi et al. 2020).

Malnutrition refers to conditions caused by an insufficient, excessive, or imbalanced intake of energy or nutrients. It is generally divided into two categories. The first is undernutrition, which includes stunting (low height for age), wasting (low weight for height), underweight (low weight for age), and deficiencies of essential vitamins and minerals. The second involves overnutrition, characterized by overweight,

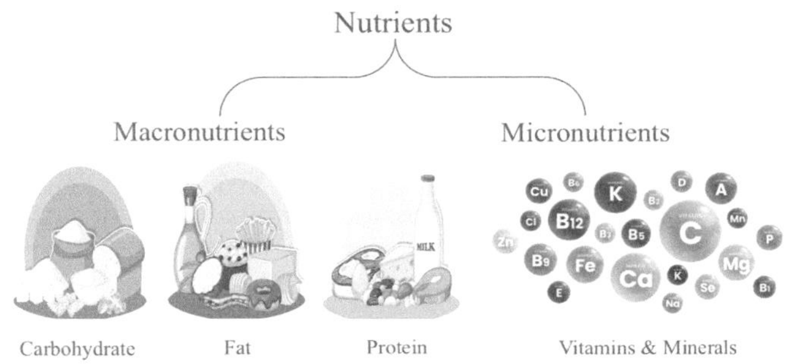

FIGURE 1.1 Macronutrients and micronutrients.

DOI: 10.1201/9781003661245-1

obesity, and diet-related noncommunicable diseases such as cardiovascular disease, diabetes, stroke, and certain cancers. Micronutrient deficiencies (MNDs) pose a major global public health and development challenge.

MNDs represent a major contributor to morbidity and mortality among children. While micronutrients are primarily obtained through the diet, they can also be provided in the form of nutritional supplements. Maternal supplementation has been shown to offer health benefits to breastfed infants by improving the micronutrient content of breast milk. The most prevalent MNDs involve insufficient intake or absorption of vitamin A, iron, iodine, zinc, folate, and vitamin B_{12}. These deficiencies lead to distinct clinical effects, such as impaired vision, anemia, developmental delays, and weakened immune responses. Iron deficiency, the most widespread globally, is linked to maternal deaths, chronic fatigue, and delayed cognitive development in children. Likewise, deficiencies in zinc and folate hinder physical growth, immune function, and proper fetal development.

Micronutrient malnutrition (MNM), often referred to as “hidden hunger,” affects over two billion people worldwide (WHO). Though not always visible, its effects are severe, particularly in low- and middle-income countries where food insecurity, poor diet diversity, and limited healthcare are common. South Asia and Sub-Saharan Africa experience the highest burden, but deficiencies also occur in high-income countries, often masked by overnutrition and processed diets. Vulnerable groups include infants, young children, pregnant and lactating women, adolescent girls, the elderly, and individuals with chronic illnesses affecting nutrient absorption.

1.2 MICRONUTRIENT DEFICIENCIES AND THEIR IMPACT

1.2.1 Iron Deficiency and Anemia

Iron deficiency (ID) and iron-deficiency anemia (IDA) are major contributors to the global disease burden. ID can occur without anemia, but still leads to significant physiological and cognitive consequences. Anemia arises when ID depletes hemoglobin concentrations to subclinical levels, impairing oxygen delivery to tissues. Absolute ID, which is more prevalent among young children (under five years of age) and premenopausal women, particularly those who are pregnant, arises when iron reserves are inadequate to satisfy bodily demands. Conversely, in individuals experiencing inflammation, iron may be sequestered from the plasma despite adequate overall iron stores, resulting in anemia and impaired erythropoiesis, a condition referred to as functional ID. Anemia is characterized by reduced hemoglobin levels in the blood. Globally, ID is estimated to account for approximately half of all anemia cases, as iron is a vital element in the formation of hemoglobin. In children, anemia poses a significant public health concern, as it can impair cognitive development, limit physical growth, and increase susceptibility to infectious diseases.

Causes and Consequences

ID arises primarily due to a sustained negative iron balance over time, in which iron intake or absorption fails to meet physiological demands (Cook 2005). The primary cause is the progressive depletion of iron stores, initially ferritin and hemosiderin, which are essential for maintaining adequate iron supply for cellular and metabolic functions (Animasahun and Itiola 2021).

Stages of Iron Deficiency (UNICEF and WHO 2001)

ID can exist in multiple stages (Table 1.1), which determine the severity and clinical manifestations of the condition.

The physiological consequences of ID vary depending on the stage. Early stages are often asymptomatic or associated with nonspecific symptoms such as fatigue and reduced exercise tolerance (Figure 1.2). As iron depletion advances to IDA, consequences become clinically apparent, including:

- Reduced oxygen-carrying capacity of blood (due to low hemoglobin).
- Impaired cognitive function and concentration.
- Decreased muscle performance and endurance.
- Increased susceptibility to infections.
- In severe cases, developmental delays in children and complications during pregnancy (WHO 2025).

These outcomes are particularly significant in vulnerable populations, such as children, pregnant women, and adolescents, where the demand for iron is high.

Impact on Cognitive Development and Work Productivity

ID in early life stages, especially between 6 months and 5 years, is strongly associated with delayed neurodevelopment. Iron is critical for myelination, neurotransmitter synthesis, and synaptic development. Its deficiency during the formative years disrupts cognitive functions, such as attention, memory, and learning capacity (Lozoff and Georgieff 2006). For instance, a notable finding from National Family Health Survey-5 (NFHS-5) data (India) reveals that 64.2% of urban and 68.3% of rural children aged 6–59 months suffer from anemia (hemoglobin <11.0 g/dL), highlighting a widespread nutritional crisis during a critical neurodevelopmental window (Figure 1.3).

TABLE 1.1
Stages of Iron Deficiency and Iron-Deficiency Anemia

Stage	Iron Status	Hemoglobin	Key Features
Normal iron status	Adequate iron stores and transport iron	Normal	Iron needs are met; normal physiological function.
Iron depletion	↓ Iron stores (ferritin and hemosiderin)	Normal	No reserve iron available if demand increases; no anemia yet.
Iron-deficient erythropoiesis	↓ Stored iron and ↓ transport iron (e.g., transferrin-bound)	Normal or slightly ↓	Insufficient iron for erythropoiesis; early functional deficiency.
Iron deficiency anemia	Severely ↓ stored and transport iron	↓ Hemoglobin significantly	Impaired production of Hb and myoglobin; systemic symptoms begin to appear.

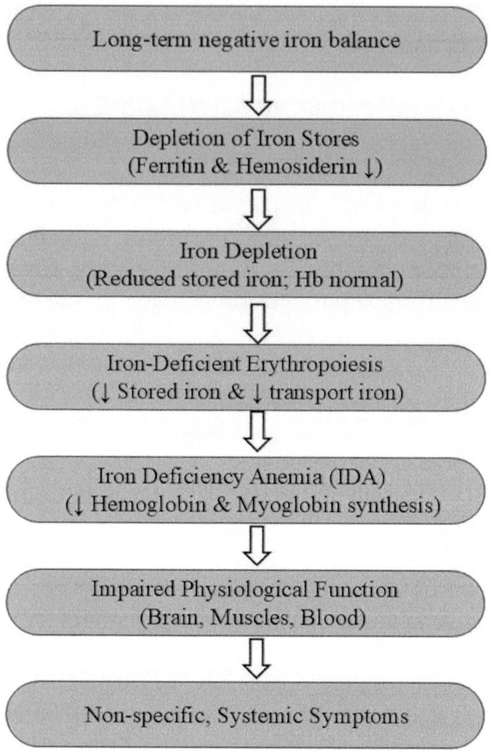

FIGURE 1.2 Stages and consequences of iron deficiency and anemia.

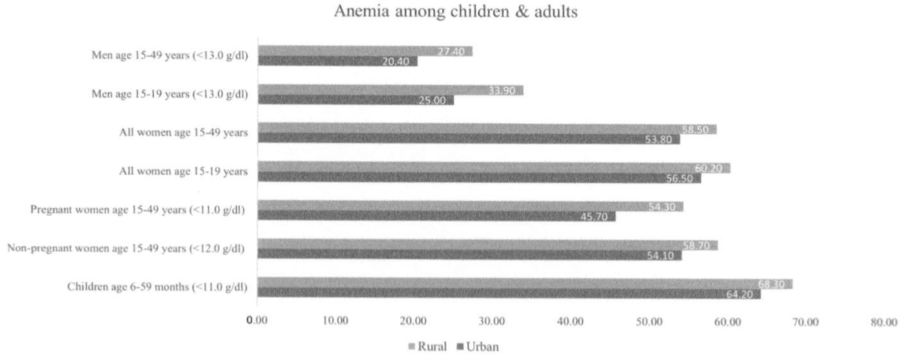

FIGURE 1.3 Comparison of anemia prevalence among different demographic groups in rural and urban settings in India. The burden is disproportionately higher among rural populations, especially children and women of reproductive age (NFHS-5 2019–2021).

ID reduces aerobic capacity and physical endurance by impairing hemoglobin-mediated oxygen transport. This has direct implications for productivity, especially in physically demanding occupations. The NFHS-5 data also highlight that more than half of reproductive-aged women and a significant proportion of men aged 15–49 years are anemic, with 58.5% of rural women and 53.8% of urban women, compared to 27.4% of rural men and 20.4% of urban men, respectively.

1.2.2 Folate and Vitamin B_{12} Deficiency

Role in Pregnancy and Neural Tube Defects

Folate (vitamin B_9) and vitamin B_{12} are critical for DNA synthesis, methylation, and cellular division, making them particularly important during pregnancy (Fenech 2012). Folate deficiency, especially in the early stages of gestation, is a well-established risk factor for neural tube defects (NTDs), including spina bifida and anencephaly (Martinez et al. 2023). When using macrocytic anemia as a hematological indication, a serum folate cut-off level of 14 nmol/L (6 ng/mL) indicates potential folate deficit, while levels below 7 nmol/L (3 ng/mL) indicate folate deficiency (WHO). A higher folate threshold (>227 nmol/L) is necessary to ensure optimal protection against NTDs, as neural tube closure demands increased folate availability to support rapid cellular division. Folate insufficiency is thus characterized by folate concentrations below this protective level. NTDs are associated with high death rates, morbidity, disability, and psychological and economic expenses. Many are preventable with folic acid, and those who are impacted can benefit from access to appropriate services, improving survival and quality of life (Blencowe et al. 2018).

Anemia and Neurological Effects

Beyond ID, deficiencies in other micronutrients including vitamin A, vitamin B_6, vitamin B_{12}, riboflavin, and folic acid are also associated with the development of anemia. Both folate and vitamin B_{12} deficiencies contribute to megaloblastic anemia, a condition characterized by the presence of abnormally large, immature red blood cells in the bone marrow. This form of anemia results in fatigue, pallor, and increased susceptibility to infections. In severe cases, folate deficiency may cause glossitis and gastrointestinal symptoms, whereas B_{12} deficiency can result in irreversible neurological damage, including peripheral neuropathy, impaired coordination, and memory decline (Asbaghi et al. 2021).

1.3 CAUSES AND RISK FACTORS OF MNM

MNM arises from a combination of immediate, underlying, and structural factors, often acting in synergy. These include poor diet quality, disease burden, social inequalities, environmental hazards, and limited healthcare access. MNDs affect individuals across all age groups; however, the impact is particularly pronounced among pregnant women and children, especially infants.

The following subsections explore these causes in depth:

1.3.1 Dietary Deficiencies and Poor Nutrition

The burden of poor nutrition, especially when occurring during critical periods such as gestation and early childhood, has long-term consequences on human capital,

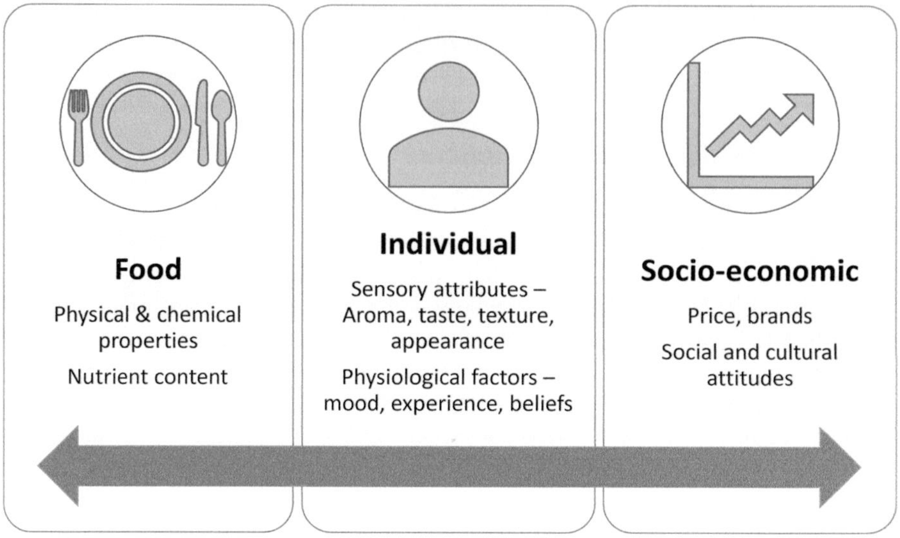

FIGURE 1.4 Determinants influencing food selection.

development, and national productivity. Poor nutrition, particularly during gestation and early childhood, has lasting impacts on individual development, human capital, and national productivity. As shown in Figure 1.4, food selection is shaped by a complex interaction of personal preferences, social norms, and environmental factors.

1.3.2 Socioeconomic and Cultural Factors

MND poses a substantial obstacle to socioeconomic development, disproportionately affecting already vulnerable populations and perpetuating a cycle of underdevelopment. Its long-term consequences include diminished productivity, impaired cognitive and learning abilities, and adverse health outcomes. The resulting burden includes significant social and economic costs, reduced labor capacity due to elevated rates of illness and disability, and a profound loss of human potential. Several interconnected factors contribute to inadequate child feeding, as shown in Figure 1.5.

In traditional and tribal communities, deeply embedded cultural practices and food beliefs significantly influence dietary habits. This issue is reflected in studies such as the one applying the PEN-3 model (James 2004), which emphasized the role of community and culture in shaping dietary behaviors among African Americans.

1.3.3 Malabsorption and Infections

Malabsorption and infections are critical, yet often underappreciated, contributors to MNM. The downstream effect of malabsorption is the development of MNDs, with profound systemic consequences. Deficiencies in iron, folate, and vitamin B_{12} often result from mucosal or ileal dysfunction (Hodgson and Epstein 2007). Clinically, this manifests as anemia, osteomalacia, coagulopathy, and immunodeficiency syndromes.

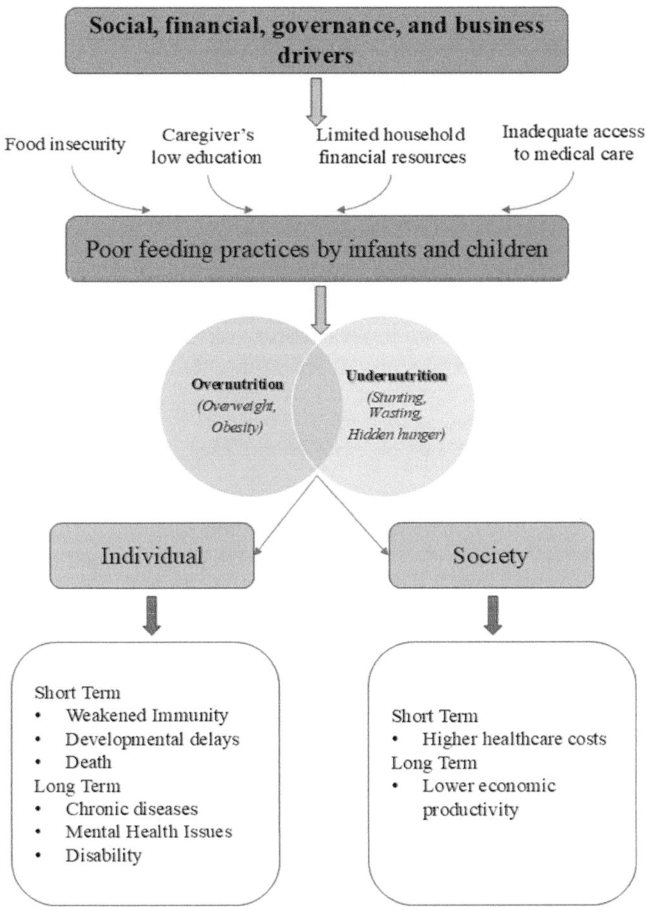

FIGURE 1.5 Determinants and consequences of inadequate infant and young child feeding practices. (Adapted from Agostoni et al. 2023.)

Infections exacerbate MNDs via multiple overlapping mechanisms. Firstly, they can directly damage the gut mucosa, reducing absorptive surface area, as seen in post-infectious malabsorption and tropical sprue (Hodgson and Epstein 2007). Moreover, systemic infections like HIV, tuberculosis, and measles can both aggravate existing malabsorption and induce a catabolic state that elevates nutrient demands while reducing intake due to anorexia (Morales et al. 2023). HIV-associated enteropathy, for instance, causes chronic diarrhea and mucosal flattening, severely impairing nutrient absorption (Hodgson and Epstein 2007).

1.3.4 Climate Change and Food Security

Climate change has a negative impact on agriculture and the nutritious value of food. Elevated CO_2 levels reduce protein and micronutrient content (e.g., iron, zinc, and vitamin A) in staple crops like wheat, rice, and maize, leading to nutritional shortages in populations who rely on these foods. Families experiencing food insecurity frequently replace nutritious foods with low-cost, energy-dense, nutrient-poor alternatives (Bouis et al. 2011). This pattern leads to diets lacking in iron, zinc, vitamin A, and folate, which primarily impact women and children.

1.4 ASSESSMENT AND MEASUREMENT OF MNDs

Accurate measurement and monitoring of micronutrient deficits is critical for both public health nutrition surveillance and clinical diagnosis. Given the wide range of physiological functions and varied bioavailability of micronutrients, a multi-method approach to assessing nutritional status is recommended.

1.4.1 Clinical Assessment (Signs and Symptoms)

Clinical assessment remains an essential method for detecting people with moderate to severe MNDs, especially in resource-constrained settings. Characteristic symptoms frequently appear after prolonged nutritional deprivation and can vary in specificity. Table 1.2 shows the clinical signs and symptoms of key MNDs.

Folate deficiency exhibits overlapping symptoms with B_{12} deficiency, most notably macrocytic anemia, but also poses specific risks such as NTDs during fetal development. Clinically, individuals may report general fatigue, pallor, and irritability.

1.4.2 Biochemical Indicators (Blood and Urine Tests)

Biochemical testing provides exact, objective assessments of micronutrient status, which is especially valuable for detecting subclinical deficiencies. These analyses include nutritional concentrations in serum or plasma, functional biomarkers, and metabolite levels.

Vitamin B_{12} deficiency can be identified using four key biological markers: total serum B_{12}, holotranscobalamin (holoTC or "active B_{12}"), and the associated metabolic indicators, total homocysteine (tHcy) and methylmalonic acid (MMA). Its status is commonly evaluated using total serum B_{12}, although this measure lacks sensitivity in early deficiency. MMA levels >260 nmol/L and tHcy >15 μmol/L are considered suggestive of deficiency (Sobczyńska-Malefora et al. 2021).

TABLE 1.2
Clinical Signs and Symptoms of Key MNDs

Micronutrient	Common Clinical Signs
Vitamin B_{12}	Megaloblastic anemia, neuropathy, glossitis, memory loss
Folate	Macrocytic anemia, neural tube defects (fetal), irritability
Zinc	Growth retardation, alopecia, immune dysfunction, dermatitis

For folate, serum concentrations below 7 nmol/L denote deficiency. Homocysteine levels also serve as a secondary indicator, with elevated levels (>15 μmol/L) indicating impaired folate metabolism (Pravst et al. 2021).

1.4.3 Dietary Surveys and Intake Assessments

Dietary intake assessment is a proactive approach used to evaluate the risk of developing MNDs. Standard methods include 24-h dietary recalls, food frequency questionnaires (FFQs), food diaries, and diet history interviews. These are often coupled with food composition tables to estimate nutrient intake.

1.4.4 Anthropometric Measurements

Anthropometric indices, though indirect, are valuable for assessing the cumulative impact of nutrient deficiencies, particularly in children. Measurements such as weight-for-age, height-for-age, and BMI-for-age help detect stunting, wasting, and underweight, all of which are often associated with chronic or multiple MNDs.

1.5 PUBLIC HEALTH INTERVENTIONS FOR MNM

A variety of strategies have been implemented to enhance micronutrient intake among mothers and children. These approaches include, individually or in combination, nutrition education, dietary modifications, food distribution programs, agricultural interventions, micronutrient supplementation, and food fortification.

1.5.1 Food-Based Approaches

Dietary Diversification

Dietary diversity is acknowledged as a viable, culturally acceptable, and long-term method for combating MNM, particularly in low- and middle-income nations. This method aims to increase the variety and quality of diets by increasing consumption of micronutrient-rich foods such as fruits, vegetables, legumes, whole grains, and animal-derived products. Unlike supplementation and fortification, dietary diversification encourages the natural absorption of numerous nutrients, lowering the likelihood of antagonistic interactions and increasing bioavailability.

Biofortification of Staple Crops

Biofortification is a sustainable approach to combat MNM by increasing the nutrient content of staple crops through conventional breeding, agronomic practices, or modern biotechnological methods (Kiran et al. 2022). The Indian Council of Agricultural Research (ICAR) has developed 87 biofortified cultivars across 16 crops, targeting deficiencies in iron, zinc, protein, provitamin A, and lysine. Biofortification is well-suited for reaching remote, vulnerable populations through seed distribution and adoption of improved crop varieties.

1.5.2 Iron and Folic Acid (IFA) Supplementation Programs

IFA supplementation is a cornerstone intervention to combat IDA and associated adverse outcomes in pregnancy and adolescence. WHO recommends daily supplementation with 30–60 mg elemental iron and 0.4 mg folic acid for at least 180 days during pregnancy to prevent maternal anemia, preterm birth, and low birth weight (WHO 2016). In Bangladesh, although 80% of pregnant women reported taking IFA, only 6% achieved the recommended ≥180 tablets during pregnancy, largely due to inadequate antenatal visits and limited counselling (Billah et al. 2022).

Among adolescents, the Weekly Iron and Folic Acid Supplementation (WIFS) program in India aims to address high anemia prevalence and implemented across all states and union territories and currently benefits over 112 million adolescents, thereby contributing to the long-term goal of interrupting the intergenerational cycle of anemia. It involves the weekly supervised administration of 100 mg elemental iron and 500 μg folic acid, along with biannual deworming using Albendazole (400 mg). In Puducherry, WIFS achieved 67.7% compliance and significantly reduced anemia in schoolgirls (Wangaskar et al. 2021). A program in Ghana also demonstrated that each additional IFA tablet taken weekly was associated with a 5% reduction in anemia risk, highlighting the importance of adherence (Gosdin et al. 2021).

1.5.3 Food Fortification Strategies

Food fortification is an important public health intervention that addresses widespread MNDs by incorporating vital vitamins and minerals into regularly consumed food items. It is a sustainable and population-wide technique that is especially beneficial in low-resource situations with little nutritional diversity.

Fortification with Iron and Folic Acid

IDA and NTDs represent major nutritional concerns, particularly in women of reproductive age and young children. The fortification of wheat and maize flours with iron and folic acid is a well-established intervention, implemented in over 80 countries. In 2023–2024, an 806 million metric tons of wheat flour were consumed worldwide (International Grains Council 2025), making wheat flour an ideal vehicle for fortification.

In Brazil, mandatory fortification of flour with 0.15 mg folic acid and 4.2 mg iron per 100 g led to a drop in NTD prevalence from 0.79 to 0.55 per 1,000 live births (Santos et al. 2015). A systematic review of 27 studies found folic acid fortification reduced NTDs by up to 70% globally (Castillo-Lancellotti et al. 2013). For anemia control, Barkley et al. (2015) showed that national flour fortification with bioavailable iron compounds reduces anemia prevalence in women by 2.4% annually.

1.5.4 Behavior Change Communication and Education

Behavior change communication and education are critical components in combating micronutrient deficiency. These strategies seek to impact individual and community behaviors through focused information, skill development, and knowledge dissemination.

Nutrition Awareness Campaigns

Nutrition awareness campaigns are run to educate the population about the importance of micronutrients, to dispel myths and misconceptions regarding food and nutrition, and to promote the consumption of locally available, micronutrient-rich foods. Details of the awareness programs are discussed in Chapter 2.

In India, Poshan Maah, observed each September under *Poshan Abhiyaan*, is a national campaign to improve nutrition among mothers and children. The program focuses on pregnant and breastfeeding women, adolescent girls, and children under six, aiming to raise awareness and change behaviors. Internationally, the UK's "Five-a-Day" program, run by the NHS, is a compelling international example of excellent nutrition marketing. To minimize the risk of heart disease, stroke, and cancer, the NHS recommends a healthy, balanced diet that includes five portions (80 g each) of fruit and vegetables per day, similar to the WHO-recommended 400 g minimum.

Maternal and Infant Nutrition Programs

Maternal and infant nutrition programs form a critical component of public health strategies to combat MNM. These programs primarily focus on promoting exclusive breastfeeding for the first six months of life, followed by the introduction of complementary feeding with micronutrient-rich foods. Additionally, they emphasize targeted supplementation and food fortification for pregnant women and infants to prevent deficiencies such as anemia and stunting. Key interventions include antenatal counselling, distribution of IFA tablets, growth monitoring, and nutrition counselling at Anganwadi Centers and health facilities. Collectively, these programs aim to build a strong foundation for child health and development during the critical first 1,000 days of life.

1.6 POLICY AND GLOBAL EFFORTS TO COMBAT MNM

Global institutions such as the WHO, the United Nations International Children's Emergency Fund (UNICEF), and the Food and Agriculture Organization (FAO) have been instrumental in developing and executing strategies to tackle MNDs.

1.6.1 Role of WHO, UNICEF, and FAO

WHO leads in public health and nutrition by formulating guidelines, establishing nutrient consumption recommendations, and coordinating global action plans. One of WHO's significant contributions is the "Global Strategy on Diet, Physical Activity, and Health" and the "Comprehensive Implementation Plan on Maternal, Infant, and Young Child Nutrition," which seek to achieve six global targets (Figure 1.6) by 2025 (WHO 2014). It plays a central role in combating MNDs through evidence-based strategies, including IFA supplementation and high-dose vitamin A for vulnerable groups like children and pregnant women.

FIGURE 1.6 The six global targets of comprehensive implementation plan on maternal, infant, and young child nutrition.

UNICEF focuses on integrating micronutrient interventions into broader child health and development frameworks. Through programs such as vitamin A supplementation, salt iodization, and micronutrient powders for home fortification, UNICEF supports national governments to enhance the nutritional status of children and women (UNICEF 2023). For example, two important programs by UNICEF are:

i. Home Fortification with Multiple Micronutrient Powders for the Prevention of IDA in Early Childhood (UNICEF 2023).
ii. Nutrition, for Every Child: UNICEF Nutrition Strategy 2020–2030: It outlines UNICEF's vision to achieve universal access to nutritious, safe, affordable, and sustainable diets for every child. The framework focuses on five key goals:

   a. Prevent all forms of malnutrition in early childhood, including stunting, wasting, anemia, and overweight.
   b. Promote breastfeeding and appropriate complementary feeding practices.
   c. Enhance maternal nutrition before, during, and after pregnancy.

d. Strengthen nutrition resilience in fragile and emergency contexts.
e. Scale up proven interventions like food fortification, MNPs, nutrition education, and social protection.

FAO promotes sustainable, food-based solutions by supporting diversified agriculture, biofortification, and nutrition-sensitive food systems to combat dietary deficiencies and encourage local food production and consumption (FAO 2017). It also aids governments in policy development and food security analysis. UNICEF partners with FAO to improve child nutrition, prevent malnutrition, and enhance access to healthy diets. This collaboration addresses links between nutrition, food access, and child development, aiming to improve children's intake of nutrient-rich foods and support their overall growth.

1.6.2 Sustainable Development Goals (SDGs) and Nutrition Targets

The SDGs, adopted by all UN Member States in 2015, underscore the interconnection between nutrition and development. The 2030 Agenda for Sustainable Development outlines 17 SDGs, with nutrition-related targets embedded in SDG 2 (Zero Hunger) and SDG 3 (Good Health and Well-being). However, nutrition is inherently linked across multiple goals, including poverty (SDG 1), education (SDG 4), gender equality (SDG 5), and sustainable agriculture (SDG 12).

Key Nutrition-Related Targets and Trends

- **SDG 2.1 (food security)**: In 2022, 735 million people were undernourished, 122 million more than in 2019. An additional 2.4 billion faced moderate to severe food insecurity.
- **SDG 2.2 (child undernutrition)**: Stunting in children under five dropped from 24.6% in 2015 to 22.3% in 2022, but remains off-track. Wasting affected 6.8%, while 5.6% were overweight, reflecting a dual burden of malnutrition.
- **SDG 3.2 (child mortality)**: The under-five mortality rate declined to 37 per 1,000 live births in 2022, a 14% drop since 2015, but undernutrition remains a key driver in low-income countries.
- **SDG 3.4 and 3.8 (NCDs and health coverage)**: Despite some progress in health service access, 4.5 billion people still lack full coverage. Diet-related NCDs are rising, especially in urban populations with shifting diets.

1.6.3 National Micronutrient Deficiency Control Programs

Several countries have initiated targeted national programs to address specific MNDs based on their population needs and dietary patterns. In India, there are several programs being run by the government to combat nutritional deficiencies.

National Iron Plus Initiative (NIPI)

NIPI, launched by the Ministry of Health and Family Welfare, targets IDA at various stages of life. It offers IFA supplements to the following target groups: Children

(6–59 months and 5–9 years), adolescents (10–19 years), pregnant and lactating women, and women of reproductive age. The program also incorporates deworming, behavior change communication, and periodic health screening to enhance iron absorption and compliance.

WIFS Program

WIFS, which was launched in 2013, focuses on adolescents aged 10–19 in both classroom and out-of-school contexts. It offers weekly IFA pills, biannual deworming, and nutritional and health education. The program, which was implemented through partnership between the health and education sectors, aims to minimize adolescent anemia, particularly among girls.

Food Fortification Initiatives

India has mandated staple food fortification under the Food Safety and Standards Authority of India (FSSAI) to treat several MNDs, including iodized salt, iron, folic acid, and vitamin B_{12}-fortified wheat flour and rice, as well as fortified milk and oil with vitamins A and D.

These initiatives are carried out through programs such as the Integrated Child Development Services (ICDS) and the Public Distribution System (PDS) to target low-income people.

1.6.4 Public–Private Partnerships in Nutrition

Public–Private Partnerships (PPPs) in nutrition play an important role in combating malnutrition and enhancing food security by combining the strengths of both sectors. These alliances seek to improve access to nutritious food, promote food fortification, and foster sustainable food systems.

i. **Global alliance for improved nutrition (GAIN)**: It is a well-known example of a PPP founded by the United Nations in 2002 to combat malnutrition. GAIN promotes government fortification initiatives, food innovation, and market-based solutions. One of its most renowned programs, the Large-Scale Food Fortification (LSFF) effort, collaborates with governments and businesses to fortify staples such as wheat flour, maize flour, oil, and salt with critical vitamins and minerals.
ii. **Food fortification initiative (FFI)**: FFI is another global PPP that aims to improve health through flour and rice fortification. It works with governments, NGOs, industry, and academics to promote mandated food fortification regulations and offers technical assistance to millers and legislators. FFI helps over 80 nations to implement flour fortification, which helps reduce anemia and neural tube abnormalities by supplementing iron and folic acid (FFI 2024).
iii. **Food fortification resource centre (FFRC)**: In India, FFRC was formed by FSSAI to promote and organize food fortification efforts. FFRC promotes PPP by connecting government activities with industry actors, development partners, and academics. It offers toolkits, technical support, and regulatory guidance to help assure quality compliance. Private players such as TATA Trusts,

DSM, and the Akshay Patra Foundation have worked with FFRC to supply fortified meals through public distribution systems and school meal programs.

iv. **HarvestPlus**: HarvestPlus is a global program led by the International Food Policy Research Institute (IFPRI) that develops and promotes biofortified crops through partnerships with national agricultural systems and private seed companies. This initiative has helped to scale up zinc-enriched wheat, iron-rich pearl millet, and vitamin A-rich sweet potato in India and Africa.

v. **World food program's nutrition improvement projects**: WFP collaborates with food companies to produce nutrient-rich Ready-to-Use Therapeutic Foods (RUTF) and complementary foods for vulnerable populations.

vi. **GAIN premix facility**: Operated by GAIN in collaboration with UNICEF and private manufacturers, this facility provides affordable, high-quality premixes of vitamins and minerals for fortification.

vii. **UNICEF–DSM partnership**: DSM, a global science-based company in nutrition, has partnered with UNICEF to develop innovative micronutrient powders and fortified foods for children and pregnant women.

1.7 CHALLENGES IN ADDRESSING MNM

1.7.1 Economic and Political Barriers

Addressing MNM in low- and middle-income countries (LMICs) faces persistent economic and political constraints that undermine program effectiveness and sustainability. Despite increased recognition of nutrition as a development priority, funding gaps remain significant. Undernutrition in LMICs has been estimated to cost economies up to 11% of GDP annually due to productivity losses and increased healthcare costs (World Bank 2019; Ramponi et al. 2021).

1.7.2 Cultural Resistance to Dietary Changes

Cultural resistance to dietary change is one of the most enduring challenges in the fight against MND (Table 1.3). Even when there are obvious health benefits, changing eating habits can be challenging since they are ingrained in social, religious, and familial institutions. While food culture includes more general eating customs, manners, and social rituals, cultural food refers to certain traditional recipes that have been passed down through the centuries (Jayasinghe et al. 2025). Family networks and cultural authority figures should be incorporated into educational efforts to increase acceptance of dietary modifications.

1.7.3 Supply Chain and Distribution Issues in Supplementation

Despite having a solid conceptual foundation, the execution of micronutrient supplementation programs frequently fails because of ongoing issues across the supply chain. Systemic inefficiencies undermine the intended health consequences of supplementing schemes by compromising coverage and quality from procurement to last-mile delivery (Table 1.4).

TABLE 1.3
Cultural Factors Contributing to Resistance Against Dietary Change

Cultural Influence	Example	Implication for Intervention	References
Religious food restrictions	Fasting during Ramadan or Lent	Timing of supplementation strategies must be adapted	Jayasinghe et al. (2025)
Traditional cooking practices	High-fat meat dishes in pastoral cultures	Conflict with low-fat diet recommendations	Yin et al. (2020)
Symbolic value of foods	Rice in Asian cultures, meat in Xinjiang	May hinder reduction in carbohydrate or meat intake	Yin et al. (2020)
Social eating norms	Communal meals, food sharing	Difficult to enforce individualized diet changes	Cowdery et al. (2010)

TABLE 1.4
Common Supply Chain Bottlenecks in Micronutrient Supplementation Programs

Supply Chain Function	Common Bottlenecks	Examples from India (AMB Program)	Source
Forecasting	Reliance on population targets alone	60% indent gap in IFA Red tablets in 2018–2019	
Procurement	Long lead times, SOP non-compliance	Average lead time: 35 weeks; up to 62 in Uttarakhand	Ahmad et al. (2023)
Warehousing	Poor inventory tracking, quality degradation	No standardized temperature control at storage facilities	
Distribution	Irregular schedules, insufficient transport	Block-level distribution delayed in hilly states	
Quality assurance	Limited testing and monitoring	Only 1 MMS brand in India close to UNIMMAP specification	Monterrosa et al. (2018)

National nutrition policies often do not align with procurement incentives, manufacturing capacity, or regulatory enforcement. As a result, even when supplements are available, their quality and timely distribution are not guaranteed.

1.7.4 Need for Integrated Multisectoral Approaches

Addressing MNDs necessitates coordinated and persistent action from all sectors. While malnutrition has generally been viewed as a health issue, its causes are intertwined with agricultural practices, educational systems, climate vulnerability, water and sanitation infrastructure, and sociopolitical structures. Siloed interventions are insufficient; instead, a holistic, cross-sectoral approach is required to address the complex and varied nature of MNDs. An integrated food-based, multi-actor strategy was drawn from global dialog at the Nutrition for Growth Summit (2021) (Figure 1.7).

Tirado et al. (2022) identified six highly feasible and effective adaptation strategies for nutrition (Figure 1.8) by linking them with climate adaptation.

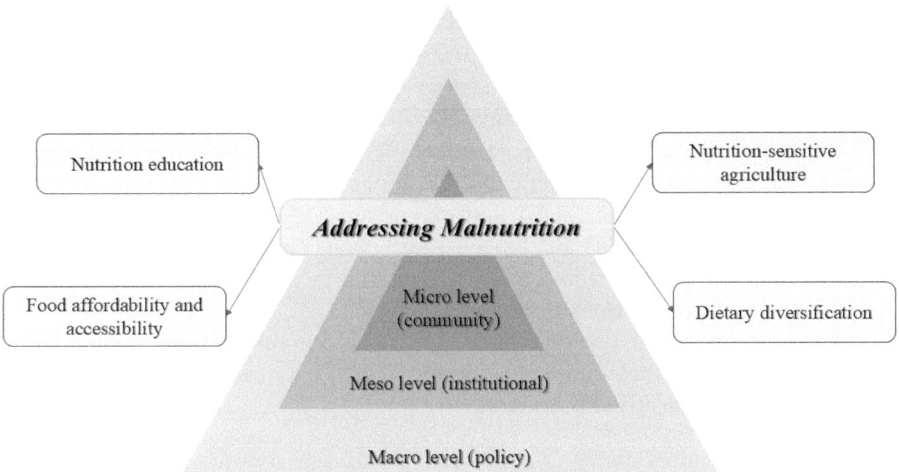

FIGURE 1.7 Collaborative food-centered strategy involving multiple stakeholders to address malnutrition.

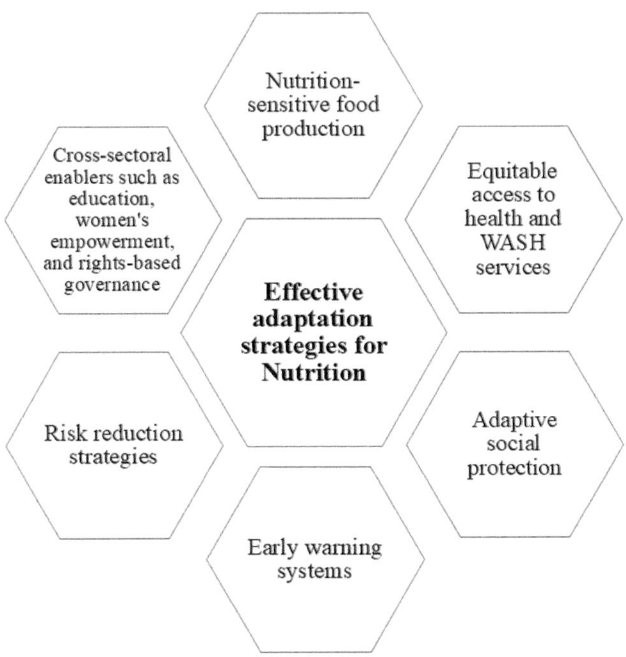

FIGURE 1.8 Six highly feasible and effective adaptation strategies for nutrition.

1.8 FUTURE DIRECTIONS AND INNOVATIONS IN MICRONUTRIENT INTERVENTIONS

1.8.1 Advances in Nutrigenomics and Personalized Nutrition

Nutrigenomics and nutrigenetics are reshaping public health nutrition by supporting precision approaches tailored to individual genetic and metabolic profiles. Nutrigenetics focuses on how genetic variation influences nutrient responses, while nutrigenomics explores how dietary components affect gene expression and metabolic pathways (Ordovas and Mooser 2004). Together, they underpin personalized nutrition strategies adapted to genetics, life stage, and environmental context. Personalized plans based on genetic and metabolic profiles can improve adherence, enhance nutrient utilization, and minimize adverse effects. Studies showed that nutrients such as polyphenols and omega-3 fatty acids can modulate genes involved in metabolism and inflammation, offering potential in managing diseases like cancer, diabetes, and cardiovascular conditions (Vyas 2022).

Despite its promise, several challenges limit the use of nutrigenomics in low-resource settings, including infrastructure gaps, costs, ethical concerns, and the need for culturally sensitive counselling.

1.8.2 Fortification with Nanotechnology and Smart Delivery Systems

Nanotechnology has emerged as a game changer in micronutrient fortification, solving fundamental issues such as low bioavailability, chemical instability, and unwanted sensory changes in fortified meals. Nanoencapsulation protects micronutrients from degradation and releases them at specific places in the gastrointestinal system, enhancing stability, absorption, and efficacy. Smart delivery systems not only mask undesirable taste and aroma but also enable nutrient release based on physiological triggers such as pH, enzyme presence, or temperature.

1.8.3 Role of Digital Technology in Nutritional Surveillance

Digital technologies have transformed nutritional surveillance by enabling real-time, personalized, and large-scale data collection. Tools such as mobile health (mHealth) apps, wearable trackers, and AI-powered analytics support continuous monitoring of dietary habits, biometric data, and health behaviors across clinical and community settings.

mHealth apps like MyFitnessPal, Snap-n-Eat, and Undermyfork exemplify how smartphones aid in dietary self-monitoring. Wearables such as smartwatches and biosensors collect continuous biometric data, including heart rate, activity, and chewing patterns, enhancing traditional assessments (Limketkai et al. 2021). Wearable devices such as smart forks, jaw-mounted EMG sensors, and sweat-analyzing patches are also being tested for detailed tracking of eating behaviors, hydration, and metabolism (Mortazavi and Gutierrez-Osuna 2023). Practical applications include the Keenoa app, which uses AI to analyze photographed meals and shares nutrient data with dietitians for remote monitoring (Salinari et al. 2023).

Despite their potential, these technologies face challenges. Key barriers include data privacy, poor interoperability across platforms, digital divides, and low digital literacy in at-risk groups. Ethical concerns, especially in emergency contexts, must also be addressed to ensure equity and build public trust.

1.8.4 Strengthening Resilience of Food Systems

Strengthening resilience of food systems is vital in addressing growing challenges such as climate change, resource scarcity, conflict, and market volatility. A resilient food system is one that can withstand, adapt to, and recover from diverse shocks, while continuing to provide sufficient, safe, and nutritious food for all populations. As seen during COVID-19 and the conflict in Ukraine, local systems and short food supply chains also improve resilience by lowering reliance on international markets (Rimhanen et al. 2023). Coherence in policy is also essential. The FAO's Twin-Track Approach provides a framework for coordinating short- and long-term food security objectives by fusing emergency assistance with long-term agricultural development (Pingali et al. 2005).

1.9 CASE STUDIES OF SUCCESSFUL MICRONUTRIENT INTERVENTIONS

1.9.1 Golden Rice and Biofortified Crops

Golden rice represents a landmark innovation in the domain of biofortification, designed to combat vitamin A deficiency (VAD). VAD is not only the leading cause of preventable childhood blindness, but it is also associated with significantly increased child mortality (Dubock 2017; WHO 2009). Early iterations, such as Golden Rice-1, contained β-carotene at levels around 1.6 μg/g (Beyer et al. 2002), while the improved Golden Rice-2 variety, utilizing a maize-derived phytoene synthase gene, achieves levels up to 35 μg/g (Tang et al. 2009).

Beyond Golden Rice, India has achieved considerable advances in biofortified crop production via conventional breeding. ICAR has introduced numerous kinds of iron- and zinc-rich pearl millet, zinc-enriched wheat, and vitamin A-rich maize, providing further weapons for combating hidden hunger. A few of the biofortified crops by ICAR are listed in Table 1.5. A total of 531 biofortified cultivars have been licensed to private seed producers and Farmers Producer Organizations (FPOs) for partnership scaling (Yadava et al. 2022).

1.9.2 Large-Scale Supplementation Programs in Developing Countries

Large-scale supplementation programs are an important part of public health nutrition initiatives, addressing essential MNDs such as iron, folic acid, vitamin A, iodine, and zinc. One of the most common interventions is IFA supplementation in countries like Nepal, Nicaragua, and Ghana. Furthermore, food fortification and biofortification are widely acknowledged as very cost-effective population-level solutions. According to systematic evaluations, iron fortification reduces the incidence of

TABLE 1.5
Biofortified Crop Varieties Developed by ICAR with Enhanced Nutritional Traits

Crop Name	Cultivar	Nutritional Benefit	Year Released	Developed By (Institute)
Finger millet	VR 929	Iron (131.8 ppm)	2020	ANGRAU, Guntur
Finger millet	CFMV1	Calcium (428 mg/100 g), Iron (58.0 ppm), Zinc (44.0 ppm)	2020	ANGRAU, Vizianagaram
Lentil	IPL 220	Iron (73.0 ppm), Zinc (51.0 ppm)	2018	ICAR-IIPR, Kanpur
Pearl millet	HHB 299	Iron (73.0 ppm), Zinc (41.0 ppm)	2017	CCS-HAU with ICRISAT, Patancheru
Rice	DRR Dhan 48	Zinc (24.0 ppm)	2018	ICAR-IIRR, Hyderabad
Rice	CR Dhan 310	Protein (10.3%)	2016	ICAR-NRRI, Cuttack

anemia by 41%, whereas flour fortification with folic acid reduces neural tube abnormalities by 40%–50%. Vitamin A fortification of sugar in countries like Guatemala has considerably reduced insufficiency rates within a year of adoption (Osendarp et al. 2018).

1.9.3 Community-Based Nutrition Programs

Community-based nutrition programs have shown significant success in addressing MNDs through locally adapted and participatory approaches. One widely recognized example is the Tamil Nadu Integrated Nutrition Project in India, which focused on reducing malnutrition among women and children through growth monitoring, nutrition education, and supplementary feeding. Its success was driven by targeted supplementary feeding, intensive supervision (1 worker per 150 households), and robust monitoring systems, setting a model later integrated into the national ICDS (Heaver 2002).

Helen Keller International's Homestead Food Production (HFP) initiative has effectively linked agricultural practices with micronutrient outcomes. Originating as a pilot in Bangladesh in 1990, the program expanded to nearly 900,000 households by 2012 and reached 4.5 million people (Helen Keller Intl. 2025). Participants consumed over double the amount of vegetables compared to non-participants and saw reductions in anemia.

In Africa, the SPRING (Strengthening Partnerships, Results, and Innovations in Nutrition Globally) project in countries like Uganda and Niger integrated agriculture and nutrition interventions, using community groups to promote the use of iron-rich foods and fortified products.

These case studies highlight that community engagement, multisectoral collaboration, and culturally sensitive education are essential components of successful micronutrient interventions. They offer replicable models for other regions facing similar nutritional challenges.

1.10 CONCLUSIONS

MNM remains a significant global health challenge, disproportionately affecting vulnerable populations such as children, pregnant women, and low-income communities. Throughout this chapter, a variety of public health strategies have been reviewed, ranging from dietary diversification, biofortification, and food fortification, to large-scale supplementation and targeted education campaigns. These interventions have shown considerable success in reducing specific deficiencies, yet their sustained effectiveness depends on holistic implementation.

Evidence-based policy, community participation, and research are all equally important. Policies based on strong surveillance and local data enable focused interventions and equitable resource distribution. Community participation increases the acceptability and sustainability of treatments, especially when they are culturally appropriate and participatory in design. Furthermore, ongoing research and innovation such as food fortification is critical for adjusting tactics to changing demographic and environmental issues. Eliminating MNM requires a coordinated, context-specific, and equity-focused approach, where science, policy, and community action converge to promote sustainable nutritional well-being.

REFERENCES

Agostoni, C., Baglioni, M., La Vecchia, A., Molari, G., and Berti, C. 2023. "Interlinkages between Climate Change and Food Systems: The Impact on Child Malnutrition—Narrative Review." *Nutrients* 15(2): 416.

Ahmad, K., Singh, J., Singh, R. A., Saxena, A., Varghese, M., Ghosh, S., … and Patel, N. 2023. "Public Health Supply Chain for Iron and Folic Acid Supplementation in India: Status, Bottlenecks and an Agenda for Corrective Action under Anemia Mukt Bharat Strategy." *PLOS ONE* 18(2): e0279827.

Animasahun, B. A., and Itiola, A. Y. 2021. "Iron Deficiency and Iron Deficiency Anemia in Children: Physiology, Epidemiology, Aetiology, Clinical Effects, Laboratory Diagnosis and Treatment: Literature Review." *Journal of Xiangya Medicine* 6: 1–14.

Asbaghi, O., Ashtary-Larky, D., Bagheri, R., Moosavian, S. P., Nazarian, B., Afrisham, R., … and Alavi Naeini, A. 2021. "Effects of Folic Acid Supplementation on Inflammatory Markers: A Grade-Assessed Systematic Review and Dose–Response Meta-Analysis of Randomized Controlled Trials." *Nutrients* 13(7): 2327.

Awuchi, C. G., Igwe, V. S., Amagwula, I. O., and Echeta, C. K. 2020. "Health Benefits of Micronutrients (Vitamins and Minerals) and Their Associated Deficiency Diseases: A Systematic Review." *International Journal of Food Sciences* 3(1): 1–32.

Barkley, J. S., Wheeler, K. S., and Pachón, H. 2015. "Anemia Prevalence May Be Reduced Among Countries That Fortify Flour." *British Journal of Nutrition* 114(2): 265–273.

Beyer, P., Al-Babili, S., Ye, X., Lucca, P., Schaub, P., Welsch, R., and Potrykus, I. 2002. "Golden Rice: Introducing the β-carotene Biosynthesis Pathway into Rice Endosperm by Genetic Engineering to Defeat Vitamin a deficiency." *The Journal of nutrition* 132(3): 506S–510S.

Billah, S. M., Raynes-Greenow, C., Ali, N. B., Karim, F., Lotus, S. U., Azad, R., … and El Arifeen, S. 2022. "Iron and Folic Acid Supplementation in Pregnancy: Findings from the Baseline Assessment of a Maternal Nutrition Service Programme in Bangladesh." *Nutrients* 14(15): 3114.

Blencowe, H., Kancherla, V., Moorthie, S., Darlison, M. W., and Modell, B. 2018. "Estimates of Global and Regional Prevalence of Neural Tube Defects for 2015: A Systematic Analysis." *Annals of the New York Academy of Sciences* 1414(1): 31–46.

Bouis, H. E., Eozenou, P., and Rahman, A. 2011. "Food Prices, Household Income, and Resource Allocation: Socioeconomic Perspectives on Their Effects on Dietary Quality and Nutritional Status." *Food and Nutrition Bulletin* 32(Supplement 1): S14–S23.

Castillo-Lancellotti, C., Tur, J. A., and Uauy, R. 2013. "Impact of Folic Acid Fortification of Flour on Neural Tube Defects: A Systematic Review." *Public Health Nutrition* 16(5): 901–911.

Cook, J. D. 2005. "Diagnosis and Management of Iron-Deficiency Anemia." *Best Practice and Research Clinical Haematology* 18(2): 319–332.

Cowdery, J. E., Parker, S., and Thompson, A. 2010. "Application of the PEN-3 Model in a Diabetes Prevention Intervention." *Journal of Health Disparities Research and Practice* 4(1): 3.

Dubock, A., 2017. "An Overview of Agriculture, Nutrition and Fortification, Supplementation And Biofortification: Golden Rice as an Example for Enhancing Micronutrient Intake." *Agriculture & Food Security* 6(1): 59. https://doi.org/10.1186/s40066-017-0135-3

Fenech, M. 2012. "Folate (Vitamin B9) and Vitamin B12 and Their Function in the Maintenance of Nuclear and Mitochondrial Genome Integrity." *Mutation Research/Fundamental and Molecular Mechanisms of Mutagenesis* 733(1–2): 21–33.

Food and Agriculture Organization (FAO). 2017. *The Future of Food and Agriculture – Trends and Challenges*. Rome.

Food Fortification Initiative. 2024. *2023 Annual Report*. Atlanta: FFI. https://ffinetwork.org/annual-reports/

Gosdin, L., Sharma, A. J., Tripp, K., Amoaful, E. F., Mahama, A. B., Selenje, L., … and Addo, O. Y. 2021. "A School-Based Weekly Iron and Folic Acid Supplementation Program Effectively Reduces Anemia in a Prospective Cohort of Ghanaian Adolescent Girls." *The Journal of Nutrition* 151(6): 1646–1655.

Heaver, R. 2002. *India's Tamil Nadu Nutrition Program: Lessons and Issues in Capacity and Management.* Washington, DC: Health Nutrition and Population Unit, Human Development Network, World Bank.

Helen Keller Intl. 2025. "Helen Keller Intl – Continuing Her Work." May 6, 2025. https://www.hki.org/working-worldwide/asia-pacific/bangladesh/

Hodgson, H. J., and Epstein, O. 2007. "Malabsorption." *Medicine* 35(4): 220–225.

International Grains Council. 2025. *Grain Market Report*. https://www.igc.int/en/gmr_summary.aspx

James, D. 2004. "Factors Influencing Food Choices, Dietary Intake, and Nutrition-Related Attitudes Among African Americans: Application of a Culturally Sensitive Model." *Ethnicity and Health* 9(4): 349–367.

Jayasinghe, S., Byrne, N. M., and Hills, A. P. 2025. "Cultural Influences on Dietary Choices." *Progress in Cardiovascular Diseases* 90: 22–26.

Kiran, A., Wakeel, A., Mahmood, K., Mubaraka, R., Hafsa, and Haefele, S. M. 2022. "Biofortification of Staple Crops to Alleviate Human Malnutrition: Contributions and Potential in Developing Countries." *Agronomy* 12(2): 452.

Limketkai, B. N., Mauldin, K., Manitius, N., Jalilian, L., and Salonen, B. R. 2021. "The Age of Artificial Intelligence: Use of Digital Technology in Clinical Nutrition." *Current Surgery Reports* 9(7): 20.

Lozoff, B., and Georgieff, M. K. 2006. "Iron Deficiency and Brain Development." *Seminars in Pediatric Neurology*, 13(3): 158–165. WB Saunders.

Martinez, H., Benavides-Lara, A., Arynchyna-Smith, A., Ghotme, K. A., Arabi, M., and Arynchyn, A. 2023. "Global Strategies for the Prevention of Neural Tube Defects Through the Improvement of Folate Status in Women of Reproductive Age." *Child's Nervous System* 39(7): 1719–1736.

Monterrosa, E. C., Beesabathuni, K., van Zutphen, K. G., Steiger, G., Kupka, R., Fleet, A., and Kraemer, K. 2018. "Situation Analysis of Procurement and Production of Multiple Micronutrient Supplements in 12 Lower and Upper Middle-Income Countries." *Maternal and Child Nutrition* 14: e12500.

Morales, F., Montserrat-De la Paz, S., Leon, M. J., and Rivero-Pino, F. 2023. "Effects of Malnutrition on the Immune System and Infection and the Role of Nutritional Strategies Regarding Improvements in Children's Health Status: A Literature Review." *Nutrients* 16(1): 1.

Mortazavi, B. J., and Gutierrez-Osuna, R. 2023. "A Review of Digital Innovations for Diet Monitoring and Precision Nutrition." *Journal of Diabetes Science and Technology* 17(1): 217–223.

Ordovas, J. M., and Mooser, V. 2004. "Nutrigenomics and Nutrigenetics." *Current Opinion in Lipidology* 15(2): 101–108.

Osendarp, S. J., Martinez, H., Garrett, G. S., Neufeld, L. M., De-Regil, L. M., Vossenaar, M., and Darnton-Hill, I. 2018. "Large-Scale Food Fortification and Biofortification in Low- and Middle-Income Countries: A Review of Programs, Trends, Challenges, and Evidence Gaps." *Food and Nutrition Bulletin* 39(2): 315–331.

Pingali, P., Alinovi, L., and Sutton, J. 2005. "Food Security in Complex Emergencies: Enhancing Food System Resilience." *Disasters* 29(S1): S5–S24.

Pravst, I., Lavriša, Ž., Hribar, M., Hristov, H., Kvarantan, N., Seljak, B. K., … and Kušar, A. 2021. "Dietary Intake of Folate and Assessment of the Folate Deficiency Prevalence in Slovenia Using Serum Biomarkers." *Nutrients* 13(11): 3860.

Ramponi, F., Tafesse, W., and Griffin, S. 2021. "Economic Evaluation of Interventions to Address Undernutrition: A Systematic Review." *Health Policy and Planning* 36(4): 533–541.

Rimhanen, K., Aakkula, J., Aro, K., and Rikkonen, P. 2023. "The Elements of Resilience in the Food System and Means to Enhance the Stability of the Food Supply." *Environment Systems and Decisions* 43(2): 143–160.

Salinari, A., Machì, M., Armas Diaz, Y., Cianciosi, D., Qi, Z., Yang, B., … and Giampieri, F. 2023. "The Application of Digital Technologies and Artificial Intelligence in Healthcare: An Overview on Nutrition Assessment." *Diseases* 11(3): 97.

Santos, L. M. P., Lecca, R. C. R., Cortez-Escalante, J. J., Sanchez, M. N., and Rodrigues, H. G. 2015. "Prevention of Neural Tube Defects by the Fortification of Flour with Folic Acid: A Population-Based Retrospective Study in Brazil." *Bulletin of the World Health Organization* 94(1): 22.

Sobczyńska-Malefora, A., Delvin, E., McCaddon, A., Ahmadi, K. R., and Harrington, D. J. 2021. "Vitamin B12 Status in Health and Disease: A Critical Review. Diagnosis of Deficiency and Insufficiency – Clinical and Laboratory Pitfalls." *Critical Reviews in Clinical Laboratory Sciences* 58(6): 399–429.

Tang, G., Qin, J., Dolnikowski, G. G., Russell, R. M., and Grusak, M. A. 2009. "Golden Rice Is an Effective Source of Vitamin A." *The American Journal of Clinical Nutrition* 89(6): 1776–1783.

Tirado, M. C., Vivero-Pol, J. L., Bezner Kerr, R., and Krishnamurthy, K. 2022. "Feasibility and Effectiveness Assessment of Multi-Sectoral Climate Change Adaptation for Food Security and Nutrition." *Current Climate Change Reports* 8(2): 35–52.

United Nations Children's Fund (UNICEF). 2023. *Home Fortification with Multiple Micronutrient Powders for the Prevention of Iron Deficiency Anaemia in Early Childhood. Brief Guidance Note*. New York: UNICEF.

UNICEF, United Nations University, and World Health Organization. 2001. *Iron Deficiency Anemia: Assessment, Prevention, and Control: A Guide for Programme Managers.* Geneva: WHO.

Vyas, S. 2022. "Advances in Nutrigenomics and Applications in Public Health: A Recent Update." *Current Research in Nutrition and Food Science Journal* 10(3): 1092–1104.

Wangaskar, S. A., Sahu, S. K., Majella, M. G., and Rajaa, S. 2021. "Prevalence of Anemia and Compliance to Weekly Iron-Folic Acid Supplementation Programme Amongst Adolescents in Selected Schools of Urban Puducherry, India." *Nigerian Postgraduate Medical Journal* 28(1): 44–50.

World Health Organization. 2009. *Global Prevalence of Vitamin A Deficiency in Populations at Risk 1995–2005*. WHO Global Database on Vitamin A Deficiency. Geneva. https://www.who.int/publications/i/item/9789241598019

World Health Organization. 2014. *Comprehensive Implementation Plan on Maternal, Infant and Young Child Nutrition*. https://www.who.int/publications/i/item/WHO-NMH-NHD-14.1

World Bank. 2019. "Nutrition. Overview." https://www.worldbank.org/en/topic/nutrition/overview

World Health Organization. 2016. *Guideline: Daily Iron Supplementation in Infants and Children*. Geneva: WHO.

World Health Organization. 2019. "Micronutrients." https://www.who.int/health-topics/micronutrients

World Health Organization. 2025. "Anaemia.". https://www.who.int/news-room/fact-sheets/detail/anaemia

Yadava, D. K., Choudhury, P. R., Hossain, F., Kumar, D., Sharma, T. R., and Mohapatra, T. 2022. *Biofortified Varieties: Sustainable Way to Alleviate Malnutrition*. 4th ed. New Delhi: Indian Council of Agricultural Research.

Yin, J., Yang, D., Zhang, X., Zhang, Y., Cai, T., Hao, Y., … and Chen, Y. 2020. "Diet Shift: Considering Environment, Health and Food Culture." *Science of the Total Environment* 719: 137484.

2 Government Policy and Programs on Rice Fortification

Gorenand Prasad Yadav and Devesh Kumar

2.1 INTRODUCTION

Rice fortification has become a vital public health concerned approach to address alarmingly high micronutrient deficiencies (MNDs), especially in country like India, where rice is consumed by almost 65% of the population as a staple food. Considering the pervasive nutritional deficits impacting population health, particularly deficiencies in bioavailable iron, folate, and vitamin B_{12}, governments across the world are increasingly turning toward fortified rice (FR) as an affordable and widespread solution to prevent the health issues. The enrichment of rice with these essential micronutrients is promoted as a cost-effective strategy to improve nutritional outcomes in communities where rice is a dietary mainstay (WHO 2018).

Governments have used rice fortification as part of social safety net programs in a number of nations worldwide. These initiatives, which include supplemental food rations and Mid-Day Meal (MDM) programs, focus on populations that are more susceptible to malnutrition or have higher nutritional demands. In fact, nations like Bangladesh, India, and others have taken the initiative to employ FR through social safety programs after consulting with the World Food Programme (WFP). Beginning in 2016, the United States government initiated the incorporation of FR into its food assistance initiatives. This deliberate action attempts to improve the nutritional content of food aid provided to vulnerable populations, both domestically and internationally.

Based on information collected during the fifth round of the National Family Health Survey (NFHS) (2019–2021), anemia was found to have a higher prevalence in India as compared to the previous NFHS-4 conducted in 2015–2016 (Muralidharan et al. 2024). In addition to iron deficiency, other nutrient shortfalls, like vitamin B_{12} and folic acid, persist and impact the population's health and efficiency. To tackle these challenges, the Indian government has introduced a number of initiatives, such as distributing FR through various Social Welfare Programs and Schemes including Pradhan Mantri Matru Vandana Yojana (PMMVY), Integrated Child Development Schemes (ICDS), and Pradhan Mantri Poshan Shakti Nirman (PM POSHAN). The Indian government has incorporated food fortification, especially rice fortification, into the National Nutrition Mission (Poshan Abhiyan) as a supplemental measure to lower the prevalence of anemia and malnutrition in India. Through rice fortification,

DOI: 10.1201/9781003661245-2

the Government of India continues to prioritize advancing the overall nutritional well-being of its population, with a strategic emphasis on combating undernutrition among vulnerable and marginalized communities. This initiative forms a critical component of the nation's broader public health and socioeconomic development agenda (Press Information Bureau, Government of India 2024). The various policies and programs undertaken by the Indian Government are major steps in addressing the malnutrition and improving the lives of millions of people, acknowledging the vital role that nutrition plays in creating a healthier country. The food fortification program launched by the Government of India stands out as few of the most effective supplementary initiative aimed at addressing nutritional anemia across the population, as the country works to fulfil Sustainable Development Goal 2, which calls for the eradication of malnutrition in all its manifestations. In order to help the poor, meet their nutritional needs, and build a stronger, healthier India, the government is fortifying the rice that is given out through these programs (Press Information Bureau, Government of India 2024).

India initiated its rice fortification program in 2019 with a pilot project across 15 districts, aiming to address MNDs by enriching rice with iron, folic acid, and vitamin B_{12}. This pilot phase targeted beneficiaries of the Public Distribution System (PDS) and other welfare schemes. Building on the pilot's outcomes, the government expanded the initiative in a phased manner. Phase II, launched in April 2022, encompassed 291 districts, including aspirational and high-burden areas, with the goal of nationwide coverage by March 2024.

2.2 EVOLUTION OF GOVERNMENT POLICIES ON RICE FORTIFICATION

2.2.1 Early Initiatives and Awareness

The Indian Government Policies on Rice Fortification started in the year 2016. In February 2016, Key stakeholders who participated in the implementation and oversight of the fortification program include Food Safety and Standards Authority of India (FSSAI), National Institute of Nutrition (NIN), Indian Council of Medical Research (ICMR), Food Corporation of India (FCI), and the Ministry of Health and Family Welfare. Mandatory rice fortification in government programs through FCI was suggested by the Ministry, and it was phased in May 2016 (Press Information Bureau, Government of India 2024). On 15 August 2021, an announcement was made by Hon'ble Prime Minister of India on a landmark initiative to combat MNDs by mandating the fortification of rice across key national food security and nutrition programs. The initiative, which builds upon global and national evidence on the effectiveness of food fortification, was launched as a Centrally Sponsored Pilot Scheme (2019–2022). This pilot, backed by a financial outlay of ₹174.64 crore, was operationalized to assess the feasibility, supply chain logistics, and community acceptance of FR. By March 2022, approximately 4.30 lakh MT of FR had been distributed across 11 states through government welfare schemes demonstrating both scalability and operational viability (Press Information Bureau, Government of India 2024; Ministry of Women and Child Development 2020).

Recognizing the success of the pilot and the critical role of FR in reducing anemia and other MNDs, the Cabinet Committee on Economic Affairs in 2022 approved the nationwide scale-up under a Central Sector Scheme, which is fully funded by the Central Government. The total financial allocation for the scale-up was ₹4,269.76 crore, earmarked to cover the period until 30 June 2024. This scale-up was meticulously structured into three implementation phases, each aligned with existing food-based welfare schemes to ensure systematic integration and coverage.

This fortification program is being implemented in compliance with the standards developed by the FSSAI, with rigorous quality assurance and monitoring mechanisms integrated at every stage of the supply chain, from production at rice mills to last-mile distribution. Through this ambitious, multiphased strategy, India aims to make significant progress toward achieving its National Nutrition Targets, aligning with the Poshan Abhiyaan, and contributing to the Sustainable Development Goals (SDG 2 and SDG 3), specifically those related to ending hunger, improving nutrition, and ensuring healthy lives for all.

As of March 2024, the initiative has achieved 100% coverage across the country. FR has completely replaced custom-milled rice in all government food schemes, ensuring the delivery of essential micronutrients to millions of beneficiaries nationwide (Press Information Bureau, Government of India 2024).

2.2.2 International Collaboration and Influences on Indian Policy

The development of large-scale food fortification (LSFF) programs around the world has been greatly aided by the involvement of international groups and organizations. LSFF has become well-known worldwide since the United States' first salt iodization program was implemented in 1920. The significance of widespread food fortification in addressing micronutrient deficiency has been acknowledged by governments across the globe. Globally accepted guidelines for extensive food fortification policies have been developed in great part through international and regional agreements, resolutions, and conferences (Regulatory governance of LSFF 2024).

India's rice fortification program, a cornerstone of its strategy to combat malnutrition, is significantly strengthened through global scientific collaboration. International Public and Private Organizations often support fortification efforts in low- and middle-income countries like India. They provide their support in designing and planning the fortification programs, which may involve recommendations on the identification of suitable foods to fortify, fortification equipment, packaging, labeling, and distribution systems. Development partners may also offer technical assistance, such as training and advice on best practices, to help stakeholders gain the abilities and know-how needed to carry out and maintain fortification projects successfully. Guidelines published by International Organizations such as the WHO-FAO and GAIN show promising influence in shaping fortification policy (Regulatory governance of LSFF 2024). These organizations also often provide financial support in the form of funding for equipment and covering other costs associated with fortification activities. They also fund or help conduct start-up activities like surveys, capability assessments, capacity building training, and standards development, therefore laying the ground for food fortification programs to be designed and implemented. A number of projects around the world have been completed with the support of International Public and Private Organizations. For instance, in India, different rounds of the National Family Health Survey (Ministry of Health and Family Welfare of India 2021), which is important for ascertaining the level of micronutrient deficiency in the population, have been conducted.

Between October 2016 and October 2017, several significant consultations and meetings of experts from different organizations were conducted to prioritize rice fortification in India. Experts focused on improving the technical side of food fortification, such as making sure that the premix (the blend of added nutrients) is of good quality and readily available. They worked in collaboration with organizations like Indian Institute of Health Management Research (IIHMR) and Global Alliance for Improved Nutrition (GAIN). As of 20 March 2025, ongoing partnerships with organizations like WHO, the GAIN, and academic institutions worldwide underscore India's commitment to evidence-based fortification. PATH's collaboration with the Akshaya Patra Foundation introduced FR in school feeding programs, benefiting approximately one million people (PATH 2024). This initiative exemplifies how international NGOs contribute to scaling up fortification efforts, leveraging global expertise to address local nutritional challenges.

2.3 NATIONAL AND STATE LEVEL POLICY FRAMEWORKS

2.3.1 Introduction to Relevant Policy Developments

India has a national policy framework for rice fortification that aims to prevent anemia and malnutrition through National Food Security Act (NFSA), 2013, particularly by mandating the distribution of FR under Targeted Public Distribution System (TPDS) and allied welfare schemes like the MDM and ICDS (NFSA 2013).

The NFSA establishes a framework of joint responsibility between the Central Government and the State/Union Territory (UT) Governments to ensure the seamless delivery of food security to eligible beneficiaries across the country. The Central Government is tasked with several key responsibilities. These include the allocation of the required quantity of foodgrains to each State and UT, as well as the transportation of these foodgrains up to the designated depots within the respective States/UTs. Additionally, the Central Government provides financial assistance to support the movement of foodgrain from the godowns of FCI to the doorstep of Fair Price Shops (FPSs). On the other hand, the State and UT Governments play a crucial role in the effective implementation of the Act at the grassroots level. Their responsibilities include identifying eligible households, issuing ration cards, and ensuring the distribution of foodgrain entitlements through the FPS network. They are also responsible for issuing licenses to FPS dealers, monitoring their operations, and establishing robust grievance redressal mechanisms. Furthermore, the States/UTs are expected to strengthen the overall functioning of the TPDS to ensure transparency, accountability, and efficient service delivery (National Food Security Portal, Department of Food and Public Distribution, Government of India). Thus, the central and state governments in India collaborate on rice fortification through a combination of central sector initiatives and state-level implementation. The central government provides funding and sets national standards, while states are responsible for procuring, distributing, and monitoring the fortification process.

2.3.2 Role of Food Safety and Standards Authority of India (FSSAI)

The FSSAI, established under the Food Safety and Standards Act, 2006, plays a pivotal regulatory and scientific role in India's rice fortification initiative. As the apex body responsible for food safety and standards in the country, FSSAI develops and enforces fortification norms to ensure micronutrient stability, bioavailability, and retention in fortified foods, particularly under the diverse post-harvest and cooking conditions prevalent across India (Institute of Competitiveness India Report 2023).

The Department of Food and Public Distribution (DFPD) ensures the implementation of Rice Fortification to counter malnutrition especially related to anemia and micronutrient deficiencies. In addition to quality certifications provided by manufacturers and millers of Fortified Rice Kernels (FRK), the Storage and Research division of the DFPD is responsible for overseeing the quality standards of the final product throughout the entire process from procurement to distribution via various social welfare programs. In March 2022, the DFPD introduced a comprehensive Standard Operating Procedure (SOP) to streamline the implementation of rice fortification across India's domestic supply chain. This SOP delineates the responsibilities of stakeholders at each stage from the production of FRK to the distribution of FR to beneficiaries, ensuring adherence to quality standards throughout the process. Complementing the DFPD's operational oversight, the FSSAI plays a pivotal regulatory role. The FSSAI has established fortification standards through the Food Safety and Standards (Fortification of Foods) Regulations, 2018, and the Food Safety and Standards (Food Products Standards and Food Additives) Regulations, 2011. To facilitate compliance and quality assurance, the FSSAI has empanelled and licensed FRK manufacturers, developed guidelines for packaging and labeling, and mapped laboratories accredited by the National Accreditation Board for Testing and Calibration Laboratories for quality testing of FRK and FR. Additionally, the Food Fortification Resource Centre (FFRC), a unit of FSSAI, serves as a resource hub, providing technical assistance, training, and capacity-building support to stakeholders involved in the fortification program. This collaborative framework between DFPD and FSSAI ensures a cohesive approach to rice fortification, aiming to address MNDs and improve public health outcomes across the nation.

The FSSAI mandates specific micronutrient levels for FR: iron (28–42.5 mg/kg as ferric pyrophosphate), folic acid (75–125 µg/kg), and vitamin B_{12} (0.75–1.25 µg/kg), with optional additions like zinc (10–15 mg/kg) and vitamin A (500–750 µg RE/kg) based on regional needs (Food Safety and Standards (Fortification of Foods) Regulations 2018). In order to support and align stakeholders, FSSAI established FFRC as a nodal point, offering technical and implementation support and educating consumers about fortification, nutrition, and food safety. The FFRC was a specialized forum that promotes and supports widespread food fortification across India. It acted as a platform, providing guidelines on technology, standards, food safety, the manufacturing and procurement of premixes and equipment, quality assurance, and quality control for fortification. It included every piece of information about food fortification, including government circulars, success stories, national and international experiences, the most recent technological advancements, and scientific proof (Institute of Competitiveness India Report 2023).

2.3.3 Role of Bureau of Indian Standards (BIS)

The BIS is the National Standards Body of India whose core actives include formulation of Indian Standards, Conformity Assessment of various products according to the relevant Indian Standards, and Laboratory services. The BIS plays a crucial role in ensuring the quality and safety of FR in India by establishing Standards and providing Certification for both the FR and the equipment used in its manufacturing. The BIS also suggest the quality and safety of FRK and mineral premix used in the manufacturing of FR. The BIS grants the certification for the use of BIS Standard Mark on the products after successful verification of manufacturing capability and testing facility with the manufacturers and ensuring conformity of the product to the requirement of relevant Indian Standard by testing from BIS recognized laboratory (Bureau of Indian Standards Act 2016).

The Indian Standards formulated by BIS are: IS 17780: 2021 (Fortified Rice-Specification), IS 17782: 2021 (Fortified Rice Kernels-Specification), IS 17854: 2022 (Equipment for Manufacture of Fortified Rice-Specification), IS 17853: 2022 (Equipment for Manufacture of Fortified Rice Kernels-Specification), and IS 17781: 2021 (Vitamin Mineral Premix for Manufacturing Fortified Rice Kernels-Specification). According to BIS Conformity Assessment portal (Available at: https://www.manakonline.in/MANAK/ApplicationLicenceRelatedrpt), around 78 manufacturers are BIS certified for using BIS Standard Mark as per IS 17782: 2021, 3 manufacturers are BIS certified as per IS 17853, and 13 manufacturers are BIS certified as per IS 17781.

The BIS plays a central role in institutionalizing the rice fortification ecosystem in India. By integrating scientific rigor, regulatory oversight, and industrial compliance, BIS not only assures quality and safety but also underpins the success of national food security interventions such as TPDS, ICDS, and PM POSHAN. Its collaboration with stakeholders from FRK manufacturers to food processors and state procurement agencies ensures that FR delivered to millions of households is both nutritionally adequate and safe for consumption. The standards developed and enforced by BIS serve as the technical backbone of India's rice fortification program, and they are instrumental in supporting the broader national agenda of eradicating hidden hunger and achieving nutritional security for all.

2.3.4 The National Nutrition Strategy (NNS) and Its Alignment with Fortification

National Institution for Transforming India (NITI) Aayog published the NNS in 2017 with the goal of ensuring that women, children, and teenage girls, especially those from vulnerable groups, receive the best possible nutrition. The NNS key targets include reducing undernutrition in children (0–3 years) by 3% annually and cutting anemia in young children, adolescent girls, and women (15–49 years) by one-third from NFHS-4 levels by 2022 (NNS 2017). The NSS is being implemented based on key principles, viz., life cycle approach focusing on critical stages of nutrition, early prevention of undernutrition across all ages, inclusive and gender-sensitive, promoting the rights of women and children, community empowerment, encouraging families to adopt better nutrition practices and participate in monitoring, support for frontline workers like Anganwadi workers and ASHAs, decentralized and flexible planning tailored to local needs, active role of local bodies like Panchayats and urban authorities, encouragement of innovation, use of science and evidence, and ensuring no conflict of interest (NNS 2017).

The Ministry suggested making rice fortification mandatory in government programs, and it started in phases from May 2016 through FCI. In March 2018, as part of the NNS, NITI Aayog, in collaboration with the FCI and the DFPD, proposed the fortification of rice and wheat flour to enhance the nutritional content of foods distributed under the ICDS and the MDM programs. The National Nutrition Mission (NNM), now known as Poshan Abhiyan, formerly aligned with the MDM Scheme, focuses on improving the nutritional outcomes of school-going children, thereby supporting both physical growth and cognitive development. The Prime Minister has given approval for the extension of the PM POSHAN scheme, aimed at providing nutritional support to schoolchildren, for a five-year duration covering the period from 2021–2022 to 2025–2026. The financial outlay for the PM POSHAN scheme over the five-year period amounts to ₹130,794.90 crore, making it one of the most significant investments in child nutrition and education in the country (Press Information Bureau, Government of India 2024).

2.4 KEY INDIAN GOVERNMENT PROGRAMS ON RICE FORTIFICATION

As part of the "Anemia Mukt Bharat" initiative, the Government of India prioritizes the nationwide distribution of FR through a number of social welfare schemes in a phase-wise manner, with full implementation targeted by March 2024. This initiative aims to combat anemia and improve the overall nutritional status of the population, particularly among vulnerable groups. In Phase I (2021–2022), FR was launched under key social welfare programs such as ICDS and PM POSHAN, formerly known as the National Programme of Mid-Day Meals in Schools. This phase ensured uniform implementation of FR distribution across India under these schemes. Phase II, completed by March 2023, expanded the distribution to aspirational districts and districts identified as having a high burden of anemia, through various welfare schemes including PDS. This was in addition to the areas already covered under Phase I. In Phase III, the remaining districts across the country were brought under the FR distribution program. By March 2024, the phased rollout was completed, ensuring nationwide coverage of FR across all social welfare schemes (Nandeep et al. 2024).

2.4.1 Pradhan Mantri Garib Kalyan Anna Yojana (PMGKAY)

Overview of the Program

The PMGKAY, or the Prime Minister's Food Security Scheme for the Poor, was introduced by the Government of India on 26 March 2020, as part of the *Atmanirbhar Bharat* (Self-Reliant India) campaign. Launched in response to the COVID-19 pandemic, this welfare scheme aims to provide free food grains to economically disadvantaged individuals, ensuring their access to essential nutrition during challenging times (myScheme Portal, 2024). In March 2020, the Government of India announced the free distribution of additional foodgrains (Rice and Wheat) to around 80 crore NFSA beneficiaries (Antyodaya Anna Yojana (AAY) households, Priority Households (PHH) beneficiaries and Other Welfare Scheme) at the scale of 5 kg per person per month under the PMGKAY, in addition to regular entitlements of their *Ration Cards* under NFSA. The scheme was initially launched for twenty-eight months and implemented in seven different phases starting from April 2020 and continued till December 2022. According to DFPD, Government of India, a total of about 1015 LMT of foodgrains were distributed under this scheme with expected financial burden of about 3.91 lakh crore to ensure food security of NFSA beneficiaries in the country. In order to ensure effective implementation, this program was further extended by the Government of India for one year from January 2023 to December 2023.

Keeping in mind of welfare of the recipients, the Government of India had decided to extend the free distribution of foodgrains under PMGKAY to approximately 81.35 crore recipients for the next five years, starting from 1 January 2024, in accordance with their entitlements.

Integration of Rice Fortification with PMGKAY

The integration of rice fortification with the PMGKAY marks a significant step toward addressing India's long-standing challenge of micronutrient deficiencies, particularly iron-deficiency anemia. The Government of India extended the FR distribution under PMGKAY and other schemes from July 2024 to December 2028. As a centrally funded initiative, this extension is covered under the previously approved ₹11.79 lakh crore allocation for PMGKAY. According to the Press Information Bureau (2024), over 81 crore beneficiaries will continue to receive free, nutrient-enriched rice, reinforcing the country's commitment to nutrition security. This integration not only ensures caloric sufficiency but also addresses the hidden hunger that disproportionately affects the poor, women, and children.

While the initiative holds transformative potential, challenges remain in terms of supply chain readiness, quality control, and public awareness. Ensuring uniform fortification standards across rice mills, regular testing, and sensitization campaigns are crucial for long-term success. Nevertheless, this effort marks a paradigm shift from mere food security to nutrition security, leveraging the existing public distribution framework for enhanced public health outcomes.

2.4.2 Public Distribution System (PDS)

Overview of the Program

The Parliament of India enacted the NFSA in September 2013 with the objective of legally guaranteeing access to affordable food for a significant portion of the population. The legislation seeks to ensure food security by providing subsidized food grains to approximately two-thirds of the country's citizens. PDS serves as a supplementary food security mechanism in India, aimed at providing essential foodgrains at subsidized prices to the population. Under this system, commodities (i.e. rice, wheat, sugar, kerosene, etc.) are distributed to beneficiaries. Additionally, some states also include items like pulses, iodized salt, and edible oil in their PDS offerings. These essential goods are made available to eligible households through designated outlets known as Fair Price Shops (FPS), also known as government-authorized distribution centers or Ration Shops (Nandeep et al. 2024).

The FCI manages procurement, storage, transport, and allocation of foodgrains to different States. State Governments have the responsibility of PDS operations, including beneficiary identification, ration card issuance, intra-state distribution, and FPS management. The current PDS allocation includes wheat, rice, sugar, and kerosene to States and UTs (National Food Security Portal, Department of Food and Public Distribution, Government of India).

Fortification of Rice in the PDS

For reducing the risk of anemia and other health issues associated with vitamin and iron deficiency, the Government of India has decided to distribute the FR through PDS. According to DFPD, the Government of India (Available at: https://sansad.in/getFile/annex/263/AU74.pdf?source=pqars) launched a Centrally Sponsored Pilot Scheme on Rice Fortification and Distribution under PDS in 2019, initially covering the period 2019–2020 to 2021–2022. Initially, fifteen states of the country agreed to implement the Pilot Scheme in their designated districts. Under the Pilot Scheme, 4.30 lakh metric tons (LMT) of FR were distributed through PDS between 2019–2020 and 2021–2022. The scheme was further extended by Government of India to further extend the FR distribution through PDS throughout India in three phases.

Challenges in Ensuring Uniformity and Quality in PDS

Analyses of the PDS reveal key implementation issues, including inaccurate beneficiary identification and significant leakages. Studies show around 61% exclusion errors and 25% inclusion errors, meaning many poor households are left out while non-poor ones are included. Additionally, food grains often leak during transport and from ration shops into the open market. There are several concerns related to trends in food grain procurement. Currently, the central government procures about one-third of the foodgrains produced, but this share is expected to rise under the new Act, raising questions about sustainability and financial viability. The food subsidy already places a heavy burden on the government, as procurement and distribution costs are nearly six times higher than the sale price. With increased procurement, the subsidy and related expenses are likely to grow.

A study conducted by Nandeep et al. (2024) analyzed the distribution and consumer perception of FR provided through the PDS. The research involved interviews with various stakeholders, including Fair Price Shop dealers, District Supply Officers of the PDS, warehouse supervisors from the State Food Corporation (SFC) and FCI, as well as the beneficiaries. The findings revealed that the majority of beneficiaries consumed the FR supplied via PDS, either fully or partially. Approximately 67.8% of beneficiaries observed that one or two FRK floated during washing. Notably, all respondents from Uttar Pradesh reported the presence of floating FRK, while none of the beneficiaries in Jharkhand observed this phenomenon. Among those who noticed floating FRK, none in Uttar Pradesh discarded them, whereas in Maharashtra, up to 90.5% (with an average of 27%) of beneficiaries discarded the floating kernels. Regarding the FR preparation, beneficiaries reported that roughly 10% of the 40 FRK added per 250 g of raw rice floated during washing. Over 90% of beneficiaries did not report any color change in FR. Texture remained unchanged for most, except in Andhra Pradesh. Approximately 3.3% observed a difference in cooking, while 18.8% perceived a variation in taste compared to unfortified rice.

2.4.3 Mid-Day Meal (MDM) Scheme

Fortification Efforts in School Meal Programs

The MDM program in India, which provides free school meals to the schoolchildren, is the largest school meal program in the world, designed to deliver adequate calories and essential nutrients to the majority of Indian children (Mohapatra et al. 2023). The MDM scheme itself started back in 1995 as the National Programme of Nutritional Support to Primary Education (NP-NSPE). The DFS program was first introduced in Tamil Nadu in 2004, aiming to address iron and iodine deficiencies among schoolchildren through the MDM scheme.

The Government of India has made a number of fortification efforts in school meal programs, including the distribution of double-fortified salt (DFS), FR, and fortified *atta*, to improve the nutritional profiles of schoolchildren. Since then, this program has expanded to other states across India. In September 2021, it was renamed as PM-POSHAN. The scheme's main goal is to improve the nutritional status of children while also encouraging greater school enrolment and regular attendance. According to Press Information Bureau, Government of India (2024), the Government of India has replaced custom-milled rice with FR in all the schemes including MDM Scheme, and by March 2024, 100% coverage of FR distribution was successfully achieved.

Impact of Rice Fortification on Child Nutrition

In India, adding FR to the MDM program improves overall health and cognitive function while also increasing child nutrition, especially in lowering anemia. According to studies, schoolchildren who frequently eat FR had higher hemoglobin levels and a lower prevalence of anemia.

A study was conducted by Mahapatra et al. (2021) to investigate the effect of FR on health of schoolchildren (6–12 years) who were provided FR through the MDM program in Gujarat, India. According to the study's findings, the schoolchildren having regular consumption of FR has an improvement of 0.4 g/dL in the average haemoglobin, 10% reduction in average anemia prevalence, and 11% improvements in average cognitive test scores. The study found no significant reduction in morbidity or improvement in school attendance attributable solely to FR. Mohapatra et al. (2023) conducted a study to assess the impact of FRK on schoolchildren across public schools in four of the eight blocks in Dhenkanal district, Odisha. The results showed a 1.63-fold improvement in anemia status and an increase of 0.18 g/dL in hemoglobin levels. The study concluded that fortified MDM can effectively improve anemia among school-aged children in real-world situations.

2.4.4 Integrated Child Development Services (ICDS)

Overview of the Program

The ICDS Scheme, initiated by the Government of India on 2 October 1975, is the largest community-based program globally targeting children below six years, pregnant and lactating women, and females aged 16–44. The scheme aims to enhance nutritional and health outcomes in children aged 0–6 years, support holistic development, reduce mortality, morbidity, malnutrition, school dropout rates, and facilitate interdepartmental coordination for effective child welfare implementation (Sachdev and Dasgupta 2001). The various services provided under ICDS Scheme include supplementary nutrition, education related to nutrition and health, pre-school education, improving the immune system, health check-up, and referral services (Integrated Child Development Services, Government of Maharashtra).

Fortified Rice and ICDS: A Strategic Partnership

The ICDS and FR are strategically linked in India's attempts to promote child nutrition and fight malnutrition. The distribution of FR through PDS, which includes ICDS, and its fortification was approved by the Government of India to prevent the malnutrition and health problems in children up to the age of 6 years, pregnant and nursing mothers, and women.

Anganwadi worker is an ICDS program's community-based voluntary frontline worker. She is selected from the community and plays an important role because she interacts with the beneficiaries frequently. The Anganwadi's profile, including her training, experience, abilities, attitude, and qualifications, greatly influences the output of the other ICDS program. The main hub for providing mothers and children with ICDS services is an Anganwadi. Typically, an Anganwadi serves 700 people in tribal areas and 1,000 people in both rural and urban areas. An Anganwadi worker provides services at the Anganwadi Centre (Joshi and Verma 2018).

Under the Wheat-based Nutrition Program of Saksham Anganwadi and Poshan 2.0 (formerly known as the ICDS), starting from the third quarter of the fiscal year 2021–2022, 100 percent FR has been distributed across all States and UTs in India. This distribution includes aspirational and high-burden districts, ensuring comprehensive nutritional support nationwide (Press Information Bureau 2022).

2.5 THE WAY FORWARD FOR RICE FORTIFICATION IN INDIA

2.5.1 Scaling Up Fortification Efforts

The Government of India decided to scale-up the uses of FR across the nation by March 2024 through a number of social welfare programs. Following the establishment of rice fortification Standards by FSSAI in 2016 and their subsequent gazetting in 2018, a number of states began to scale up rice fortification through various safety net programs.

At present, there is no specific scheme to connect rice millers with banks. However, a letter has been sent to the States asking them to help link millers with banks under the Atma Nirbhar Bharat Package for MSMEs. The extra cost for fortifying rice has been fixed at ₹0.73 per kg. This cost includes expenses like buying and transporting FRK, running costs, equipment wears and tear, annual interest on investments, interest on working capital, and quality checks (like lab tests and sample collection). Under the PMGKAY, an annual procurement target of 520 LMT of FR has been set. Currently, India has 1023 FRK manufacturers with a combined production capacity of 111 LMT per year, significantly exceeding the program's requirement of 5.20 LMT. Additionally, there are 232 premix suppliers having a production capacity of 75 LMT per annum, which is well above the 0.104 LMT needed for the scheme (Ministry of Consumer Affairs, Food and Public Distribution 2024). India's rice fortification ecosystem has grown considerably. Over 21,000 of the 30,000 rice mills that are now in service have blending machinery installed, allowing them to produce 223 LMT of FR per month. With multiple laboratories accredited by National Accreditation Board for Testing and Calibration Laboratories across India doing stringent quality testing on FR, testing infrastructure has also expanded.

PATH South Asia is helping scale up rice fortification in 11 Indian states and 1 UT to support the government's goal of reaching 800 million people with FR. Working with partners, PATH focuses on increasing awareness, access, and affordability to improve health and nutrition. This effort also supports people's ability to make informed, healthy choices. The approach can be used in other countries too. Fortifying rice in social programs is a big opportunity for improving nutrition in India, where many face MNDs. However, India's large population, complex governance, and diverse rice industry make it challenging. To address this, the WFP worked with national and state governments to run four major pilot projects, proving that FR can be successfully added to social programs. WFP also helped to set standards, support local production, and spread awareness. After years of work, the government decided in 2021 to include FR in all major food programs. Today, it is reaching over 400 million people, and India now produces enough fortified kernels on its own. This effort offers valuable lessons for improving nutrition worldwide (World Food Programme 2022).

2.5.2 Public Engagement and Education

Consumer education and awareness is an important part of any food fortification program. It is important to create consumer demand for fortified products, and education can help achieve this. If people are unaware of the health benefits of FR, they might not consume it in an open market, even if there is a plentiful supply. Therefore, in order to integrate FR into communities' diets, large-scale rice fortification programs must prioritize these groups, whether they be schoolchildren, pregnant women, community health workers, or others. In a similar vein, PATH, an international non-profit global health organization, raises knowledge of FR by dispelling myths, emphasizing its health advantages, and generating demand for its consumption through social behavior change initiatives. For instance, in India, PATH carries out efforts aimed at changing social behaviors, education, and information regarding FR. The purpose of these programs is to raise awareness among government representatives, community health workers of food SSNPs, teachers, cooks, owners of fair price stores, expectant and nursing mothers, and schoolchildren about the advantages of FR. Additionally, PATH operates across the whole supply chain for FR and provides training to private sector participants, such as rice millers, producers of FR kernels, and makers of blending machines, on how to maximize the production of FR that can be distributed to communities through SSNPs (PATH 2022).

In order to increase awareness of the advantages of FR, particularly among tribal populations, DFPD is collaborating with Gujarat, Maharashtra, Chhattisgarh, Madhya Pradesh, Jharkhand, Telangana, Rajasthan, and Kerala to conduct various types of workshops and seminars in districts and tribal areas with a high risk of Thalassemia and Sickle Cell Anemia (Press Information Bureau, Ministry of Consumer Affairs, Food and Public Distribution 2022).

In order to support and align stakeholders, FSSAI established the "'Food Fortification Resource Centre" (FFRC) as a nodal point, offering technical and implementation support and educating consumers about fortification, nutrition, and food safety. The FFRC was a specialized forum that promotes and supports widespread food fortification across India. It included guidelines on standards, food safety, technology, the manufacturing and procurement of premixes and equipment, quality assurance, and quality control for fortification. It included all of the information about food fortification, including government circulars, success stories, national and international experiences, the most recent technological advancements, and scientific proof (Institute of Competitiveness Report 2023).

On 20 November 2021, the Department of Food and Civil Supplies, Nagaland, along with the Centre for Health Research Innovation (CHRI) and the National Stock Exchange (NSE) Foundation, held an awareness campaign on FR under "Azadi Ka Amrit Mahotsav". The chief guest highlighted rice fortification as a key step to fight iron-deficiency anemia, now part of National Programs like POSHAN Abhiyaan and Anemia Mukt Bharat. The representative of CHRI explained that fortifying rice with nutrients like iron and folic acid improves health, especially since rice is a staple in India. The event encouraged people to consume FR, available through the public distribution system, and also included distribution of FR and food packets to beneficiaries (Nagaland State Portal 2021).

Similarly, FCI has launched an awareness campaign in Kakinada district of Andhra Pradesh, India, to educate the public about FR and address misconceptions surrounding it. Some groups of people are worried because they think that the FR that is given out by the PDS contains plastic. Because of this, some people have been taking out the fortified kernels when cleaning the rice before cooking (The New Indian Express 2025).

2.6 CONCLUSIONS

Rice fortification has emerged as a strategic public health intervention to address MNDs, particularly in countries like India where rice is consumed as a staple food. Government policies and programs play an important role in driving this initiative, from establishing regulatory frameworks and quality standards to ensuring equitable distribution through social welfare programs. The durability of rice fortification initiatives clearly depends on political will, strong monitoring systems, and community involvement, even though technical viability and cost-effectiveness are crucial. Governments must continue to scale fortification programs in accordance with national nutrition targets, invest in awareness efforts to boost public acceptance, and modify policies to solve operational issues. Policymakers can significantly improve population health outcomes and reduce hidden hunger by integrating rice fortification into larger food and nutrition security programs.

REFERENCES

Bureau of Indian Standards. 2016. *Bureau of Indian Standards Act, 2016*. https://www.bis.gov.in/the-bureau/bis-act-rules-and-regulations/.

Food Safety and Standards Authority of India. 2018. *Food Safety and Standards (Fortification of Foods) Regulations, 2018*. https://www.fssai.gov.in/upload/uploadfiles/files/Compendium_Food_Fortification_Regulations_30_09_2021.pdf.

Institute of Competitiveness India. 2023. "Report: Rice Fortification." https://www.competitiveness.in/wpcontent/uploads/2023/10/Report_Rice_Fortifcation_Web_version.pdf.

Joshi, K., and K. Verma. 2018. "Knowledge of Anganwadi Workers and Their Problems in Rural ICDS Block." *IP Journal of Paediatrics and Nursing Science* 1 (1): 8–14.

Mahapatra, S., S. K. Mishra, A. Patel, S. Patel, and R. Patel. 2021. "Micronutrient-Fortified Rice Improves Haemoglobin, Anemia Prevalence and Cognitive Performance among Schoolchildren in Gujarat, India: A Case-Control Study." *International Journal of Food Sciences and Nutrition* 72 (5): 690–703.

Ministry of Consumer Affairs, Food and Public Distribution. 2024. "Fortified Rice: Centre's Ambitious Initiative to Combat Micronutrient Deficiencies." Press release, October 15. https://pib.gov.in/PressReleasePage.aspx?PRID=2065835.

Ministry of Health and Family Welfare of India. 2021. *National Family Health Survey (NFHS-5) 2019–21*. https://ruralindiaonline.org/en/library/resource/national-family-health-survey-nfhs-5-2019-21-haryana/.

Ministry of Women and Child Development. 2020. "Distribution of Fortified Rice to Curb Malnutrition." Press release, March 3. https://www.pib.gov.in/PressReleasePage.aspx?PRID=1606112.

Mohapatra, M. K., S. K. Panda, S. Sahoo, and B. K. Mishra. 2023. "Effectiveness of Fortified Mid-Day Meal in Reducing Anemia among School Children in Dhenkanal, Odisha: A Quasi-Experimental Study." *Food and Nutrition Bulletin* 44 (2): 79–87.

Muralidharan, J., S. Seshadri, and S. Swaminathan. 2024. "Mandatory Fortification of Rice in the Public Distribution System in India: An Ethics Perspective." *Indian Journal of Medical Ethics* 9 (1): 26–30.

myScheme Portal. 2024. "Pradhan Mantri Garib Kalyan Anna Yojana (PM-GKAY)." https://www.myscheme.gov.in/schemes/pm-gkay.

Nagaland State Portal. 2021. "Awareness Campaign on Fortified Rice Conducted in Dimapur." https://ipr.nagaland.gov.in/awareness-campaign-fortified-rice-conducted-dimapur.

Nandeep, E. R., S. K. Mishra, S. Sahoo, and M. K. Mohapatra. 2024. "Implementation, Delivery, and Utilization of Iron Fortified Rice Supplied through Public Distribution System across Different States in India: An Exploratory Mixed-Method Study." *PLOS Global Public Health* 4 (8): e0003533.

PATH. 2022. "Four Principles for Large-Scale Rice Fortification." https://www.path.org/our-impact/articles/four-principles-for-large-scale-rice-fortification/.

Press Information Bureau, Government of India. 2024. "Nutritious Boost: Free Fortified Rice for a Healthier India." Press note, August 15. https://pib.gov.in/PressNoteDetails.aspx?NoteId=153272.

Press Information Bureau, Government of India. 2022. "Under Wheat Based Nutrition Programme..." Press release, July 27. https://pib.gov.in/PressReleasePage.aspx?PRID=1846133.

Sachdev, Y., and J. Dasgupta. 2001. "Integrated Child Development Services (ICDS) Scheme." *Medical Journal Armed Forces India* 57 (2): 139–43.

The New Indian Express. 2025. "FCI Launches Awareness Campaign on Fortified Rice." November 21. https://www.newindianexpress.com/states/andhra-pradesh/2024/Nov/21/fci-launches-awareness-campaign-to-educate-public-about-fortified-rice.

World Food Programme. 2022. "Proof Pilot: 9 Insights into India's Rice Fortification Pilot Scale Approach." https://www.wfp.org/publications/2022-proof-pilot-9-insights-indias-rice-fortification-pilot-scale-approach.

World Health Organization. 2018. *Guideline: Fortification of Rice with Vitamins and Minerals as a Public Health Strategy*. Geneva: WHO. https://apps.who.int/iris/bitstream/handle/10665/272535/9789241550291-eng.pdf.

3 Rice Fortification Technology
Issues, Challenges, and Possible Solutions

Hari Niwas Mishra, Chandrakant Genu Dalbhagat, and Shubham Mandliya

3.1 INTRODUCTION

Micronutrient malnutrition (MNM) is a global health challenge affecting a large population segment due to inadequate intake of the essential vitamins and minerals. It is also called ‘hidden hunger’ as the visual symptoms cannot be identified in an individual by the naked eye in the early stages. It is a major challenge in underdeveloped and developing countries because these deficiencies hinder physical growth and cognitive development, affecting health, learning capabilities, and productivity. As a result, they contribute to increased morbidity and mortality across the lifespan and hurt social and economic development. Around 2 billion of the global population is deficient in key micronutrients such as vitamin A, iodine, iron, and zinc (WHO 2025). Iron deficiency is a notable concern leading to anemia due to reduced hemoglobin levels and diminished oxygen-carrying capacity in the blood (Vishwakarma, Dalbhagat, and Mishra 2021). It is often referred to as Iron-Deficiency Anemia (IDA). As per the National Family Health Survey-5 (NFHS-5), the prevalence of anemia in India is at an alarming stage. The anemia in children (67.1%) aged 6–59 months and women (57.0%) has increased by 14.50% and 7.34%, respectively, compared to the anemia status of Children (58.6%) and women (53.1%) in NFHS-4% (International Institute for Population Sciences (IIPS) 2021). Even the World Health Organization (WHO) portrays a concerning picture, estimating that 37% of pregnant women and 40% of children under five are anemic on a global scale (WHO 2025). This issue needs urgent intervention and strategic planning to alleviate it and improve the nutritional status of the affected population.

Food fortification has emerged as a sustainable and cost-effective intervention that involves the deliberate addition of nutrients (macro and micronutrients) to commonly consumed staple foods. The staple foods include rice, wheat flour, salt, edible oil, and milk, which are consumed daily by the household. Staple food fortification is a scientifically supported intervention to prevent, reduce, and manage MNM and IDA. Rice, a staple food for half of the global population, provides a fortification opportunity as most families incorporate it into their daily diet at least once daily

DOI: 10.1201/9781003661245-3

(Dalbhagat and Mishra 2019). Brown rice (after shelling paddy) is a rich source of macro and micronutrients; however, in the milling process, especially during the polishing stage, the removal of the bran layer results in the loss of most vitamins and minerals. Also, most of the population preferred white milled rice over the brown or parboiled rice in their diet, which lacks the essential micronutrients. Therefore, fortifying rice with essential micronutrients (iron, folic acid, and vitamin B_{12}) offers a sustainable strategy to alleviate the prevalence of anemia (Dalbhagat and Mishra 2021b). Rice fortification can be achieved using different techniques such as coating, dusting, and extrusion. However, extrusion is advantageous as it retains more micronutrients during washing and cooking than coating and dusting (Dalbhagat and Mishra 2021a; Yusufali et al. 2015). Moreover, in dusting and coating, the whole rice or head rice is used for fortification, while in the extrusion processing, the broken rice is used to manufacture the fortified rice kernels (FRK). FRK is an extruded product with a high concentration of desired micronutrients such as iron, folic acid, and vitamin B_{12} (Dalbhagat and Mishra 2021a; Dalbhagat and Mishra 2019). This approach not only uses the broken rice, a byproduct of the rice milling industry, but also adds value to the product and lowers the cost of rice fortification (Mishra, Mishra, and Rao 2012; Ganachari et al. 2022).

The Government of India has initiated a centrally sponsored pilot scheme 'Fortification of Rice and its Distribution under Public Distribution System', mainly to supply the fortified rice throughout the Targeted Public Distribution System (TPDS) under the National Food Security Act (NFSA), Pradhan Mantri Poshan Shakti Nirman (PM POSHAN), Integrated Child Development Services (ICDS), and any other welfare schemes. The rice fortification scheme was initiated in 2019–2020 as a pilot scheme with a Rs. 174.64 Crore budget, supplying 4.30 lakh metric tonnes of fortified rice in 11 states till March 2022. In August 2021, the Prime Minister of India, announced rice fortification in the government schemes, boosting the rice fortification ecosystem in the country. Currently, 100% coverage of the distribution of fortified rice is achieved, and the milled rice is replaced by the fortified rice in the government schemes (Ministry of Consumer Affairs, Food and Public Distribution 2024). This visionary scheme addresses long-standing nutritional disorders such as MNM and IDA. It shows the commitment of the government to establish an infrastructure and supply chain network to effectively deliver the fortified rice to the 80 crore beneficiaries. The rice fortification program still has specific issues and challenges that must be identified and tackled for successful implementation. All the stakeholders, viz., Government, academia, industry, and non-government organizations, should work together to navigate through identified challenges deeply and optimize the potential of the program, fostering a more resilient and impactful approach in combating the MNM and IDA.

3.2 RICE FORTIFICATION

As rice is a staple food, it is a good vehicle to carry vitamins and minerals to the affected population. These micronutrients are infused into the FRK, which is blended into the normal rice, to make the fortified rice. Rice is economically accessible to low-income or below-poverty households; therefore, for targeting the mass population, rice fortification always has the potential to tackle and prevent MNM.

Rice fortification is a process in which the essential micronutrients, such as vitamins (folic acid and vitamin B_{12}) and minerals (iron, zinc), are added to the rice to address nutritional deficiencies (WHO 2025). It is done by blending the FRK into the normal rice in a specified ratio, such as 1:100 or 1:200 (BIS 2021a). FRK is the main product, which may be produced using dusting, coating, extrusion, or any other suitable process. In dusting and coating, the whole rice kernels are used to coat the micronutrients on the surface of the rice, while the extrusion process utilizes the broken rice flour and vitamin-mineral premix (VMP) to produce FRK. VMP is formulated so that the micronutrient used targets the specific nutritional deficiencies prevalent in the target population. Besides these two primary ingredients, water and additives are added to the formulations before the extrusion processes. The additives include stabilizers and emulsifiers to achieve specific quality improvement in the FRK, such as textural properties and cooking characteristics.

3.3 FORMULATION AND SPECIFICATION

3.3.1 Rice/Broken Rice

Rice or broken rice serves as the primary base material and significantly decides the properties of FRK. Broken rice is generally used for making rice flour as it is the byproduct of the rice mill industry, which generally contributes 10%–12% (Dalbhagat and Mishra 2021a). The variety of rice is also essential while preparing the FRK, as the varietal composition significantly affects the textural and cooking quality of FRK (Nithya, Dalbhagat, and Mishra 2022). The clean and impurity-free rice or broken rice (moisture content 11%–12% w.b.) is milled into rice flour with particle size generally less than 250 μm, preferably less than 150 μm (BIS 2021b). Rice flour is hygroscopic and very susceptible to insect infestation; therefore, it is handled carefully to control the moisture migration and contamination. The contaminated or infested broken rice should be discarded from raw material storage and prohibited from being used in manufacturing FRK. While designing the formulation, the broken rice should be characterized, especially for the mineral and vitamin contents. The initial composition of broken rice flour is used to calculate the amount of VMP.

3.3.2 Vitamin-Mineral Premix

VMP is one of the most essential ingredients on which the success of the rice fortification program is based. If the VMP is not of good quality, the entire FRK batch and fortified rice production would be wasted, and no health benefits would be observed for the beneficiary. The VMP is a formulated mixture of micronutrients such as vitamins, minerals, carrier (base) material, and flow-improving ingredients. The selection, suitability, and amount of micronutrients in VMP depend on the stability during processing, cooking and storage, interaction with the rice flour and other nutrients, bioavailability, sensory acceptability, target group, and price. The well-prepared VMP shows a homogeneous distribution of micronutrients within it. The FRK manufacturer should buy the VMP from a vendor who has a valid FSSAI license of 99.5 category. Moreover, the VMP manufacturer should declare that the VMP is per the FSSAI guidelines and for FRK manufacturing.

The formulation of VMP should use food-grade ingredients and comply with the specifications laid by the FSSAI. The certificate of analysis (COA) should be provided to the FRK manufacturer, indicating the source of materials, particle size of the iron source, overages of minerals and vitamins, expiry date, and all other information suggested in the FSSAI guideline. The COA of VMP for FRK manufacturing should also be shared with the State Food Safety Authorities and District Officers for final approval (Department of Food and Public Distribution 2020). The VMP is required to produce FRK, and its micronutrient composition is well described in the IS 17781: 2021 Vitamin and Mineral Premix for Manufacturing Fortified Rice Kernels—Specification (BIS 2021c). The VMP formulation includes carrier material (mostly corn starch, rice starch), micronized ferric pyrophosphate, folic acid, cyanocobalamin, and permitted additives such as antioxidants and flow agents. The concentration of vitamins and minerals is important for calculating the quantity of VMP required for FRK manufacturing. The amount VMP (g) is calculated (Eq. 3.1) based on the amount of concentration of micronutrients in VMP (M_{VMP}) and the final concentration of the micronutrients in the FRK (M_{FRK}).

$$\text{VMP} = \frac{M_{\text{FRK}}}{M_{\text{VMP}}} \tag{3.1}$$

where VMP is the quantity of VMP (g) per 100 g of rice flour, M_{FRK} is the desired concentration of micronutrient in the FRK (mg/100 g), and M_{VMP} is the actual concentration of micronutrient in 1 g of VMP (mg/g).

3.3.3 Water

Water serves as a solute that penetrates the rice flour, i.e., starch structure, and helps gelatinization during extrusion. Potable water should be used to manufacture the FRK as set by the standard IS 10500:1991 (BIS 1991). Water quality should be checked regularly, and detailed reports should be maintained, complying with the specified standard. This ensures the quality and safety in the FRK manufacturing process. The quantity of water required for conditioning the rice flour and VMP mix can be calculated using the Eq. 3.2:

$$\text{W} = \frac{\text{F}_{\text{RFM}}\left(x_f - x_i\right)}{\left(100 - x_f\right)} \tag{3.2}$$

where W is the amount of water to be added to the rice flour–VMP mix (kg), F_{RFM} is the quantity of rice flour–VMP mix (kg), x_i is the initial moisture content of rice flour–VMP mix (% w.b.), and x_f is the final (desired) moisture content (% w.b.) in the wet and conditioned rice flour–VMP mix.

3.3.4 Additives

Permitted additives, such as emulsifier/acid regulators/antioxidants/hydrocolloids, can be added to the FRK formulation to improve its quality, such as cooking and textural properties (BIS 2021b).

3.4 TECHNOLOGIES FOR RICE FORTIFICATION

There are three potential techniques for rice fortification by which the micronutrient status in rice can be improved. Vitamins and minerals can be placed on the surface of the rice, or they may be infused into the newly produced kernels, such as FRK. The methods are coating, dusting, and extrusion processing.

3.4.1 Coating

The coating process refers to applying a high concentration of micronutrients on the surface of the rice kernel with a water-resistant edible coating. The coated kernels are mixed with the normal rice in a specific ratio of 1:50 to 1:200 to make fortified rice (WHO 2018). The coating material includes the edible waxes and gums where the micronutrients are dissolved, and the spray solution is prepared (Yusufali et al. 2015; Mridula and Pooja 2014). The gums and waxes are crucial in fixing the micronutrient layer on the rice surface. The spraying of micronutrient mixes on the rice surface is done in a large rotational drum or pan coating machine (Mridula and Pooja 2014; Shrestha, Arcot, and Paterson 2003). After this coating phase, the kernels undergo careful drying, producing fortified kernels (Steiger et al. 2014). The innovative aspect of this technology lies in its capacity to concentrate micronutrients on the surface of the rice grains. The major problem associated with the coating technique is that the color, taste, and appearance of the fortified kernels change, and consumers can easily distinguish them in the fortified rice during cleaning. Moreover, the loss of micronutrients also occurs during the washing and cooking process (WHO 2018).

3.4.2 Dusting

The dusting process involves blending polished rice grains in the dry powder form of micronutrient premix. The micronutrients adhere to the surface of the rice grains because of the electrostatic force between the micronutrients and the rice. The problem of nutrient loss is significant in washing and cooking; therefore, a remark ‘Do not wash before cooking’ on the fortified rice package should be added (WHO 2018; Steiger et al. 2014). Dusting is not recommended in countries where washing of fortified rice before cooking and draining off excess water during cooking is followed. Moreover, there is a possibility of segregation of fortified kernels in the fortified rice (Steiger et al. 2014).

3.4.3 Extrusion

Extrusion is a versatile process that involves mixing, kneading, shearing, cooking, shaping, and forming (Dalbhagat, Mahato, and Mishra 2019). It is carried out using equipment called an extruder. The extruder has barrel, screw, heaters, feeders, and die-cutter assembly (Gujral et al. 2012; Xu et al. 2016). During extrusion, starch undergoes gelatinization, proteins denature, and lipids, starch, and proteins form complexes, leading to changes in microstructure, chemical properties, and the overall shape of the product (Hagenimana, Ding, and Fang 2006; Chen et al. 2022).

For FRK manufacturing, mainly twin-screw extruders (TSEs) are used (Nithya et al. 2024; Dalbhagat and Mishra 2021). The extrusion process may be hot, warm, or cold. This method effectively produces FRK from the conditioned rice flour–VMP mix, which embeds the vitamins and minerals inside the FRK. Therefore, when blended in normal rice, the FRK produced through the extrusion process becomes visibly indistinguishable. Extrusion also provides surface appearance to FRK, such as sheen and transparency in hot extrusion due to full or partial cooking, while opaque in warm and cold extrusion due to partial to no cooking (Yusufali et al. 2015; WHO 2018). In FRK manufacturing, the critical parameters are moisture content, mechanical energy input, thermal energy input, and residence time of material inside the barrel, which decides the degree of gelatinization in FRK (Mishra, Mishra, and Rao 2012). The extrusion process is, therefore, a sophisticated and effective technique to produce FRK.

3.5 MANUFACTURING FORTIFIED RICE KERNELS THROUGH EXTRUSION TECHNOLOGY

The manufacturing process for FRK through the extrusion process is illustrated in Figure 3.1.

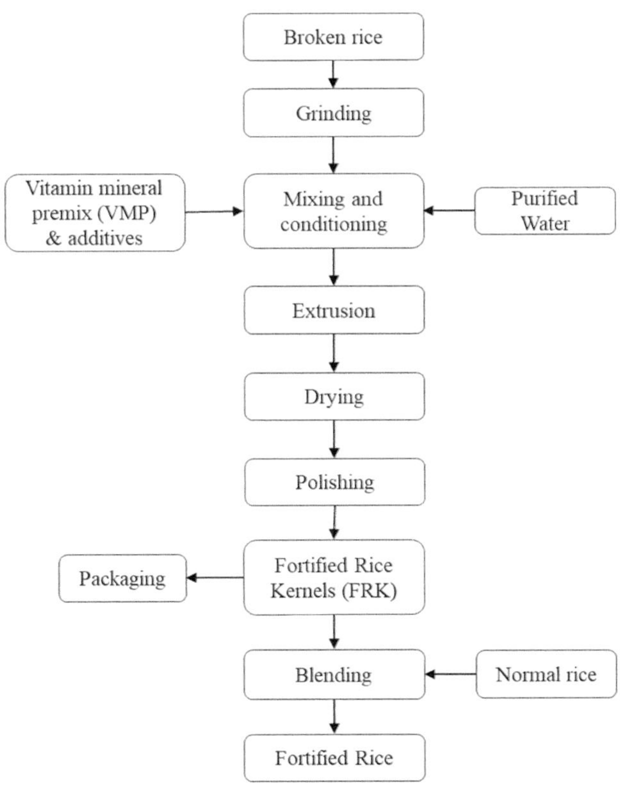

FIGURE 3.1 Process flowchart for the manufacture of FRK and fortified rice.

The broken rice, procured from the local rice mill, is first ground into flour, and then desired amounts of water and VMP are uniformly mixed and conditioned to prepare the rice flour and VMP mixture. The mixing is done to distribute the micronutrient and water evenly throughout the conditioned mix. This mixture is fed into a TSE, specially designed and fabricated for the purpose, with optimized production parameters. At the end of the extruder, a rice-shaped die (specially designed) is used to give rice shape to FRK. After extrusion, the FRK are dried to achieve the desired storage moisture content. Finally, the FRK are blended with the normal rice to make fortified rice. The pilot-scale facility (capacity of 100 kg per day) for manufacturing FRK and fortified rice was established by IIT Kharagpur, West Bengal, India, with financial support from the Department of Biotechnology, Government of India (Figure 3.2). The facility was named the 'Iron Fortified Rice Manufacturing Unit' because it was established to conduct research and development trials to manufacture iron fortified rice. Later, the same was used to develop the process technology for FRK, and it is now being used to improve the process technology and quality of FRK and fortified rice. The schematic of the continuous FRK manufacturing plant designed by IIT Kharagpur can be seen in Figure 12.1.

3.5.1 Grinding

Broken rice or rice (raw/parboiled) is ground in a micropulverizer, i.e., hammer mill, to obtain the fine rice flour, as suggested in IS 17782:2021 (BIS 2021b). Before

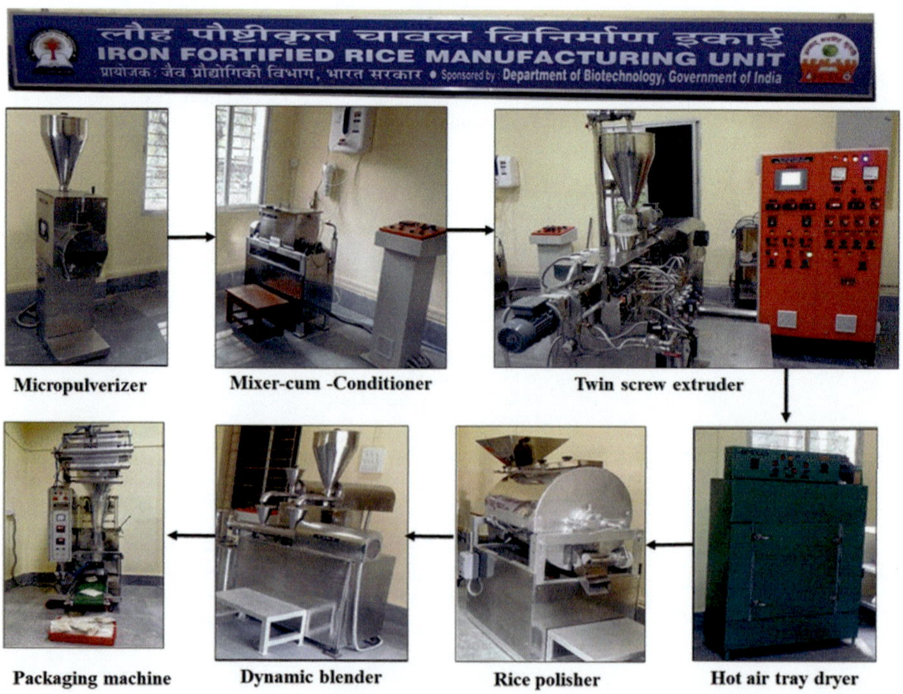

FIGURE 3.2 Iron Fortified Rice Manufacturing Unit, IIT Kharagpur, West Bengal, India.

grinding, the broken rice should be cleaned for foreign matters such as metal, stone, chaff, and plastic. The hammer mill works on the principle of impact to reduce the size of the material. The mill consists of a high-speed rotating hammer inside the grinding chamber with a screen at the bottom. The hammers hit the grains at very high speeds due to the impact on the wall of the grinding chamber. The feed rate of broken rice/rice is controlled to obtain good quality of rice flour. The fineness of the rice flour is regulated by using different-sized stainless steel screens. All the contact parts of the hammer mill should be food-grade stainless steel (SS304).

3.5.2 Mixing and Conditioning

Mixing of the rice flour and VMP is performed in the paddle/ribbon mixer-cum-conditioner to distribute the micronutrient uniformly throughout the rice flour–VMP blend. Mixing VMP with rice flour is important as the micronutrient content in the FRK is dependent on the mixing process. Mixing is done by dry mixing rice flour and VMP, followed by adding water to moisten the rice flour–VMP blend, generally called 'wet mixing'. The reverse osmosis ultraviolet-treated water is used to condition the rice flour–VMP blend. Adding moisture to the blend is helpful in the extrusion process as it promotes the starch gelatinization process. The paddle/ribbon mixer-cum-conditioners are made up of food-grade stainless steel (SS304). It has a 'U' shape mixing chamber for single shaft and twin 'U' or 'W' shape chambers for twin shafts. The spray pipe is installed at the top to sprinkle water during conditioning. The top cover is provided to close the chamber during operation, while the discharge of the blend can be done by tilting the mixer or using a discharge gate provided at the bottom, which is operated using pneumatic valves.

3.5.3 Extrusion

The extrusion process is the heart of the FRK manufacturing process, in which the conditioned rice flour–VMP blend is fed to the extruder. For FRK manufacturing, the choice of extruder is mostly the TSE, because it provides better conveying and mixing abilities during processing. As heat-labile vitamins are added in FRK, the warm extrusion process is used with a gradually increasing temperature profile from feeding to the die head section. The heat is supplied to the material inside the barrel through the electric heater. At the same time, frictional heat is also generated due to interparticle friction within the material, material-screw, and material-barrel friction. Due to this heating mechanism, the starchy material melts and turns into a viscoelastic dough. This viscoelastic dough is extruded through the die with rice-shaped holes and cut to produce the FRK. The extrusion process is controlled by the operating parameters such as barrel and die head temperature, extruder screw speed, feed rate, cutter speed and feed material characteristics such as moisture content, particle size, variety of rice, and amylose-amylopectin content (Dalbhagat and Mishra 2019).

3.5.4 Drying

The drying process in FRK production is carried out to reduce the moisture content of extruded FRK to 12% w.b. and prevent microbial growth (BIS 2021b). It makes the FRK shelf stable and safe for storage and transportation. On a laboratory scale, the drying of FRK is done in a tray or microwave dryer; however, in commercial production, the continuous belt dryer is used to dry large quantities. The drying process needs to be adequately controlled to preserve the quality of FRK; otherwise, changes in surface color and fissure formation in the FRK occur. In drying, the hot air is generally used as a drying medium, which passes over the bed or thin layer of FRK either in a batch or continuous dryer. Monitoring air process parameters such as air temperature, relative humidity, flow rate, and drying time is essential to achieve the consistent quality of FRK. Properly dried FRK shows good surface appearance, color, no fissure formation, and minimum or nil nutrient losses.

3.5.5 Polishing

During cutting at the extruder die face, the FRK surface may carry fines attached, which required to be removed. The rough and dull FRK surface may attract the consumer's attention and may get identified in fortified rice. Therefore, the polishing process is employed to enhance the surface smoothness and appearance, and make FRK compatible with the normal rice. The polishing process is optional and carried out using a rice polisher in which the FRK are fed. It has an emery stone roller and a stainless steel screen at the bottom. The FRK are gently rubbed between the roller and screen to remove the fines and make the surface shiny and smooth. The polished FRK exhibits a similar appearance and texture to normal rice.

3.6 SPECIFICATIONS FOR FORTIFIED RICE KERNELS

The specifications of FRK are detailed in the standard IS 17782:2021 Fortified Rice Kernels Specification. The FRK quality is expected to closely match that of the normal rice in which it is intended to be blended. The physical dimensions (length, width, and thickness), density (bulk and true), thousand kernel weight, and color should be similar to normal rice. Moreover, the FRK should not have any off odor. The allowed variation in these physical attributes is ±10%. It also mentioned that the cooking characteristics should match normal rice, and solid losses should be less than 10%. The cooking time deviation should be within 10% of the normal rice in which it is to be blended. The micronutrient requirement for level fortification (mandatory micronutrients) is presented in Table 3.1.

The optional micronutrients include zinc, vitamin A, thiamine, riboflavin, niacin, and pyridoxin, which can also be added to the FRK. Compliance with the regulations ensures the production of high-quality fortified rice meeting the nutritional requirements expected.

3.7 BLENDING OPERATION TO PRODUCE FORTIFIED RICE

Fortified rice is a blend of FRK and normal rice so that the micronutrient content would be as per the specifications laid out by the regulatory bodies. Blending is done by mixing the FRK with rice in a ratio between 1% and 2%, depending on the concentration of the added micronutrients to the FRK (BIS 2021a). Uniform

TABLE 3.1
Requirements for the Level of Fortification for Mandatory Micronutrients

Micronutrient	Chemical Form of Micronutrient	Fortification Level for 1: 100 Blending Ratio	Fortification Level for 2: 100 Blending Ratio
Iron, mg/kg	Micronized ferric pyrophosphate	280–425	140–212.5
	or		
	Sodium iron (III) ethylenediamine tetraacetate trihydrate (Sodium Feredetate—NaFeEDTA)	140–212.5	70–106.25
Folic acid, μg/kg	Folic acid	750–1250	375–625
Vitamin B_{12}, μg/kg	Cyanocobalamin or hydroxocobalamin	7.5–12.5	3.75–6.25

Source: BIS (2021b).

distribution of FRK in the fortified rice is desired during the blending process. The blending ratio and homogeneity of FRK need to be checked during the production of fortified rice. This process is carried out either in batch or continuous mode. The selection of a blending system depends upon factors such as blending capacity, suitability, gentleness of blending, low maintenance, cost of blender, time of blending, and homogeneity. The paddle, ribbon, and fluidized bed blenders are used in batch blending. The preweighed FRK and normal rice are fed into the batch blenders and operated at the optimized blending speed and time to obtain the desired blending ratio and homogeneity. A continuous (dynamic) blending system is used to meet the large demand for fortified rice. The dynamic blender, i.e., FRK and normal rice blending unit installed at the Iron Fortified Rice Manufacturing Unit, IIT Kharagpur, West Bengal, can be seen in Figure 10.3. The automatic feeders also employed in both batch and dynamic blenders, which include vibratory, belt conveyor, and screw feeders.

3.8 PACKAGING AND STORAGE OF FRK AND FORTIFIED RICE

FRK and fortified rice packaging can be done manually or using automatic packaging machines. The FRK are filled and sealed manually in polypropylene or woven bags, maintaining durability and moisture resistance. The form, fill, and seal (FFS) packaging machines may also be used to pack the FRK pouches. The FFS packaging machine provides an automatic dispensing, forming, and sealing mechanism for multilayer laminates and metallized films, which provide a robust barrier against moisture and oxygen. The non-transparent packaging film provides a barrier against the light. Stringent personal safety and hygiene practices are adhered to during the packaging process. Specific storage conditions (temperature, relative humidity, and air ventilation) are required to store the packaged FRK in a designated space. The BIS (2021b) suggested the storage of FRK at a temperature of 25°C or less and a relative humidity of less than or equal to 60%. Regular quality evaluation of FRK for shelf-life assessment is necessary to guarantee consumers sustained nutritional efficacy and safety of the fortified rice.

BIS (2021a) suggested packing the fortified rice in food-grade bags such as food-grade jute (IS 12650) and high-density polyethylene (HDPE)/polypropylene (PP) woven sacks (IS 16208). The bags shall have a heat-cut mouth to prevent fibrillation, and a sewn, single-fold bottom enhances durability. The bags should be double-stitched with strong thread to prevent microbial contamination and insect infestation (BIS 2021a). The fortified rice blender should follow the Legal Metrology (Packaged Commodities) Rules 2011 and the Food Safety and Standards Act, 2006, to pack the fortified rice in a stipulated quantity. The storage of the fortified rice should be done in a dedicated space where care has been taken to control rodents and pest attacks. The storage environment may be similar to FRK; however, maintaining environmental conditions for such large quantities of fortified rice is quite difficult.

3.9 QUALITY ASSURANCE

Stringent good manufacturing practices (GMP) and good hygienic practices (GHP) are involved in manufacturing FRK and fortified rice, ensuring safe and high-quality products. The entire process is aimed at maintaining the nutritional integrity and safety of the final product. The manufacturing facility design is critical to implement the GMP and GHP during the production process, as the facility should have dedicated space for different activities in the plant, such as production, storage of raw ingredients and finished products, storage rooms, washrooms, storerooms, laboratories, and offices. It should have proper cleaning, maintenance, and clear demarcation to prevent cross-contamination. Central to this ethos is a focus on personal hygiene among workers, with measures like regular handwashing and protective clothing to mitigate contamination risks.

Cleaning and sanitization scheduling ensure thorough cleansing of equipment and contact surfaces, instruments and utensils used in ingredient preparation with food-grade cleaning, and sanitization agents to eliminate potential threats from bacteria, viruses, or contaminants. The quality of raw materials such as broken rice, VMP, additives, and water needs to be rigorously checked to meet the specified standards. Maintenance of equipment and instruments, process control, and adherence to critical production process parameters ensure the consistent quality of the FRK and fortified rice. Storage and handling protocols prevent spoilage, contamination, or degradation by following the first-in, first-out (FIFO) approach for the raw ingredients. It manages inventory and reduces the risk of using expired or outdated raw ingredients.

Quality assurance involves routine testing of raw ingredients, finished products, equipment calibration, employee training on GMP and GHP, personnel hygiene practices, and different regulations and standards. A standardized traceability system enables tracing raw materials throughout production to identify quality and safety issues. The integrity of the packaging material is ensured to check for contamination and preventing damage. These practices help to produce high-quality products and ensure the safety and nutritional efficacy of the FRK and fortified rice. Audits and inspections improve the process and quality management system and maintain high-quality products.

3.10 SUPPLY CHAIN OF FRK AND FORTIFIED RICE

The supply chain is important in delivering fortified rice to the consumer. Different approaches may be adopted to develop the supply linkages for fortified rice. The supply chain begins with procuring raw materials, including rice/broken rice, VMP, additives, and other ingredients, to manufacture FRK and fortified rice and its distribution (Muthayya et al. 2014). The rice fortification involves two major processes: manufacturing FRK and blending FRK in normal rice. Therefore, two supply chain linkages are exhibited, one for FRK and the other for fortified rice. In the FRK supply chain, broken rice might be procured from a rice mill, or the rice mill may have its manufacturing unit for FRK, while the VMP and additives may be procured from vendors who comply with the set of government regulations. The FRK is then supplied or used to produce fortified rice, which is distributed to the consumer. The fortified rice supply may have different approaches like (a) large mills that has its own FRK manufacturing and blending unit, (b) medium/small mills that procure FRK from other sources and blend at their unit, (c) safety net distribution unit, and (d) distribution point for retail sale where the FRK and normal rice are procured and blended. The blended fortified rice is then supplied to different government schemes or for retail sales. During the distribution, adequate storage and handling practices are necessary to maintain the integrity of fortified rice. At retail points, consumer awareness campaigns should be conducted to raise awareness about the health benefits of the fortified rice. A continuous monitoring and evaluation system should be established to track the effectiveness of fortified rice (Muthayya et al. 2012). Effective collaboration among stakeholders, including government agencies, manufacturers, distributors, and NGOs, is key to the success of the supply chain of fortified rice (Pee et al. 2018).

3.11 ISSUES, CHALLENGES, AND POSSIBLE SOLUTIONS

On a commercial scale, the FRK is manufactured using extrusion technology. Although the process technology is proven and validated, the rice fortification programs face various technological and implementation challenges due to operational difficulties, an inability to understand the process and equipment, and the lack of training for the operators, quality personnel, and government officials. Moreover, a comprehensive and cumulative approach is needed to implement this program across different regions. Active participation of all stakeholders is required to implement and comply with the regulatory standards while producing the FRK and fortified rice.

Technological issues arise for the manufacturing of FRK and fortified rice, which may include matching the different properties of FRK with normal rice, operation, repair, and maintenance of the production line of FRK, achieving homogeneity of FRK in fortified rice, quality analysis and control, etc. Manufacturers must understand that the physical, cooking, and textural properties of FRK must match those of normal rice. Moreover, the desired micronutrient content in FRK and fortified rice should follow government regulations. During FRK and fortified rice production, implementing GMP, GHP, and HACCP-like prerequisite programs is essential to prevent potential safety risks. The technological issues during the FRK and fortified rice manufacturing process, together with possible solutions, are listed in Table 3.2. The implementation issues are related to the stakeholders, such as FRK and the fortified rice manufacturer, machinery manufacturer, quality assurance and control (QA/QC), technology provider (academia), regulators (government officials), and consumer. In a manufacturing unit, the operators may not know the machinery

TABLE 3.2
Issues, Causes and Possible Solutions in the FRK Manufacturing Process

Issues	Causes	Possible Solution
• Non-uniform micronutrient (Fe, folic acid and vitamin B_{12}) contents in FRK	• Inefficient mixing process • Uneven distribution of micronutrients in VMP • Significant variation in the particle size of the flour and VMP • Flour clumps formation due to improper water dosing	• Optimize mixing process parameters for dry blending (rice flour and VMP) • Ensure quality and uniformity of micronutrients in VMP • Minimum deviation should be in the particle size of rice flour and VMP • A fine water spraying system in the blender should be used
• Non-uniform physical dimensions	• Improperly designed die • Poorly controlled extrusion process parameters (barrel and die head temperature, screw speed, moisture content and feed rate) • Die openings blockages	• Design a die according to the rice variety to be fortified • Optimize the extrusion process and maintain the process parameters during production • Ensure uniform moisture distribution in rice flour- VMP mixture
• Differences in the density of FRK and raw rice • Floating FRK in water	• Severe extrusion process (high temperature and low moisture) causes expansion of FRK • Cutting speed at die (High speed – thin FRK and low speed – thick FRK)	• Use optimum extrusion conditions (avoid high temperature and low moisture extrusion) • Optimize extrusion process parameters
• Colour of FRK (yellowish tint/ browning)	• Improper extrusion and drying process • Colour of ingredients, i.e., mostly VMP • Presence of foreign matter (organic/inorganic)	• Maintain the extrusion temperature below 100°C • Use low temperature drying (<40°C) • Use VMP with white/off-white in color and ensure the iron source is micronized
• Fissure development in FRK	• Improper drying process	• Use low temperature drying (<40°C)
• Segregation of FRK in fortified rice	• Variation in physical dimensions, density, and individual kernel weight of FRK and normal rice	• Use a properly designed die for FRK manufacturing • Maintain optimum extrusion process and cutter speed

and lack the skill to operate it. Without understanding the process and operational parameters, if the production continued, the quality of both products may be compromised, causing the product to be rejected in further processes. Additionally, the scarcity of QA/QC personnel, particularly for analyzing the micronutrient and operating the high-end instruments such as High-Performance Liquid Chromatography (HPLC), Inductively Coupled Plasma-Mass Spectrometry (ICP-MS)/optical emission spectroscopy (ICP-OES), and Atomic Absorption Spectrophotometer (AAS), causes problems in the supply chain of FRK and fortified rice. Moreover, the limited access to these facilities for small manufacturers also hampers the production of both products.

Sometimes, FRK manufacturers compromise on the quality of FRK to curtail production costs by using low-quality ingredients. Such products then fail in the quality checks at the procurement point, causing economic losses. The complication further increases due to non-standard machinery, which may not align with the process and specific requirements of FRK and fortified rice manufacturing according to set regulations. Moreover, the ongoing rice fortification initiative needs continuous monitoring, evaluation, and long-term commitment from various stakeholders. Comprehensive skill development training for all stakeholders, accessibility to analytical instruments and equipment, implementation of strict regulations, and consumer awareness campaigns are essential to address these technological and implementation challenges. Addressing these challenges, the success of the rice fortification scheme can be ensured by promoting the health and well-being of the affected population on a larger scale.

3.12 CONCLUSIONS

Food fortification has proven to be a successful and widely accepted strategy for enhancing the nutritional status of staple foods and combating the prevalence of MNM globally. Notably, rice is a staple food for nearly half the world's population, particularly in countries like India, where it is a part of their tradition and culture. It presents a unique opportunity to become a vehicle for delivering the essential vitamins and minerals to the affected population. Using innovative technologies such as extrusion has shown excellent outcomes in manufacturing the FRK, providing an opportunity to produce a large quantity of fortified rice. The selection of raw ingredients (broken rice, VMP, additives, and water) and technology (extrusion, dusting, and coating) is crucial in manufacturing the FRK. Different issues and challenges arise in implementing the rice fortification program and underscore the complexities involved, which indicates the need for a nuanced and strategic approach. On the way forward, the successful integration of fortified rice into global nutritional initiatives needs a concerted effort to address these challenges. This involves leveraging advanced technology, implementing GMP, GHP, quality control, and assurance protocols, establishing resilient supply chain practices, and devising targeted solutions. By doing this, the complexities of the rice fortification program can be resolved to meet the global nutritional goals. The journey of rice fortification encompasses continuous innovation and collaboration to ensure the nutritional requirements of the diverse population worldwide.

REFERENCES

BIS. 1991. *Indian Standard Drinking Water — Specification (First Revision)*. Vol. 25. Bureau of Indian Standards, New Delhi.

BIS. 2021a. *Fortified Rice — Specification (IS 17780 :2021)*. Bureau of Indian Standards, Government of India, New Delhi.

BIS. 2021b. *Fortified Rice Kernels — Specification (IS 17782:2021)*. Bureau of Indian Standards, Government of India, New Delhi. www.standardsbis.in.

BIS. 2021c. *Vitamin and Mineral Premix for Manufacturing Fortified Rice Kernels — Specification (IS 17781:2021)*. Bureau of Indian Standards, Government of India, New Delhi.

Chen, Fei, Yan Lu, Lixu Pan, Xiaolei Fan, Qianfeng Li, Lichun Huang, Dongsheng Zhao, Changquan Zhang, and Qiaoquan Liu. 2022. "The Underlying Physicochemical Properties and Starch Structures of Indica Rice Grains with Translucent Endosperms under Low-Moisture Conditions." *Foods* 11 (10). https://doi.org/10.3390/foods11101378.

Dalbhagat, Chandrakant Genu, and Hari Niwas Mishra. 2019. "Effects of Extrusion Process Conditions on System Parameters; Physicochemical Properties and Cooking Characteristics of Extruded Fortified Rice Kernels." *Journal of Cereal Science* 89: 102782. https://doi.org/10.1016/j.jcs.2019.05.016.

Dalbhagat, Chandrakant Genu, Dipendra Kumar Mahato, and Hari Niwas Mishra. 2019. "Effect of Extrusion Processing on Physicochemical, Functional and Nutritional Characteristics of Rice and Rice-Based Products: A Review." *Trends in Food Science and Technology*: 226–40. https://doi.org/10.1016/j.tifs.2019.01.001.

Dalbhagat, Chandrakant Genu, and Hari Niwas Mishra. 2021a. "Drying Modeling, Cooking Characteristics, Pasting Properties, and Crystallinity of Fortified Rice Kernels." *Journal of Food Processing and Preservation* 45 (6): 1–9. https://doi.org/10.1111/jfpp.15579.

Dalbhagat, Chandrakant Genu, and Hari Niwas Mishra. 2021b. "Effect of the Drying Process on the Color Change, Fissure Development, and Morphology of Fortified Rice Kernels." *Journal of Food Process Engineering* 44 (7): e13719. https://doi.org/10.1111/jfpe.13719.

Department of Food and Public Distribution. 2020. "Fortification of Rice and Its Distribution under Public Distribution System: Operational Guidelines." https://cdnbbsr.s3waas.gov.in/s3e21e4e58ad9ab56e8a4634046da90113/uploads/2023/06/2023060963.pdf

Ganachari, Ambrish, Udaykumar Nidoni, Sharanagouda Hiregoudar, K. T. Ramappa, Nagaraj Naik, S. Vanishree, and P. F. Mathad. 2022. "Development of Rice Analogues Fortified with Iron, Folic Acid and Vitamin A." *Journal of Food Science and Technology* 59 (9): 3474–81. https://doi.org/10.1007/s13197-021-05337-5.

Gujral, Hardeep Singh, Paras Sharma, Arvind Kumar, and Baljeet Singh. 2012. "Total Phenolic Content and Antioxidant Activity of Extruded Brown Rice." *International Journal of Food Properties* 15 (2): 301–11. https://doi.org/10.1080/10942912.2010.483617.

Hagenimana, Anastase, Xiaolin Ding, and Tao Fang. 2006. "Evaluation of Rice Flour Modified by Extrusion Cooking." *Journal of Cereal Science* 43 (1): 38–46. https://doi.org/10.1016/j.jcs.2005.09.003.

International Institute for Population Sciences (IIPS). 2021. "National Family Health Survey (NFHS-5) India 2019–21." *Mumbai: International Institute for Population Sciences* I:1–714.

Ministry of Consumer Affairs, Food and Public Distribution. 2024. *Nutritious Boost: Free Fortified Rice for a Healthier India* https://www.pib.gov.in/PressNoteDetails.aspx?NoteId=153272.

Mishra, Abhinav, Hari Niwas Mishra, and Pavuluri Srinivasa Rao. 2012. "Preparation of Rice Analogues Using Extrusion Technology." *International Journal of Food Science and Technology* 47 (9): 1789–97. https://doi.org/10.1111/j.1365-2621.2012.03035.x.

Mridula, D., and Jha Pooja. 2014. "Preparation of Iron-fortified Rice Using Edible Coating Materials." *International Journal of Food Science and Technology* 49 (1): 246–52. https://doi.org/10.1111/ijfs.12305.

Muthayya, Sumithra, Jessica Hall, Jack Bagriansky, Jonathan Sugimoto, Daniel Gundry, Dipika Matthias, Shane Prigge, Peter Hindle, Regina Moench-Pfanner, and Glen Maberly. 2012. "Rice Fortification: An Emerging Opportunity to Contribute to the Elimination of Vitamin and Mineral Deficiency Worldwide." *Food and Nutrition Bulletin* 33 (4): 296–307. https://doi.org/10.1177/156482651203300410.

Muthayya, Sumithra, Jonathan D. Sugimoto, Scott Montgomery, and Glen F. Maberly. 2014. "An Overview of Global Rice Production, Supply, Trade, and Consumption." *Annals of the New York Academy of Sciences* 1324 (1): 7–14. https://doi.org/10.1111/nyas.12540.

Nithya, A., Chandrakant Genu Dalbhagat, and Hari Niwas Mishra. 2022. "A Comparative Study on the Physicochemical, Cooking and Textural Properties of Fortified Rice Kernels Prepared from Raw and Parboiled Rice." *International Journal of Food Science and Technology* 57 (2): 1325–32. https://doi.org/10.1111/ijfs.15529.

Nithya, A., Siddharth Vishwakarma, Chandrakant Genu Dalbhagat, and Hari Niwas Mishra. 2024. "Apparent Amylose Content Positively Influences the Quality of Extruded Fortified Rice Kernels." *Carbohydrate Polymers* 338: 122213. https://doi.org/10.1016/j.carbpol.2024.122213.

Pee, Saskia de, Becky L. Tsang, Sarah Zimmerman, and Scott J. Montgomery. 2018. "Rice Fortification." In *Food Fortification in a Globalized World*, edited by M.g. Venkatesh Mannar and Richard F. Hurrell, 131–141. Academic Press, Elsevier. https://doi.org/10.1016/B978-0-12-802861-2.00013-4.

Shrestha, Ashok K., Jayashree Arcot, and Janet L. Paterson. 2003. "Edible Coating Materials—Their Properties and Use in the Fortification of Rice with Folic Acid." *Food Research International* 36 (9–10): 921–28. https://doi.org/10.1016/S0963-9969(03)00101-7.

Steiger, Georg, Nadina Müller-Fischer, Hector Cori, and Béatrice Conde-Petit. 2014. "Fortification of Rice: Technologies and Nutrients." *Annals of the New York Academy of Sciences* 1324 (1): 29–39. https://doi.org/10.1111/nyas.12418.

Vishwakarma, Siddharth, Chandrakant Genu Dalbhagat, and Hari Niwas Mishra. 2021. "Food-to-Food Fortification of Rice Flour (Swarna Cv.) Using Dried Basil, Marjoram, and Spearmint Leaves Powders: A Physicochemical and Nutritional Study." *Foods 2021*, 6:15. https://doi.org/10.3390/Foods2021-10947.

WHO. 2018. *Guideline: Fortification of Rice with Vitamins and Minerals as a Public Health Strategy.* Geneva: World Health Organization. https://apps.who.int/iris.

WHO. 2025. "Anaemia." World Health Organisation. 2025. https://www.who.int/news-room/fact-sheets/detail/anaemia.

Xu, Enbo, Xiaowei Pan, Zhengzong Wu, Jie Long, Jingpeng Li, Xueming Xu, Zhengyu Jin, and Aiquan Jiao. 2016. "Response Surface Methodology for Evaluation and Optimization of Process Parameter and Antioxidant Capacity of Rice Flour Modified by Enzymatic Extrusion." *Food Chemistry* 212:146–54. https://doi.org/10.1016/j.foodchem.2016.05.171.

Yusufali, Rizwan, Katrien Ghoos, United Nations, World Food, and Peiman Milani. 2015. "Scaling Up Rice Fortification in Asia Workshop Bangkok 2014." *Sight and Life Magazine: Scaling up Rice Fortification in Asia* 2015 (3). https://www.researchgate.net/publication/279829295_Scaling_up_Rice_Fortification_in_Asia

4 Rice Milling and Parboiling

Gayatri Mishra, Ayushi Chatterjee, Raju Kumar Saha, Payal Karmakar, Koppada Sunanda, and Brajesh Kumar Panda

4.1 INTRODUCTION

Rice is a staple food for billions of people worldwide, overcoming geographical borders and cultural variations to provide a key source of nourishment. While there is still debate over the origin of rice, one hypothesis claims that an ancient Gondwanan grass gave rise to African rice (*Oryza glaberrima*) in West Africa and Asian rice (*Oryza sativa*) in South and Southeast Asia as a result of post-continental drift (Chang 2003). With the Jeypore tract in Odisha mentioned as a secondary location, India is also acknowledged as an independent origin of Indica rice (Gross and Zhao 2014). Only two of the 24 Oryza species, *O. sativa* and *O. glaberrima*, are grown. India mainly produces Indica subspecies cultivars (DRD 2014).

Building on its heritage and geographic significance in rice domestication, India is a key player in global rice production, with constant growth in yield and planted area. India produced 100 million metric tons of rice in 2015–2016, and it is expected to rise to 147 million metric tons by 2024–2025 with harvested area from 43 to 51 million hectares and yield (tons per hectare) from 4.19 in 2023–2024 to 4.32 in 2024–2025 (USDA 2025).

The caryopsis and husk of the rice grain are separated during the husking process. Husk (5%–6%), endosperm (92%), and embryo (2%–3%) make up brown rice (Kim et al. 2012), as shown in Figure 4.1. Rich in nutrients, the bran layer is composed of the pericarp, seed coat, and aleurone layer. These layers are eliminated by milling; the amount of bran removed is indicated by the degree of milling (DOM). About 20% husk, 8%–12% bran, and 68%–72% milled rice, including whole and broken grains, are the results of ideal milling (IRRI 2024).

In order to maintain nutritional quality, post-harvest techniques like milling and parboiling must be optimized in light of India's growing rice output. Grain texture, appearance, and nutrition are all impacted by milling; while eliminating bran layers enhances visual appeal, it lowers fiber, vitamins, and minerals (Müller et al. 2022). The pre-milling treatment, like parboiling, uses steps like soaking, steaming, and drying, which help in pruning nutrient loss, intensifying milling yield, transferring nutrients from bran to endosperm, and reducing grain breakage (Heinemann et al. 2005). Parboiling helps in micronutrient fortification by augmenting the absorption of additional vitamins and minerals. Research conducted by Tulyathan et al. (2007)

 DOI: 10.1201/9781003661245-4

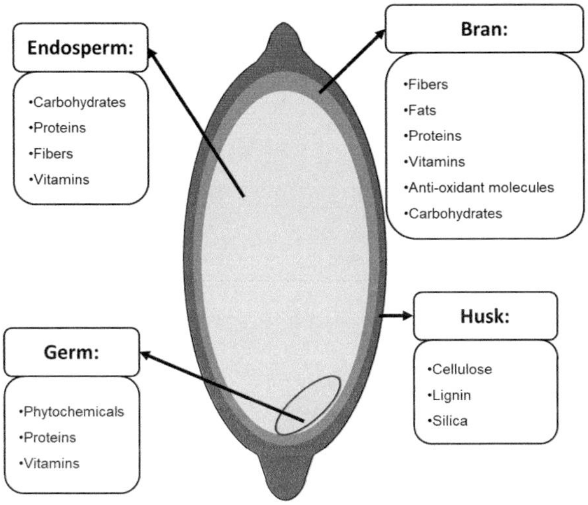

FIGURE 4.1 Anatomy of paddy grain.

reported that up to 19% iodine was absorbed into the grain by soaking brown rice in 70°C iodine solution during parboiling. Following this, another study was conducted to enrich rice with iron and zinc in a similar manner. They showed that parboiling is an efficient way to add essential minerals to rice; parboiling would, therefore, be a relevant strategy for fortification. According to Kam et al. (2012), the fortification of rice with folic acid (0.15–1.2 g/300 g brown rice) reached a maximum uptake of 299 μg/g and 90% retention after cooking. Notably, while maintaining head rice yield (65%) and kernel size, parboiling is cost-effective and scalable and has a positive impact on the nutrient improvement of rice.

Brown rice retains most of the nutrients because of the presence of bran and germ. On the other hand, 90% of the vitamins and minerals are lost due to heavy milling (Kaur et al. 2020). Through multiple studies, it has been observed that parboiling helps in improving metabolic function because of the resistant starch along with 18% more nutrient content (Sarangapani et al. 2016; Heinemann et al. 2005).

4.2 PARBOILING PROCESS: PRINCIPLES AND TECHNIQUES

The paddy undergoes a hydrothermal pre-treatment before milling, known as parboiling, which helps to improve the different qualities of the grain, like the structure, nutrition, and function. It is a conventional but scientifically proven method that involves steps like soaking, steaming, and drying the paddy, which not only helps in nutritional retention but also helps in improving milling efficiency and cooking performance. Parboiling helps to reduce breakage of paddy grains during milling, improving head rice yield. Table 4.1 presents a comparison of the physical, chemical, and cooking properties of raw and parboiled rice. This comparative study concludes that parboiled rice is superior to raw rice based on milling, nutritional quality, and higher shelf life.

TABLE 4.1
Comparative Study of Physical, Chemical, and Cooking Properties of Raw and Parboiled Rice

Properties	Raw Rice	Parboiled Rice
Grain hardness	The grains are softer	Harder due to gelatinization of starch during parboiling
Head rice yield	Lower (higher breakage during milling)	Higher (less breakage due to stronger grain structure)
Color	White or slightly translucent	Yellowish or amber color due to Maillard reaction and starch gelatinization
Nutrient retention	Loss of B-complex vitamins and minerals during milling	Improved retention as nutrients migrate from bran to endosperm during parboiling
Amylose content	Slightly higher (but may vary with variety)	Slightly lower or similar; changes depending on parboiling process
Starch structure	Native crystalline structure	Gelatinized and retrograded, making it more resistant to digestion
Cooking time	Shorter	Longer due to hardened starch structure
Water absorption	Higher	Lower; more compact grain resists water uptake
Stickiness	Higher (more clumping)	Lower; grains remain separate
Shelf life (cooked rice)	Shorter (spoils quickly)	Longer due to lower moisture migration and microbial resistance
Digestibility	Rapidly digestible	Slower digestibility; lower glycemic index

During parboiling, the paddy is first soaked in water before steaming and drying. During soaking, the kernel hydrates, followed by steaming, in which starch gelatinizes and covers the cracks present in the kernels; post steaming, the kernel is dried, and the kernel recovers its hardness. The reverse of gelatinization, known as retrogradation, takes place during drying and storage. Crystallites of starch are broken down via gelatinization, which also uses glycosidic linkages to hold onto extra water. Water is released and starch crystallites are reformed during retrogradation. Because the starch in the grain changes its energy state during parboiling, the grain's physicochemical, functional, and nutritional properties are significantly impacted. Notwithstanding the differences among the researchers, it was feasible to see that the hydration stage of paddy was frequently carried out at 60°C–70°C for 4–6h, and the starch was gelatinized by 10–20min of steam at 100°C–110°C (Bakshi et al. 1980).

The process diagram of the parboiling process has been depicted in Figure 4.2. The process of submerging rice in water, with or without husk, is known as hydration. In this phase, the grain should be hydrated at a temperature and time that allows it to reach a gelatinization moisture of at least 30% on a wet basis without degrading. However, raising the water's temperature speeds up the hydration process, which reduces the amount of time (Miah et al. 2002).

The water is drained after the paddy has been hydrated, and the starch is then exposed to conditions that ensure gelatinization.

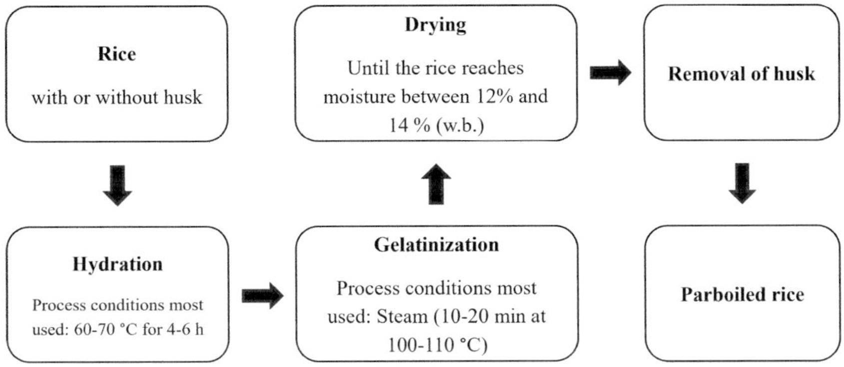

FIGURE 4.2 Diagram of the rice parboiling process.

When starch granules are subjected to heat in the presence of water, their crystalline structure is disrupted (Figure 4.3). Previous research on starch suspensions (starch/water: 30/70 w/w) has shown that amylose and amylopectin tend to partially separate during gelatinization, primarily due to their mutual incompatibility (Keetels et al. 1996). However, in thermoplastic starch systems with insufficient water content, this separation can be hindered. Shear forces encountered during extrusion can facilitate this separation, allowing a portion of the amylose to leach out from the amylopectin matrix. In systems containing 10% starch, most of the amylose remains

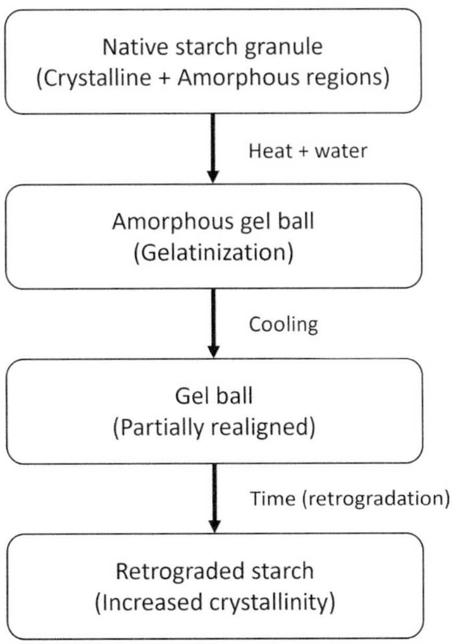

FIGURE 4.3 Diagrammatic illustration of the starch phase change during retrogradation and gelatinization.

embedded within the amylopectin network (Svegmark and Hermansson 1991). While minute amounts of ungelatinized amylopectin persist, the gelatinized form of amylopectin initially retains an amorphous state. Despite the breakdown of its double-helical crystalline structure during gelatinization, the short-branched chains of amylopectin maintain a level of structural order or "memory." These branches are believed to form gel-like spheres (Yu et al. 2005), composed predominantly of chains stemming from the same sub-main chain (Figure 4.3b). A single amylopectin molecule may form a distinct super-globular structure. Owing to the short chain lengths (only 4–6 glucose units), there is significantly less molecular entanglement among these gel-balls and super-globules compared to that found in linear polymer chains. It has also been noted that V-type single-helix crystals (Figure 4.3c) form almost instantly during extrusion, initially enhancing the modulus and yield stress in amylose-rich materials. Over time, these V-type crystals grow in crystallinity (Figure 4.3d).

According to Lanning and Siebenmorgen (2011), different cultivars have varying gelatinization temperatures, which impact the pace of hydration and the degree of gelatinization. But when it comes to traditional parboiling, the soaking process uses up to 1–1.2 L of water per kg of parboiled rice, resulting in large amounts of effluent; also, the soaking, steaming, and drying phases need a lot of energy (0.8–13 4 MJ/kg) (Srisang et al. 2021).

When paddy is parboiled, rice nutritional qualities also improve because more resistant starch (RS) accumulates and vitamins and minerals migrate toward the endosperm. By modifying the parboiling method, inorganic arsenic can be decreased (Chowdhury et al. 2024), calcium in rice can be fortified, and vitamin B_6 can be increased (Bhar et al. 2022). However, this change of the parboiling technique also lowered potassium contents by 40% (Rahman 2023), because vitamin B is soluble in water, changing the parboiling processes that may have an effect on this nutritional class. Different parboiling techniques have been depicted in Table 4.2.

4.2.1 Other Parboiling Methods

- **Avorio technique**: This technique, created in Italy, involves soaking paddy in a hot water tank while it is contained in a perforated basket. It is then steamed under pressure in a revolving cylinder and dried in hot air dryers (Kumar et al. 2018).
- **Conversion process**: The USA designed the procedure. The rice is soaked in cold water, the air is extracted using a vacuum, the rice is then steam-cooked under pressure, and it is dried in a vacuum dryer (Kumar et al. 2018).
- **Malek process**: Another technique, developed by the USA, involves soaking paddy for 3–6 h at 100°F, steaming it for 15 min, and then drying it at a low temperature in a hot air dryer (Kumar et al. 2018).
- **Fernandez process**: The procedure is comparable to hot soaking techniques and is widely utilized in Latin America (Kumar et al. 2018).
- **Schule process developed** in Germany and rice growers' association of California process developed in USA are also some of the modern parboiling methods (Kumar et al. 2018).

TABLE 4.2
Different Modern Parboiling Methods

Parameter	CFTRI Method	Jadavpur University Method	Pressure parboiling Method
Location/ developer	CSIR-CFTRI (India)	Jadavpur University	Tiruvarur, Tamil Nadu
Soaking temperature	~70°C (heated water at 85°C)	60°C–70°C	85°C–90°C
Soaking time	3–3.5 h	1–3 h	40 min
Steaming method	Open steam coils inside tank	Saturated steam	Pressurized steam
Steaming pressure / time	Approximately atmospheric pressure, time not specified	3.5 min	18 min
Soaking and Steaming	In the same tank	Option 1) Same tank Option 2) Separate horizontal apparatus	Soaking and steaming performed sequentially
Cooling step	Specifically, not mentioned	Rapid cooling with cold air draft	Specifically, not mentioned
Drying method	Sun drying or mechanical dryer	Rotary steam jacketed air dryer	Specifically, not mentioned
Automation level	Semi-automatic manual	Fully automatic	Semi-automatic
Processing time	Specifically, not mentioned	5–6 h	1–1.5 h
Output characteristics	Standard Parboiled rice	Improved Efficiency	Yellowish Rice, better shelling, higher bran fat, longer storage life
Unique features	Steam passed through coils; open bottom door	Rapid cooling; rotary dryer; optional apparatus configuration	Gelatinization under pressure; quick processing

Source: Kumar et al. (2018)

4.3 RICE MILLING: PRINCIPLES AND FUNDAMENTALS

In the rice business, "milling" refers to a number of processes, including size separation, bran removal, cleaning, shelling, and polishing. This improves cooking quality, texture, appearance, and shelf life (Tie et al. 2024). It may also particularly refer to the process of removing the bran from brown rice in order to produce white rice. A well-run milling operation guarantees that rice satisfies market and consumer quality criteria by being clean, well-polished, and little broken (IRRI 2024).

4.3.1 Stages of Rice Milling

Small farmers use traditional rice milling techniques like hand pounding and single-pass hullers (e.g., Engelberg) even though they are less efficient, labor-intensive, and produce more broken grains (IRRI 2024). In contrast, multi-stage procedures including dehusking, washing, and grading are used in contemporary milling to

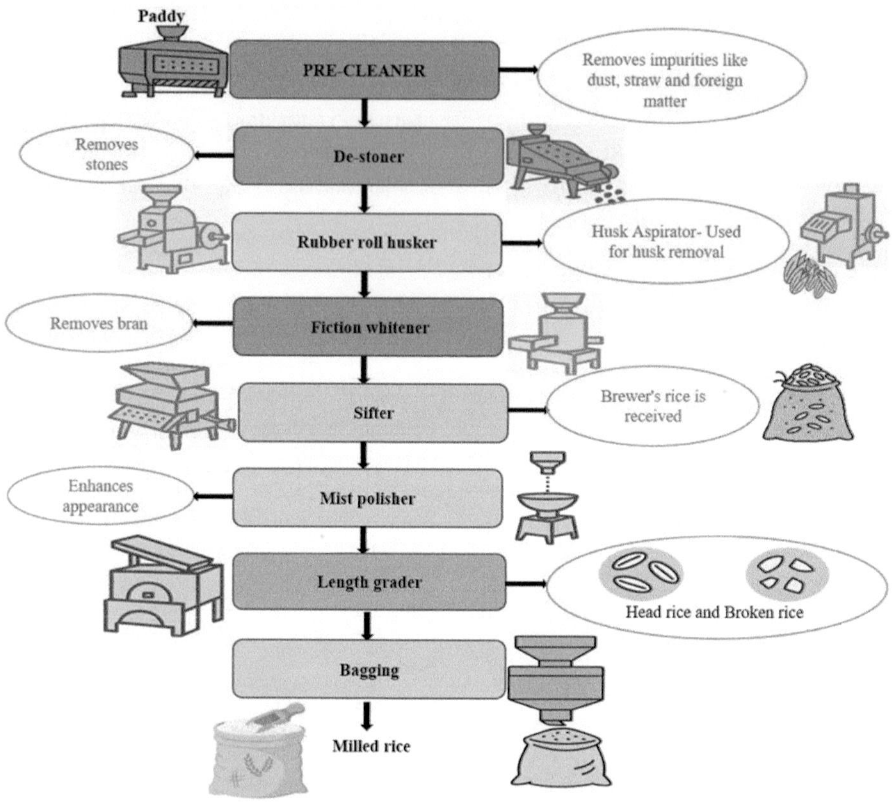

FIGURE 4.4 Stages of rice processing.

increase yield, decrease labor, and improve quality using automated systems (Bodie et al. 2019). The different stages of rice processing are shown in Figure 4.4.

Stage 1: Storage and Paddy Drying

Rice quality may be improved, and post-harvest losses can be reduced with the use of efficient drying and storage methods, which promote food security. Traditional storage solutions like gunny sacks suffer challenges including fungus development. Modern drying techniques like infrared and ultrasound-assisted system along with storage in hermetic bags like GrainPro and AgroZ retain the quality of the rice paddy (Dibagar et al. 2019).

Stage 2: Cleaning and Destoning

Paddy after harvesting contains multiple foreign substances like weed seeds, straw, and dirt, which enter the mills and reduce the efficiency and productivity; therefore, they should be removed before milling. The two phases involved in pre-cleaning are de-stoning and cleaning. A paddy cleaner aids in eliminating impurities, removing lighter matter like chaff, following heavy ones like stones with the help of sieves and air blasts. Destoning equipment have a high efficiency ranging from 40.8% to 99.75%, with a capacity of 1.8–7,500 kg/h (Adetola and Akindahunsi 2020).

Stage 3: Dehusking and Paddy Separation

A critical phase in rice milling is dehusking. In this phase, the husk is removed from the rough rice in order to produce brown rice. In order to reduce kernel breakage and yield optimization, the dehusking operation should be carefully administered. The rubber roll huller is used for husking, followed by the resulting mixture being aspirated in order to eliminate the husk, bran, and broken rice. Based on density, the left-over paddy is separated from brown rice, which again goes through the same process for complete dehusking.

Stage 4: Whitening and Polishing

In order to improve the texture, shelf life, and appearance of rice, polishing and whitening becomes an important step after dehusking. The conventional process to remove the bran layer is abrasive, but the contemporary approaches help in the application of enzymes such as xylanase and cellulase to remove the bran layer (Das et al. 2008). In case of both low and high amylose rice starches, the duration and rate of polishing have a significant effect on its rheological and physicochemical properties.

Stage 5: Rice Grading

Rice grading is a predominant post-milling activity that is directly linked to customer preferences, rice quality, and market value (FAO 2004). It also helps in the classification of rice, according to different characteristics such as shape, color, size, chalkiness, and the percentage of damaged and broken grains. Conventional equipment such as indented cylinders, length graders, and optical sorters are generally used (IRRI 2013).

4.3.2 Milling Efficiency

Milling efficiency is a prime factor influencing the quantity, quality, and market value of rice. It can be estimated using different methods such as head rice recovery (HRR), milled rice recovery (MRR), and brown rice recovery (BRR). The efficiency of milling is influenced by variables like equipment type, moisture level in paddy, and machine settings. Inadequacy in milling reduces production along with diminished nutritional properties of rice and esthetic appeal, which further impacts the consumers choice.

With the help of oscillating sieves, polished white rice is split into four categories that are head rice, big broken, tiny broken, and fine broken (brewer's). According to the definition of head rice, the kernels should be undamaged, accounting for 75% or more of the entire kernel length. They are of high value in the consumer market because of their quality. As the rice is broken into big or fine fragments, it loses its market value. A study states that a good mill generally produces 50%–60% head rice, 5%–10% big rice, and 10%–15% little broken rice (Mutungi et al. 2020).

Direct Measurement Methods

Estimation of the quality of kernel and processing effectiveness can be determined at different stages of milling with the help of quantitative techniques such as direct physical measurements. These techniques employ simple yet accurate equations to determine important parameters like brown rice percentage (Eq. 4.1), milled rice recovery (Eq. 4.2), head rice recovery (Eq. 4.3), broken rice percentage (Eq. 4.4), etc.

$$\text{Percentage brown rice} = \frac{\text{Weight of brown rice } (\text{g})}{\text{Weight of paddy } (\text{g})} \times 100 \tag{4.1}$$

Equation 4.1 presented above demonstrates yield of brown rice without damaging the germ and bran layers. More percentage will indicate superior quality paddy and effective husk removal (Lapis et al. 2019).

$$\text{Percentage milled rice} = \frac{\text{Weight of milled rice } (\text{g})}{\text{Weight of paddy } (\text{g})} \times 100 \tag{4.2}$$

Equation 4.2 provided above is an example of milled rice recovery, i.e., total yield of rice, after dehusking and polishing. It represents the overall milling effectiveness since it accounts for head as well as broken rice (Lapis et al. 2019).

$$\text{Head rice yield} = \frac{\text{Weight of whole milled rice } (\text{kg})}{\text{Weight of clean milled rice } (\text{kg})} \times 100 \tag{4.3}$$

The above equation is the quantity of intact, unbroken rice grains after milling. Greater percentage of head rice indicates higher kernel quality and market worth, which is valuable in high-quality rice markets (Tikuneh 2025).

$$\text{Percentage broken rice} = \frac{\text{Weight of broken rice } (\text{kg})}{\text{Weight of clean milled rice } (\text{kg})} \times 100 \tag{4.4}$$

The above equation represents the percentage of broken rice from the total weight of milled rice. It holds a significance when the real weight of broken rice and raw paddy is known post milling (Tikuneh 2025).

4.4 TYPES OF RICE MILLING PROCESSES

Different types of rice milling systems are enlisted in Figure 4.5.

4.4.1 Traditional Milling Methods

The term "traditional rice milling" describes the ancient techniques used for processing paddy into edible rice in rural areas. These methods are community-based, manual, inexpensive, and need little to no power or mechanical equipment. Various manual and semi-mechanized processes are used in traditional rice milling technologies, which are common in rural areas. In order to dehusk rice, manual stone milling uses instruments like the dhenki, a traditional wooden device with a lever. This method has long been used in regions of India and Nepal. A more effective method of dehusking was introduced by hand-powered hullers, including the Engelberg huller,

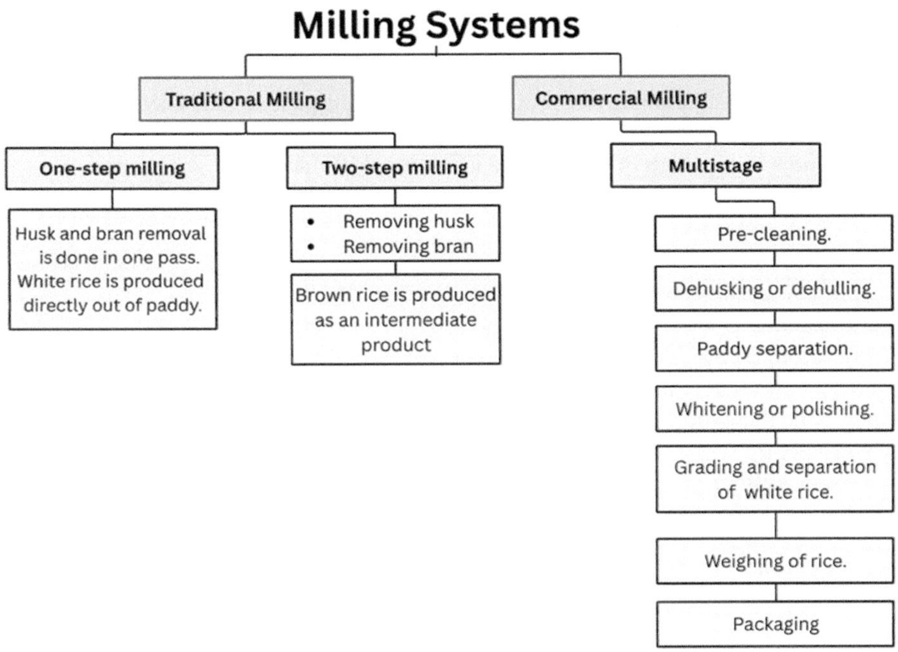

FIGURE 4.5 Types of rice milling processes.

which was created in the late 19th century and used a spinning mechanism to remove husks. Although they need a lot of attention and work, animal-powered mills, where bullocks or horses rotate millstones, have been utilized to alleviate manual labor. Water-powered mills provide an environmentally beneficial way to grind stones in hilly regions by using the energy of running streams. Traditional hand polishing methods are used after milling to improve the appearance of rice by getting rid of any remaining bran. Compared to contemporary mechanical milling technologies, these techniques frequently provide lesser efficiency, uneven quality, and more grain breakage, despite being more economical and ecologically friendly.

Single stage mill/Micro mill: A micro mill is a single-stage, low-capacity device that is made from locally fabricated parts and is intended for remote, local use. It provides low head and milling rice recovery, making it primarily appropriate for domestic use (IRRI 2024).

Two-stage compact mill: This technique, which consists of a friction whitener and a rubber roller husker, is utilized in custom milling services in rural areas. Compared to single-stage hullers, it offers higher recovery rates (over 60%) and superior milling performance (IRRI 2024).

Mobile rice mill: These mills, which are mounted on self-propelled trucks, eliminate the need for transportation by delivering milling services straight to farmers. They are popular in Southeast Asia and combine polishing and husking for ease of use in rural areas (IRRI 2024).

4.4.2 Modern Milling Techniques

Modern rice milling, which includes mechanical dryers, automated mills, and enhanced packaging, is revolutionizing rice production by lowering post-harvest losses and improving grain quality. These innovations boost local rice competitiveness, but implementation is hampered by high prices, limited technical skills, and insufficient assistance.

Rubber Roll Huller

The rubber roll huller is a novel, more efficient, and gentler process of dehusking in comparison to the conventional Engelberg hullers. It follows the concept of abrasion and differential speed. The fundamental principle used in dehulling is to advance the paddy through two contra-rotating rubber rollers and remove the husk by shear and pressure (Tikuneh et al. 2025). Rubber roll huller uses two revolving rubber rollers, which contributes to low kernel breakage and high head rice recovery. In addition, they efficiently segregate bran and husk, improvising the quality of the rice kernels. Engelberg hullers are more abrasive and less efficient compared to the rubber roll hullers, but they are easy and less costly.

Mist Polisher

Mist polisher is an advanced rice milling equipment, which uses thin water mist during rice polishing in order to reduce abrasion and heat, and it also helps in reducing breakage of rice kernels and increasing shine. According to research by Rachmat et al. (2022), the mist polisher enhances milled rice yield and quality. The mist does this by softening the bran layer, allowing easy removal without damaging the rice kernel, later obtaining shinier kernels and high milling recovery.

Grader

Rice grading is a critical post-milling procedure as it helps in segregating head rice from broken rice. In order to improve market value and rice quality, the contemporary rice graders play a vital role. Ojediran et al. (2020) developed a mechanized rice grader that used multi-screen system with reducing mesh sizes, tilted at multiple angles (6°, 8°, and 16°) to help in efficient sorting. Furthermore, a centrifugal blower was used to eliminate light contaminants, along with an eccentric motor creating oscillatory motion to move the rice kernels over the screens. The grading efficacy of 91.67% along with a product purity of 90.97% was achieved at a feed gate aperture of 5 mm. It assists in terms of modernized imaging system, lowers human errors that occur during grading, and ensures cost-effective solution for small- and medium-scale producers.

4.4.3 Comparison between Traditional and Modern Rice Milling

There has been prominent evolution in rice milling, from primitive human-intensive approaches to extensively mechanized technologies. Conventional milling techniques, generally witnessed in rural areas, use rustic tools and human labor for production of rice mainly for local use. On the other hand, modern milling uses advanced equipment to increase the efficacy, quality, and scalability, making it suitable for both local and international markets. Table 4.3 illustrates the different features of traditional and modern rice milling.

TABLE 4.3
Difference between Traditional and Modern Rice Milling

Feature	Traditional Rice Milling	Modern Rice Milling	References
Technology used	Manual tools like mortar and pestle, winnowing trays; small-scale mechanical (village-type) mills	Automated systems: rubber roll hullers, de-stoners, polishers, graders, elevators, and silos	IRRI (2024)
Power source	Human labor, sometimes animal or water power	Electricity, diesel engines, or solar-powered modern systems	Nabavi-Pelesaraei et al. (2019)
Labor requirement	Highly labor-intensive; manual effort needed at each step	Low labor; machine-operated with skilled operators and technicians	IRRI (2024)
Processing steps	Hand pounding → winnowing → manual separation; mostly single-step dehusking	Multi-stage: cleaning → dehusking → separating → polishing → grading → packaging	IRRI (2024)
Efficiency	Low yield (55%–65%); high grain breakage; dependent on worker skill	High efficiency (70%–75% head rice yield); minimal breakage due to automation	Mutungi et al. (2020)
Quality of Rice	Inconsistent; more broken grains and impurities	Uniform, polished grains; consistent quality suitable for retail/export	Kim and Lee (2012)
Nutritional value	Often retains bran and germ (brown rice), offering higher fiber and micronutrients	Polished white rice with reduced nutrition; fortified and parboiled rice available	Kalita et al. (2021)
Environmental impact	Low—biodegradable tools, minimal emissions	Moderate to high—requires energy; some systems are energy-efficient and use waste-to-value	Nabavi-Pelesaraei et al. (2019)
Maintenance	Simple; tools easily repaired or replaced	Requires skilled technicians and regular maintenance	IRRI (2024)
Waste management	Husk and bran reused as compost, fuel, or fodder	Bran and husk are commercialized (e.g., oil extraction, feed industry)	IRRI (2024)
Product types	Mainly brown rice or semi-milled rice	Brown, white, parboiled, fortified, broken rice, etc.	IRRI (2024)
Market suitability	Home or local consumption; unsuitable for regulated trade	Meets export, packaging, food safety standards; global trade-ready	IRRI (2024)
Flexibility	Can process multiple varieties but lacks uniformity	High standardization with adjustable configurations for variety	IRRI (2024)

4.5 ADVANCES IN MILLING AND PARBOILING TECHNOLOGIES

As posed by Harini et al. (2023), a machine learning approach has been proposed for the correct classification of Indian rice varieties according to the DOM, which is a critical quality feature affecting the nutritional value, consumer preference, and export potential. For this, digital photographs of rice grains at different milling levels were taken using image processing techniques. Several morphological and textural properties were computed from these images, which include area, perimeter, length of major and minor axes, eccentricity, convex area, and solidity, and these collectively represent the metamorphosis that the rice grains undergo during milling.

Baqui et al. (2008) documented the design and field implementation of two energy-efficient rice parboiling systems in Bangladesh as part of the Sustainable Energy for Development (SED) initiative. They developed a semi-cylindrical, pure husk-fired reactor (Type 1) and a cylindrical reactor based on a bran-husk mix (Type 2) for replacing conventional husk-fired systems with poor thermal efficiency (20%–30%) and extremely hazardous carbon monoxide (CO) emissions (>10,000 ppm). The design enhancements, like better insulation, effective chimney design, and steam-tight rooms, kept energy loss at a minimum while enhancing thermal efficiency to 42.75% in Type 1 and 50.3% in Type 2, while CO emissions were significantly lowered to a lower safety limit of 100–500 ppm. Economically, the units realized a benefit–cost ratio ranging from 7.24 to 11.48, utilizing 50% less husk than conventional units, with surplus husk having the potential to be reused in briquetting or power production.

The study by Jamal (2021) and Jan et al. (2017) conveys a portable rice mill machine that tackles post-production issues associated with rice processing, offering farmers an alternative to traditional rice mill industries. When integrating a small huller mill for rice whitening and a pair of leather rollers for husk removal, the suggested technique allows the farmers to efficiently process their own rice.

In an effort to combat global malnutrition, the World Intellectual Property Organization (WIPO 2025) pointed out a novel rice fortification technology created in Sri Lanka on its IP advantage platform. The patented method, designed by Seneviratne et al. (2019), involves adding fortification to the parboiling process via vacuum-assisted paddy rice soaking in a micronutrient solution.

The process facilitates the penetration of nutrients deep into the grain beyond the bran layer. The whole process is outlined in Figure 4.6. When the rice is steamed and dried, the nutrients become trapped within the starch matrix so that the resultant product will preserve more than 90% of its nutrients after washing and cooking. This innovation is a cost-effective, scalable, and sustainable approach to enhancing food and nutrition security in rice-consuming areas.

4.6 QUALITY CONTROL AND REGULATORY STANDARDS

The parboiling and rice milling operations should meet national and international standards, such as the FSSAI regulations and Codex (FSSAI 2021; Codex Alimentarius Commission 2020).

FIGURE 4.6 Diagram of the whole process of Fortigrains Lanka (Fortified Rice).

4.7 ECONOMIC AND ENVIRONMENTAL CONSIDERATIONS

4.7.1 Cost–Benefit Analysis of Milling and Parboiling Technologies

Milling and parboiling are key stages that shape rice quality, nutrition, and profitability. Their cost–benefit depends on capital, raw material, labor, processing time, technology, scale, and local resources. Traditional methods are cheaper but yield lower quality and market access, limiting long-term profit. Semi-automatic mills improve efficiency but still compromise head rice recovery and byproducts. Modern multi-stage mills deliver higher yields, superior quality, and valuable byproducts, though they require high investment. Similarly, basic parboiling methods often produce inconsistent quality, while improved industrial systems ensure uniform, nutrient-rich rice with reduced losses. An integrated cost–benefit analysis must consider resource availability, paddy quality, market demand, capital limits, byproduct use, and regulations.

4.7.2 Water and Energy Consumption Concerns

Parboiling and milling technologies raise several concerns relating to water and energy consumption, which affect challenges in the sustainability of rice processing. These processes involve significant resource inputs and are responsible for enhancing quality of grain by reducing post-harvest losses. The adopted technology highly influences water and energy consumption, which affects the rice processing. Nutrient-rich wastewater is generated during traditional parboiling, which consumes 3–5 L of water per kilogram of paddy, posing serious risk to environment. This immense water usage can be drastically cut to 75% using innovative reduced-water methods, through water recycling and vacuum-sealed soaking, which can offer a feasible solution for water-deprived regions (Vijaya et al. 2025). Modern multi-stage milling shows increased efficiency in comparison to smaller huller mills, as, despite using higher total energy, it exhibits lower specific energy consumption. Higher energy is required for parboiled rice as compared to raw rice milling, which is because the grain is harder and more energy is required for steaming in the parboiling stage. Advanced parboiling and milling machinery need to be implemented to optimize operational efficiency and energy utilization, while integrating sustainable practices such as water conservation and biomass energy generation using rice husk. In order to develop a more sustainable and environment-friendly rice processing industry, such efficient technologies along with by-product valorization need to be embraced.

4.7.3 Waste Management and Byproduct Utilization

Rice parboiling and milling produce a huge amount of waste matters and byproducts, creating several opportunities as well as challenges. The waste generated needs to be managed effectively, and the byproduct should be utilized in order to minimize environmental impact, thereby establishing economic viability, besides promoting sustainability in the rice-based industries. The main byproducts that are produced are rice bran, rice husk, and broken rice, with wastewater being a significant concern, mainly in case of parboiling. The efficient management of waste and the utilization of byproducts in rice parboiling and milling technologies can aid in reducing environmental impact and strengthening economic returns. In order to attain sustainability goals, strategic handling of husk, bran, broken grains, and wastewater produced in huge amount during these processes is necessary. In major rice-producing regions like India, these by-products are transformed into resources using various innovations in treatment methods by developing value-added products. Economic enhancement and environmental stewardship both call for effective waste management and byproduct utilization, which comprises a significant section of rice processing. Parboiling generates a huge amount of wastewater with high biological oxygen (BOD) and chemical oxygen demand (COD), thereby requiring physical, chemical, and biological treatment using anaerobic digestion to produce biogas (Kandagatla et al. 2023). Rice husk and other solid byproducts can be utilized by performing gasification or cogeneration to generate energy, and the resulting ash is used in material and cement production. Rice bran finds application in oil, animal feed, and valuable nutraceuticals for being rich in oil and nutrients.

An undervalued byproduct of rice processing is broken rice, which finds itself useful in serving as a base for fortified foods that can provide a solution to micronutrient deficiencies and also finds application in brewing and starch production. It is a byproduct of milling, which proves itself as a practical solution to address micronutrient deficiencies mostly prevalent in rice-consuming populations by being used in fortified rice. Broken rice is ground into flour and blended with essential nutrients to create fortified rice kernels, which are used in combination with regular rice. The technologies used in this process include extrusion, coating, or dusting, which create a product with similar taste and cooking properties to regular rice, without compromising consumer acceptance (Sarfaraz 2023). By putting byproducts to efficient use and by managing wastes strategically, the rice industries can minimize its environmental footprint, which will contribute to a more sustainable and economically viable future and also generate additional revenue streams.

4.8 FUTURE TRENDS AND INNOVATIONS

The industry is fueled by rising global demand and supported by technological advancements, with a focus on sustainability and nutritional enhancement, promoted by a shift to digital involvement including AI, IoT, and automation to optimize processing and ensure food safety, going hand in hand with eco-friendly practices and circular economy models. Potential research areas include developing advanced parboiling and milling technologies to gain better efficiency and nutrient retention. Key advancements in rice quality assessment, fortification, byproduct management, climate-resilient varieties, energy efficiency, AI integration, and supply chain optimization require joint efforts from research, industry, and government to enhance nutrition, ensure food security, and support sustainability in the rice sector (Nabavi-Pelesaraei et al. 2019; Sarfaraz 2023; Muchlisyiyah et al. 2023). However, challenges like high capital costs and gaps in expert skill still persists, which calls for policy support and efficient training.

4.9 CONCLUSIONS

The rice parboiling and milling sectors are witnessing enhanced efficiency through several technological transformations employing automation, IoT, and by implementing innovative parboiling techniques involving nutritional retention and enhanced resource utilization with immense emphasis on water recycling and solar energy adoption in order to maintain sustainability. However, high initial investment and need for skilled labor are among few of the persistent challenges. The path of rice fortification in a large scale can be paved by addressing these challenges and optimizing the processes. The nutrient-enriched paddy can be soaked during parboiling, and gentle milling techniques can be employed to maximize the retention of essential micronutrients. To maximize the nutritional benefits of rice parboiling and milling, a coordinated strategy encompassing innovation, intensive research, consumer awareness, policies, and capacity development is essential. Thus, technological advancements and fortification goals need to be aligned strategically, which can unlock the potential of rice to prove itself as an important factor contributing toward global health and nutrition security, thereby ensuring a better and nourished future for the population that relies on rice as a staple.

REFERENCES

Adetola, O. A., and D. L. Akindahunsi. 2020. "A Review on Performance of Rice De-Stoning Machines." *Journal of Engineering Research and Reports* 13 (1): 1–11.

Bakshi, Amarjit S., and R. Paul Singh. 1980. "Kinetics of Water Diffusion and Starch Gelatinization during Rice Parboiling." *Journal of Food Science* 45 (5): 1387–92.

Baqui, M. A., M. Ahiduzamman, and M. M. Rahman. 2008. "Experiences in Installation and Operation of Energy Efficient Rice Parboiling Systems." In *Short Course on Energy Efficiency,* edited by Shahidul I. Khan and M. Obaidullah, 108–24. Dhaka: Bangladesh Rice Research Institute (BRRI).

Bhar, S., T.Bose, A. Dutta, and S. S. Mande. 2022. "A Perspective on the Benefits of Consumption of Parboiled Rice over Brown Rice for Glycaemic Control." *European Journal of Nutrition* 61 (2): 615–24.

Bodie, A. R., A. C. Micciche, G. G. Atungulu, M. J. Rothrock Jr, and S. C. Ricke. 2019. "Current Trends of Rice Milling Byproducts for Agricultural Applications and Alternative Food Production Systems." *Frontiers in Sustainable Food Systems* 3: 47.

Chang, T. T. 2003. "Origin, History, and Domestication." In *Rice: Origin, History, Technology, and Production*, edited by C. Wayne Smith and Robert H. Dilday, 3–26. Hoboken, NJ: Wiley.

Chowdhury, N. R., M. Joardar, A. Das, P. Bhattacharya, and T. Roychowdhury. 2024. "Current Opinion on the Role of Post-Harvesting and Cooking on Arsenic Mobility in Rice Grain, Its Surmounting Risk towards Human and Domestic Livestock with Sustained Management." *Current Opinion in Environmental Science and Health* 38: 100535.

Codex Alimentarius Commission. 2020. *General Standard for Rice (CXS 198–1995).* https://www.fao.org/fao-who-codexalimentarius/sh-proxy/zh/?lnk=1&url=https%253A%252F%252Fworkspace.fao.org%252Fsites%252Fcodex%252FStandards%252FCXS%2B198-1995%252FCXS_198e.pdf

Das, M., S. Gupta, V. Kapoor, R. Banerjee, and S. Bal. 2008. "Enzymatic Polishing of Rice–A New Processing Technology." *LWT – Food Science and Technology* 41 (10): 2079–84.

Dibagar, N., R. Amiri Chayjan, S. Jan Kowalski, and S. Hossain Peyman. 2019. "Deep Bed Rough Rice Air-Drying Assisted with Airborne Ultrasound Set at 21 kHz Frequency: A Physicochemical Investigation and Optimization." *Ultrasonics Sonochemistry* 53: 25–43.

DRD (Directorate of Rice Development). 2014. *Status Paper on Rice*. Patna: Govt. of India.

FAO (Food and Agriculture Organization). 2004. Rice*: Post-Harvest Operations.* Rome: Food and Agriculture Organization of the United Nations.

FSSAI (Food Safety and Standards Authority of India). 2021. *Chapter 2.4: Cereals and Cereal Products*. New Delhi: Food Safety and Standards Authority of India.

Gross, B. L., and Z. Zhao. 2014. "Archaeological and Genetic Insights into the Origins of Domesticated Rice." *Proceedings of the National Academy of Sciences* 111 (17): 6190–97.

Harini, S., S. Chakrasali, and G. N. Krishnamurthy. 2023. "Machine Learning-Based Approach for Degree of Milling Analysis of Indian Rice Variety." *International Journal of Agriculture Innovation, Technology and Globalisation* 3 (2): 177–92.

Heinemann, R. J. B., P. de Lara Fagundes, E. Aparecida Pinto, M. de Vuono Camargo Penteado, and U. M. Lanfer-Marquez. 2005. "Comparative Study of Nutrient Composition of Commercial Brown, Parboiled and Milled Rice from Brazil." *Journal of Food Composition and Analysis* 18 (4): 287–96.

IRRI (International Rice Research Institute). 2024. *Milling systems*. Los Baños: International Rice Research Institute.

IRRI (International Rice Research Institute). 2013. *Rice Quality and Processing*. Los Baños: International Rice Research Institute.

Jamal, A. 2021. "A Literature Review on Design of Mini Rice Milling Machine." *International Journal of Scientific Research in Science and Technology* 8 (4): 1–4.

Jan, S., R. P. Kingsly Ambrose, and D. C. Saxena. 2017. "Effect of Grinding Action on the Flowability of Rice Flour." *Journal of Food Measurement and Characterization* 11: 801–11.

Kalita, T., U. P. Gohain, and J. Hazarika. 2021. "Effect of Different Processing Methods on the Nutritional Value of Rice." *Current Research in Nutrition and Food Science Journal* 9 (2): 683–91.

Kam, K., J. Arcot, and A. A. Adesina. 2012. "Folic Acid Fortification of Parboiled Rice: Multifactorial Analysis and Kinetic Investigation." *Journal of Food Engineering* 108 (1): 238–43.

Kandagatla, N., B. Kunnoth, P. Sridhar, Vinay Tyagi, P. V. Rao, and R. D. Tyagi. 2023. "Rice Mill Wastewater Management in the Era of Circular Economy." *Journal of Environmental Management* 348: 119248.

Kaur, A., S. Bhise, and M. Kaur. 2020. "Hydrothermal Treatments for Paddy to Improve Physicochemical Quality of Brown Rice." *Journal of Microbiology, Biotechnology and Food Sciences* 9 (5): 913–26.

Keetels, C. J. A. M., T. Van Vliet, and P. Walstra. 1996. "Gelation and Retrogradation of Concentrated Starch Systems: 2. Retrogradation." *Food Hydrocolloids* 10 (3): 355–62.

Kim, H.-W., S.-K. Oh, D.-J. Kim, M.-R. Yoon, J.-H. Lee, I.-S. Choi, Y.-G. Kim, and K.-N. Cha. 2012. "Changes in Contents of Nutritional Components and Eating Quality of Brown Rice by Pericarp Milling." *Korean Journal of Crop Science* 57 (1): 35–40.

Kim, S. Y., and H. Lee. 2012. "Effects of Quality Characteristics on Milled Rice Produced under Different Milling Conditions." *Journal of The Korean Society for Applied Biological Chemistry* 55: 643–49.

Kumar, V., J. Singh, N. Chauhan, S. Chandra, and M. Yadav. 2018. "Process of Paddy Parboiling and Their Effects on Rice—A Review." *Journal of Pharmacognosy and Phytochemistry* 1 (1): 1727–34.

Lanning, S. B., and T. J. Siebenmorgen. 2011. "Comparison of Milling Characteristics of Hybrid and Pureline Rice Cultivars." *Applied Engineering in Agriculture* 27 (5): 787–95.

Lapis, J. R., R. P. O. Cuevas, N. Sreenivasulu, and L. Molina. 2019. "Measuring Head Rice Recovery in Rice." In *Rice Grain Quality: Methods and Protocols*, edited by Nese Sreenivasulu, 89–98. New York: Humana Press (an imprint of Springer Science+Business Media).

Miah, M. A. K., A. Haque, M. Paul Douglass, and B. Clarke. 2002. "Parboiling of Rice. Part I: Effect of Hot Soaking Time on Quality of Milled Rice." *International Journal of Food Science and Technology* 37 (5): 527–37.

Muchlisyiyah, J., R. Shamsudin, R. K. Basha, R. Shukri, S. How, K. Niranjan, and D. Onwude. 2023. "Parboiled Rice Processing Method, Rice Quality, Health Benefits, Environment, and Future Perspectives: A Review." *Agriculture* 13 (7): 1390.

Müller, A., M. Trojahn Nunes, V. Maldaner, P. Carteri Coradi, R. Santos de Moraes, S. Martens, A. Fernandes Leal, V. Fogliato Pereira, and C. König Marin. 2022. "Rice Drying, Storage and Processing: Effects of Post-Harvest Operations on Grain Quality." *Rice Science* 29 (1): 16–30.

Mutungi, C., A. Gaspar, and A. Abass. 2020. *Postharvest Operations and Quality Specifications for Rice: A Trainer's Manual for Smallholder Farmers in Tanzania*. Ibadan: IITA.

Nabavi-Pelesaraei, A., S. Rafiee, S. Saeid Mohtasebi, H. Hosseinzadeh-Bandbafha, and Kwok-wing Chau. 2019. "Comprehensive Model of Energy, Environmental Impacts and Economic in Rice Milling Factories by Coupling Adaptive Neuro-Fuzzy Inference System and Life Cycle Assessment." *Journal of Cleaner Production* 217: 742–56.

Ojediran, J. O., C. E. Okonkwo, S. A. Alake, E. A. Alhassan, and A. T. Olayanju. 2020. "Design, Development, and Evaluation of a Motorized Rice Grader." *Journal of Food Process Engineering* 43 (2): e13336.

Rachmat, R., E. Rahayu, S. Darniadi, and M. Hadipernata. 2022. "Enhancement of Quality and Milled Rice Yield Recovery Through the Improvement of Polishing Process and Configuration." *IOP Conference Series: Earth and Environmental Science* 1024 (1): 012008. IOP Publishing.

Rahman, A. 2023. *Rice Processing Industry in* Bangladesh*: An Analysis of Rice Milling,* edited by Arifur Rahman. Bangladesh: Emerging Credit Rating Limited.

Sarangapani, C., R. Thirumdas, Y. Devi, A. Trimukhe, R. R. Deshmukh, and U. S. Annapure. 2016. "Effect of Low Pressure Plasma on Physico–Chemical and Functional Properties of Parboiled Rice Flour." *LWT – Food Science and Technology* 69: 482–89.

Sarfaraz, H. 2023. "Rice Production, Quality, and Nutrition: A Comprehensive Review on Challenges, Opportunities, and Global Perspectives." *International Journal of Agriculture and Sustainable Development* 5 (3): 90–103.

Seneviratne, H., M. Gunawardana, A. Hettiarachchy, and N. Rajapaksha. 2019. "Development of a Diffusion Based Fortification System and Process to Improve the Iron and Zinc Content in Rice." In *2019 From Innovation to Impact (FITI).* Piscataway, NJ: IEEE.

Srisang, N., S. Prachayawarakorn, S. Soponronnarit, and T. Chungcharoen. 2021. "An Innovative Hybrid Drying Technique for Parboiled Rice Production without Steaming: An Appraisement of the Drying Kinetics, Attributes, Energy Consumption, and Microstructure." *Food and Bioprocess Technology* 14: 2347–67.

Svegmark, K., and A. M. Hermansson. 1991. "Distribution of Amylose and Amylopectin in Potato Starch Pastes: Effects of Heating and Shearing." *Food Structure* 10 (2): 2.

Tie, X.-Y., N.-N. An, D. Li, L.-J. Wang, and Y. Wang. 2024. "Recent Advances in Rice Milling Technology: Effect on Nutrient, Physical Properties and Digestibility." *Food Reviews International* 40 (8): 2256–82.

Tikuneh, D. B., M. A. Mihiretu, A. D. Molla, T. A. Haile, and M. A. Fetene. 2025. "Evaluation of a Rubber Roller One-Pass Rice Milling Machine for Improving Milled Rice Quality." *Engineering Reports* 7 (4): e70101.

Tulyathan, V., T. Laokuldilok, and S. Jongkaewwattana. 2007. "Retention of Iodine in Fortified Parboiled Rice and Its Pasting Characteristics during Storage." *Journal of Food Biochemistry* 31 (2): 217–29.

USDA. 2025. "India Rice: Record Production Forecasted." *USDA Foreign Agricultural Service,* April 10, 2025. https://apps.fas.usda.gov/newgainapi/api/Report/DownloadReportByFileName?fileName=Grain%20and%20Feed%20Annual_New%20Delhi_India_IN2025-0023.pdf

Vijaya, R. I., S. Singh, and M. Islam. 2025. "Processing of Rice Husk and Its Applications." In *Rice Husk Biomass: Processing, Properties and Applications*, edited by Mohammad Jawaid and Brajendra Parmar, 1–26. Singapore: Springer Nature Singapore.

World Intellectual Property Organization. 2025. *Fortigrains Lanka Fortified Rice — A Solution to World Hunger.* April 23. https://www.wipo.int/en/web/ip-advantage/w/stories/fortigrains-lanka-fortified-rice-a-solution-to-world-hunger?utm_source=chatgpt.com

Yu, L., and G. Christie. 2005. "Microstructure and Mechanical Properties of Orientated Thermoplastic Starches." *Journal of Materials Science* 40: 111–16.

5 Vitamin-Mineral Premix
Formulation, Processing, and QA/QC

Rakesh Kumar Raigar, Indrajit Dalai, Sandip Sanjay Gite, Chandrakant Genu Dalbhagat, and Hari Niwas Mishra

5.1 INTRODUCTION

Vitamin-mineral premixes (VMP) are a combined blend of desired and precise doses of vitamins, minerals, and carrier or filler agents, along with suitable preservatives. These premixes are extensively used to prevent micronutrient deficiencies and improve public health, mainly in vulnerable populations. Staple foods such as rice, wheat flour, maize flour, salt, and edible oils are fortified with VMP to deliver essential nutrients like iron, zinc, vitamin A, vitamin D, folic acid, and vitamin B_{12} through dusting, coating, and extrusion (Steiger et al. 2014). For instance, in rice fortification, VMP are incorporated into fortified rice kernels (FRK), which are then blended with regular rice to prevent anemia and neural tube defects. Furthermore, in flour fortification, iron and folic acid are added to prevent iron-deficiency anaemia, particularly significant among women and children. Fortification of milk and oil with vitamin A and vitamin D enhances bone health and vision, while fortified salt with iodine prevents goitre and cognitive impairments. The use of VMP ensures uniform distribution, stability during processing and storage, and compliance with regulatory standards such as those set by the Food Safety and Standards Authority of India (FSSAI) and the Bureau of Indian Standards (BIS) (BIS 2021; FSSAI 2022). The premix may contain various essential vitamins without minerals, but the inclusion of a carrier agent and preservative is mandatory for the mixture to ensure the stability and consistency. Despite the lower quantity of individual or mixed VMP, accurate measurements, uniform dispersion, and controlled manufacturing process are highly essential for food fortifications. Meanwhile, to achieve consistent nutrient delivery and avoid uneven distribution, an appropriate formulation of ingredients and their blending precision are vital.

5.2 RAW INGREDIENTS

5.2.1 Minerals

Minerals play an important role in food fortification, which is a key strategy used to address micronutrient deficiencies in populations, especially in developing countries.

DOI: 10.1201/9781003661245-5

Minerals such as iron, zinc, calcium, iodine, and selenium are commonly added to staple foods like rice, wheat flour, salt, and milk to enhance their micronutrient values. The selection of the source of ingredients is an important part of any VMP, and their suitability is based on the bioavailability, digestibility, solubility in water, storability, and cost of manufacturing. Generally used ingredients for the fortification of iron in the staple foods are as follows.

Ferrous Sulphate

Ferrous sulphate ($FeSO_4$) is a blue-green, odourless, water-soluble iron salt containing 20% iron and is commonly known as green vitriol. Ferrous sulphate has a molecular weight and density of 151.908 g/mol and 2.84 g/cm^3, respectively, making it one of the cheapest sources of food supplements. This compound is produced by reacting scrap iron, iron ore, and other forms of iron with sulphuric acid, resulting in ferrous sulphate and hydrogen gas. Impurities and unreacted materials are filtered out, and the filtered materials undergo cooling, drying, and milling into a powder. This ingredient has good bioavailability compared to other forms of iron, but its disadvantage is that it is prone to oxidation, which changes its colour (Sachdeva et al. 2015).

Ferric Pyrophosphate

Ferric pyrophosphate ($Fe_4(P_2O_7)_3$) is a yellow to white powder, odourless, lower solubility, better stability, and less metallic flavour base powder. It is significantly used for the fortification of rice, cereals, and dairy products. In its anhydrous form, it has an approximate molecular weight of 745.2 g/mol. Additionally, it also contains around 25%–30% elemental iron and is produced by reacting ferric salts with pyrophosphate, followed by filtration, drying, and milling. It has lower bioavailability compared to ferrous sulphate. This element reduces the metallic flavour due to ferric phosphate and have oxidation-resistant property of the powder. Additionally, this element enhances vitamin C absorption in the body (Roe et al. 2009).

Ferrous Lactate

Ferrous lactate ($C_6H_{10}FeO_6$) is a green, odourless, with good solubility, mild test, and highly bioavailable iron (II) salt of lactic acid. This element is particularly used in beverages, baby foods, and other fortifications. It contains approximately 19%–20% of elemental iron. It is produced by neutralizing lactic acid with ferrous compounds under controlled conditions. This element is very sensitive and has a high oxidation rate, which produces the bowing off flavour in fortified storage samples if not properly stabilized and packaged (Scheers et al. 2016).

Electrolytic Iron

Electrolytic iron is a pure powdered form of elemental iron (Fe^{o}) produced by the electrolysis of iron salts. It contains approximately 98%–99% elemental iron. It is used in dry base fortification, such as flour and cereal powder. It has excellent stability, is non-reactive, and has no undesirable change such as color, test, and shelf life, but it has lower bioavailability compared to iron salts. Due to lower bioavailability, this element slowly dissolves in the gastrointestinal tract (Ziegler et al. 2011).

Ferrous Fumarate

Ferrous fumarate is a reddish-brown, odourless iron (II) salt of fumaric acid, containing about 33% elemental iron, making it one of the most iron-dense oral supplements. It is commonly used in tablets and fortified foods due to its good bioavailability and lower cost compared to chelated forms. Although it is better tolerated than ferrous sulphate, it may still cause gastrointestinal discomfort in some individuals, such as constipation or nausea. It has moderate solubility, which helps to control its reactivity in food systems but may also slightly reduce absorption in individuals with low stomach acid. In rice fortification, it is not recommended due to its negative impact on colour and taste (Steiger et al. 2014; Yusuf et al. 2022).

Ferrous Bisglycinate (Iron Bisglycinate Chelate)

Ferrous bisglycinate is a chelated form of iron, where ferrous iron (Fe^{2+}) is bound to two glycine molecules. It contains approximately 20% elemental iron. It has more bioavailability and greater ability of gastrointestinal tolerance. In a low acid condition, the chelated form of iron forms stable complexes in the gut. These complexes protect themselves from inhibitory interaction with phytates and enhance the overall absorption and bioavailable capacity of iron. It is more expensive than inorganic iron salt, which is used for fortification mainly in sensitive populations such as pregnant women, children, and anaemic patients (Fischer et al. 2023).

Sodium Iron (III) Ethylene Diamine Tetra Acetate Trihydrate (Sodium Feredetate—NaFeEDTA)

Sodium feredetate, chemically known as sodium iron (III) ethylene diamine tetra acetate trihydrate, is a chelated iron compound used in food fortification. It also contains 13% elemental iron. The EDTA (ethylene diamine tetraacetic acid) forms a stable chelated complex with ferric iron (Fe^{3+}), protecting it from forming insoluble compounds with dietary inhibitors like phytates. It is especially used in food fortification for public health programmes targeting iron-deficiency anaemia. This element is safe and effective as per the World Health Organization (WHO) and the Food and Agriculture Organization of the United Nations (FAO), and it is recommended for food fortification (Hurrell et al. 2000). Meanwhile, in the nutrient-loaded rice fortification, there is still a colour issue that might need to be solved. Additionally, its impact on Vitamin A stability needs to be considered when using it in the formulation.

5.2.2 Vitamins

Vitamins are essential organic compounds that support various physiological functions, and their fortification in food is widely adopted in concern with public health to prevent vitamin deficiencies. Commonly fortified vitamins include vitamin A, D, E, C, folic acid, and vitamin B complex. Among these vitamins, vitamin B_{12} (which supports immunity and disorders) and vitamin D (which is important for calcium absorption and bone health) are extensively used as ingredients for VMP preparation and food fortification purposes.

Folic Acid

Folic acid ($C_{19}H_{19}N_7O_6$) has a molecular weight of 441.4 g/mol. It is a yellow to orange coloured crystalline powder, odourless, and moderately soluble in water and its solubility increases under alkaline conditions. This powder is more sensitive to light, heat, and oxidation, but it is stable in dry conditions. Structurally, folic acid consists of three distinct parts like a pteridine ring, para-aminobenzoic acid (PABA), and glutamic acid linked together via a methylene bridge. This folic acid has different functional groups such as –NH, –COOH, and –C=O. It helps to address folate deficiency, especially in women of reproductive age (Wusigale and Liang 2020).

Vitamin B_{12}

Vitamin B_{12}, also known as cobalamin ($C_{63}H_{88}CoN_{14}O_{14}P$), has an approximate molecular weight of 1355.4 g/mol, which vary depending on its specific form such as cyanocobalamin, methyl cobalamin, or hydroxocobalamin. It is a red crystalline compound with a complex corrin ring structure containing a central cobalt (Co) atom base powder. Vitamin B_{12} in the cobalamin form is more stable and soluble in water and ethanol. While it is more sensitive to light, heat, and acidic or alkaline pH. The cobalt atom in the centre of the molecule can form different ligands (CN^-, OH^-, CH_3^-, etc.), which influence both its biological activity and stability. Its sensitivity to degradation requires protective packaging, especially in fortified food systems.

5.3 FORMULATION

5.3.1 Principles of VMP Formulation

The design of VMP formulation for any food fortification is a delicate step guided by several core principles, as it directly influences the total micronutrient requirement in the target food matrix. The formulation is based on the target nutrient needs, the target population (such as children or pregnant women), age, recommended dietary allowances (RDA), nutrient bioavailability, storage stability, safety, organoleptic properties, cost of fortificant, ingredient compatibility, and interactions with main food ingredients.

The selection and level of each micronutrient are based on the specific nutritional deficiencies of the target population and their RDAs. Moreover, it is important to note that the selection of a particular micronutrient depends not only on sensory acceptability, price, legal status, stability, and expected bioavailability but also on product form fitting and applied fortification techniques (Steiger et al. 2014). For the rice fortification, a premix typically contains the iron, folic acid, and vitamin B_{12} in the ratio of 3,400:100:1 to get the desired level of 28–42.5 mg, 75–125 μg, and 0.75–1.25 μg/kg in fortified rice, respectively (FSSAI 2021). The quantities are considered to be higher level because of the bioavailability and losses of premix during the storage of products. For example, ferrous fumarate or sodium iron EDTA is preferred over elemental iron due to better bioavailability.

For the formulation to prepare, a lot of things need to be considered, such as micronutrient content in carrier material, final micronutrient content required, its efficiency, and losses during processing and storage. Following are the equations (5.1–5.3) through which VMP can be formulated.

$$Y_{AF} = X_A - X_B \tag{5.1}$$

$$Y_{BF} = Y_{AF} + Y_{AF} \times (z_A + z_B + z_C + \ldots) \tag{5.2}$$

$$Y_{SF} = \frac{Y_{BF}}{M_{YB}} \tag{5.3}$$

Where X_A is the final amount of fortificant in the VMP, X_B is the available amount of fortificant present in the carrier material, Y_{AF} is the requirement of fortificants in VMP in particular weight; Y_{BF} is the actual requirement of fortificants in VMP after the addition of fortificant loss during various stages, $(z_A + z_B + z_C + \ldots)$ are the losses during processing, operation, storage, etc., Y_{SF} is the quantity of source of fortificants, and M_{YB} is the fraction of the pure element/nutrient present in the source of fortificant.

5.3.2 Ingredients Compatibility and Interaction

In food fortification, ingredient compatibility and interaction are important factors that influence the stability, efficacy, and safety of fortified products. Some vitamins and minerals can interact chemically or physically with each other or with components of the food matrix, leading to degradation and reduced bioavailability (Bett-Garber et al. 2004). In a coating method, the ferrous sulphate produced the off flavour due to the reaction between iron and hydrogenated oils or ethyl cellulose. For example, iron can catalyse the oxidation of vitamin A or C, reducing their potency and causing rancidity or discoloration. Similarly, calcium can interfere with the absorption of zinc or iron when not properly balanced. To address these challenges, fortification strategies often involve selecting nutrient forms that are more stable or using protective techniques such as microencapsulation or coating. Carriers, flowing agents, and antioxidants may also be added to improve compatibility and prevent unwanted reactions during processing and storage. Additionally, processing conditions like pH, moisture, temperature, and light exposure need to be carefully controlled. Ensuring good ingredient compatibility helps to maintain the nutritional value, shelf stability, and sensory quality of fortified foods, making them more acceptable and effective to consumers.

5.3.3 Bioavailability

Bioavailability refers to the proportion of a nutrient, such as a vitamin or mineral, that is absorbed and utilized by the body after ingestion. In the context of fortified foods or VMP formulations, high bioavailability is essential to ensure that the intended health benefits are effectively delivered. Several factors influence bioavailability, including the chemical form of the nutrient, particle size, source (synthetic vs. natural), presence of enhancers (like vitamin C for iron absorption), and interactions with other food components (e.g., phytates or calcium inhibiting iron absorption). Processing conditions' techniques can also impact nutrient stability and absorption. The bioavailability depends on the solubility in water; water-soluble sources of iron contain higher bioavailability.

5.4 PROCESSING TECHNIQUES

The preparation of VMP follows certain unit operations like cleaning or filtration, weighing, size reduction, dosing and mixing, packaging, and storage (Figure 5.1). This process involves the proper receiving, inspection, storage, and transfer of vitamins, minerals, carriers, and other additives before they enter the production line.

5.4.1 Preprocessing of VMP Mixture

Preprocessing steps are essential to ensure its stability, bioavailability, and proper mixing into the final product. The cleaning steps are used to separate foreign impurities, lumps, agglomerates, oversized agglomerates, and finest particles by using the large-scale sieving or screening (using vibrating screens with specific mesh sizes). The metallic contaminants from equipment wear in VMP premix can be caught through metal detectors to ensure the safety of VMP mix. The 20 mesh (0.84 μm) size is commonly used for the screening of water-soluble vitamins. Similarly, for the different iron sources (like ferrous sulphate) in wheat-based snacks, the 325 mesh (0.44 μm) size sieve is used. BIS also recommended that a particle size of ferric pyrophosphate should be 1–3 μm. Meanwhile, to maintain specific particle size ranges, appropriate screening methods are essential. For instance, rotary vibrating screen consisting of two or more layers of mesh mounted on a circular frame is used for the separation of particles in the VMP premix. This screen vibrates at high frequency and effectively separates material with the desired particle size. The gyratory screen is the circular screen (wire mesh or perforated/slotted sheet) operating with a special vibro motor imparting vibration in multiple directions to the screen assembly, resulting in increased sieving efficiency.

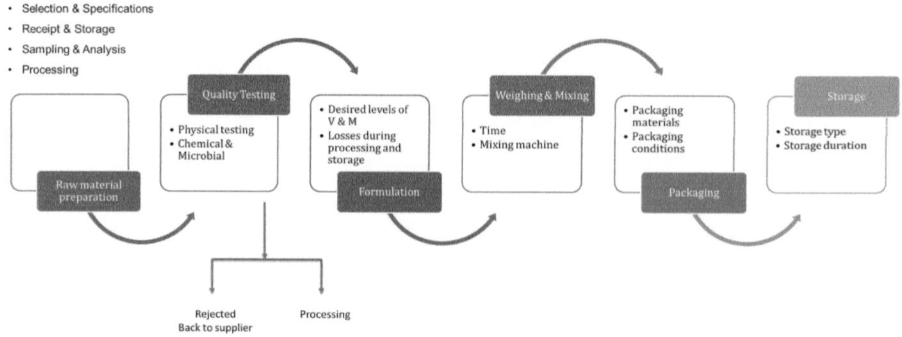

FIGURE 5.1 Unit operation in VMP manufacturing.

5.4.2 Micronization

The process of lowering iron compound particle sizes to the micron scale, usually between 1 and 10 micrometres (μm), is known as "micronization" of iron salts. The micronization can be produced by using the jet milling, ball milling, bead milling, hammer milling, and spray drying with encapsulations. The micronization of iron salts is a vital processing step that enhances the functional properties of these compounds, particularly their bioavailability and sensory characteristics of VMP premix.

5.4.3 Weighing System

A correct and accurate weight of an ingredient is paramount in the manufacturing of VMP, especially for food fortification (i.e., rice fortification). The constant weight of each ingredient is needed to maintain the quality, consistency, and nutritional characteristics in food production (Wine et al. 1997). Usually, the use of separate scales to weigh individual ingredients is prevented as it is costly and leads to inaccuracies in the final batch due to scale variations. Instead, the automatic weighing system is commonly used for precise measurements. It consists of the weighing hopper, load cell sensor, and PLC controlled systems.

In an automated weighing system, the loss feeder is installed over the weighing hopper to ensure precision and continuous dosing of the ingredients into the system. Subsequently, the weighing hopper is mounted on the load cell sensor. The "loss in weight feeder" controls the coarse and fine flow of material during the weighing operation and also controls the dispensing of material by weight at a precise rate in a continuous manner. Engisch (2014) reported that the hopper fill level is the most significant factor that can be used in mitigating the deviation effects during feeding. The feeding accuracy of weighing equipment is dependent on the properties of the food materials (such as bulk density, angle of repose, and internal angle of friction) and mechanical design of the equipment (such as internal surface of hoppers, shock, and vibration). Small weighing and dosing of VMP, especially vitamins, is performed using micro dosing system. It consists of highly sensitive load cell and precise feeders, which are highly suitable for the weighing of trace minerals.

5.4.4 Mixing of VMP

Mixing is an essential unit operation, and its main target is to achieve desired homogeneity, which means that every single particle of the VMP premix, regardless of where it's tested, has the same ratio of all ingredients as specified in the formulation. Both the micro- and macro-mixing are used to mix VMP. Where less than 1% capacity of mix is considered to be micro-mixing. However, the selection of mixing technology depends on the type of food product (such as flour, oil, or premix), ratio of the products, physical properties of the ingredients (such as particle size, density, hygroscopicity, flowability, electrostatic charge, and moisture content), and the desired homogeneity. The dry mixing of VMP is achieved by various types of mixers, i.e., ribbon blender, paddle mixer with single shaft or twin shaft, V-blenders, etc. Table 5.1 depicts the types of mixers and their purposes in the mixing of VMP premix.

TABLE 5.1
Types of Mixers and their usefulness in VMP Premix Manufacturing

Mixer	Usefulness
Ribbon blenders	Used for low-shear mixing of powders to achieve uniform blending
Paddle mixer	Suitable for handling sticky or cohesive materials like liquid vitamins
Double cone blenders	Gentle mixing, suitable for fragile or heat-sensitive nutrients
V-blenders	Effective for free-flowing powders, widely used in premix manufacturing
High-shear mixers	Used when micronutrient particles need to be dispersed into a cohesive mass (e.g., when binding agents are added)

Paddle mixer is suitable for sticky or cohesive materials, has a rapid, high-intensity mixing, and can handle some liquid vitamins. Twin-shaft designs offer even faster and more thorough mixing with lower power consumption. The V-blender is appropriate for the free-flowing powder, but it will take more time for mixing as compared to a ribbon blender. The desirable characteristics of mixing devices are short time, variable degree of filling, high efficiency, good clean ability, breakup of the lumps, smooth surface for easy emptying, no heat production during mixing, lower energy requirement, least maintenance, and cost-effectiveness. The mixing speed (45–60 rpm) and time (2–5 min) are common variables needed to optimize desired homogeneity. Most manufacturers are using the less than 5% of coefficient of variation for homogeneity of mixing.

For the better performance for homogeneity of mixing, the particle size and density of particles should be matching. As the larger size of particles is floated on the surface and the finest in the bottom, too long mixing time will create segregation of materials, while premix particles are not properly dispersed in a short period of time. The selection of carrier agent also affects the homogeneity of mixing, so the same characteristics of the materials should be taken into consideration. Rice starch should be used for the rice fortifications. Carrier should be chemically inert, non-hygroscopic, and have physical properties that should match with food. The appropriate surface characteristics, i.e., static charges and hygroscopicity of powder particles, also influence the mixing of VMP because static charges promote clinging to the wall of the mixer. The mixing index of VMP premix is also influenced by the degree of filling. The ideal degree of filling is used up to 50%–75% of their capacity for better mixing (Zimmermann et al. 2005).

5.4.5 Packaging and Storage of VMP

Packaging plays an important role in preserving the quality, safety, and shelf life of VMP premix during the storage. Selection of an inappropriate packaging material led to the production of undesirable changes. Packaging materials for VMP premix must be low-cost, compatible, protective against environmental factors, provide gas and water barriers, meet regulations, and ensure machinability with good printability. The ink should be food grade and should not react with VMP premix during storage. Good design of package and packaging should enhance product visibility, provide necessary information, and ensure convenience while maintaining the integrity of the contents throughout storage and distribution.

The multilayer packaging film consists of polypropylene for strength, a middle layer of aluminum foil or metallic sheet for protection from light and gas, and an inner layer of polyethylene for a moisture barrier that is used for VMP packaging. HDPE containers are good for transportation and robustness. Multilayered bags with excellent barrier attributes, particularly those using aluminum foil or polyethylene liners, are prevalent for VMP premixes due to their efficacy and practicality. The appropriate foil for lipid-based products has an inner layer suitable for an oily product that ensures a resistant seal, followed by a thin metalized film that creates a barrier for oxygen, moisture, and ultraviolet transmission and is often used as an additional outer layer. The printing process must be designed in such a manner that the ink should not be ingested, if the product is eaten directly from the sachet. Sachets are packed in low oxygen atmosphere with flushing of nitrogen gas to reduce the oxidation reactions of the product inside the pouches.

The main objective is to ensure precise dosage, maintain product integrity by preventing contamination and deterioration, and offer efficient packaging for diverse end-use formats. The vertical form fill seal machines equipped with auger filler mechanisms are used for the packaging of pouches from 500 g to 5 kg. For bulk packaging of the VMP, i.e., 20–50 kg, the bulk bagging machines are employed. These machines consist of a multi-head weighing load cell system for maintaining the accuracy of the VMP premix in packets.

Proper storage involves maintaining optimal environmental conditions such as temperature, humidity, and light exposure to prevent spoilage, contamination, or nutrient loss. Similarly, careful handling during transportation, loading, and unloading helps to avoid physical damage, microbial contamination, or chemical changes. Adhering to good storage and handling practices ensures product integrity from production to consumption and is essential for regulatory compliance and consumer satisfaction. The shelf life of a particular food product depends on factors such as product formulation, processing technique, storage, and type of distribution (Kilcast and Subramaniam 2000). The surrounding atmospheric relative humidity penetrating inside the product through pore spaces of pouch increases the water activity of the product, and this is highly favourable for the growth of microbes that produce toxic and undesirable nutritional characteristics.

5.4.5 GMP/GHP and Quality Management in Building & Equipment Layout and Production

The quality assurance systems are used to improve quality and hygiene and reduce costs, while hazard analysis critical control point (HACCP) programmes are specifically deployed to assure food safety. Based on quality management principles, HACCP, GHP, and GMP are implemented to help plants to maintain high levels of hygiene and quality (Sampers et al. 2012). Implementation of QA and HACCP in concert improves production efficiency as well as product safety, as it is a popular worthwhile method in the food industry for the generation of uncontaminated and sanitized manufacturing practices that produce safe-to-consume food items. Considering the above factors, an area that needs attention in the manufacture of the VMP premix is the relationship with the hygiene facility and process equipment layout. One of the crucial elements in current food processing is the safety of the products (Brennan 2006). Most consumers equate food safety with adherence to the performance of the ultimate hygienic practices during VMP production. However, failure in maintaining high hygienic standards has a direct influence on production efficiency through loss of production and requirement of the product recalls, but an even greater effect on loss of consumer confidence (Pendergast 2021). Furthermore, the cross-contamination is a major contributing factor for the growth of *Salmonella* and *Listeria monocytogenes*, and it causes food borne diseases in humans (Carrasco, Morales-Rueda, and García-Gimeno 2012).

The poor design, cracks, and crevices in floors and walls, the interface between the floor and equipment, and hollow structures in the building or in equipment cause the growth of pathogenic bacteria in food products. The building floors and walls should be easy to clean, free of particulate matter, and unable to harbour bacteria. The corners of the building should have covered corners where the walls, floors, and ceilings meet. The eroding of tile grouting and cracked tiles serves bacterial growth, so it should not be used in high-care areas of production. The polyurethane (PU) flooring should be used on the top of a cemented, tiled, or stone surface. Surfaces and sealing that are not easy to clean and act as a vector of contamination or dust accumulation in the processing areas should be avoided. Window sills must be avoided and flush/double glazing should be used. Doors shall be made of smooth and non-absorbent surfaces so that they are easy to clean and disinfect wherever necessary. Ceilings and overhead fixtures shall be designed, constructed, finished, and maintained so as to minimize the accumulation of dirt. Properly lighted, ventilated, and painted with proper drainage and waste disposal systems, as well as smoke/steam outlets, are requirements of a production premises.

To meet both ends of desired quantity and quality output, plant layout is designed in such a way to obtain an ideal spacing and equipment arrangement of different entities most economically and minimize overall operating costs. The three main classes of layouts are product, process, and group layout. The main target of layout is the location of facilities, equipment, and staffs that are primarily based on their mutual relationship. The layout of the food establishment shall be such that the food preparation/manufacturing process is not subject to cross-contamination. Raw materials and packing materials entering a high care area need to pass through a clean-down airlock to limit the ingress of contaminants into the high care area by effective cleaning of the materials. Staff flow and the layouts and locations of change rooms and ablutions are equally important. Pathogens can also enter with fresh products from the environment and equipment. Poorly designed equipment can contain niches where sanitizers are not able to penetrate and pathogens are protected. Common machine defects, improperly or poorly welded surfaces, and poor junctions can accumulate water and lead to the formation of biofilm of pathogens.

Additionally, equipment and containers to be handled in food processing operations should be made from corrosion-resistant and safe materials. For adequate protection against foreign materials, each utensil or container should be properly covered with clean gauze net or closed with a lid or any other material of suitable surface. The personnel facilities, including changing room and wash rooms, are the prerequisites and shall be located in such a way that these are not directly concerned into food processing, handling, or storage areas.

5.5 QUALITY ASSURANCE AND CONTROL (QA/QC) IN VMP PRODUCTION

5.5.1 Quality Characteristics

Physical Characterization

a. **Particle size distribution**: It is essential for powder to promote uniform mixing and is measured using sieve analysis technique.
b. **Bulk and tapped density:** It affects packaging and mixing, measured by filling a graduated cylinder with VMP and recording weight per unit volume, both before and after tapping.
c. **Flowability:** It is measured by the angle of repose or flow rate of powder to assess ease of handling and uniform blending. Poor flow rate of powder can cause segregation.
d. **Moisture content:** It is a key factor in preventing microbial growth and is determined using oven drying at 105°C until constant weight reached.

Chemical Analysis

a. **Vitamin assay:** Quantitative analysis of vitamins such as vitamin B_9 and B_{12} is carried out using high-performance liquid chromatography and hyperspectral imaging.
b. **Mineral content analysis:** Minerals like iron are analysed using atomic absorption spectroscopy (AAS).
c. **pH and acidity:** They determine chemical stability and potential interactions of the premix. They are measured by using a pH meter on an aqueous premix suspension.

Microbiological Testing

a. **Total viable count (TVC):** The microbial load should be within acceptable limits to prevent spoilage or health risks. TVC is measured by using the plate count method on plate count agar, with incubation at 30°C–37°C for 48 h.
b. **Yeast and mould count:** Fungal contamination is determined using selective media like potato dextrose agar, incubated at 28°C for 72 h.
c. **Pathogen testing:** It confirms the presence and absence of pathogens such as *Salmonella* spp. and *Escherichia coli* by using media and plating techniques.
d. **Shelf-life evaluation:** It involves real-time monitoring of nutrient stability under recommended storage conditions until the end of shelf life.

5.5.2 Quality Assurance

Good Manufacturing Practices (GMP)

QA in food fortification ensures that all processes involved in the production of fortified foods and premixes consistently meet established safety, quality, and regulatory standards. A key component of QA is adherence to GMP, which is a set of internationally recognized guidelines that provide a framework for the systematic control of production, ensuring products are consistently produced and controlled according to quality standards.

GMP includes important protocols for facility hygiene, equipment maintenance, staff training, standard operating procedures (SOPs), and record-keeping. It also mandates traceability from raw material sourcing to final packaging, ensuring that the micronutrients used in fortification are of high quality and are safely handled. GMP minimizes the risks of contamination, mix-ups, or errors during production, which is especially critical in vitamin and mineral fortification, where nutrients are added in small, precise amounts. By implementing QA through GMP, manufacturers not only comply with regulatory requirements (such as FSSAI in India or Codex Alimentarius globally) but also build consumer trust by delivering safe, effective, and high-quality fortified products (Tola 2025).

Hazard Analysis and Critical Control Points (HACCP)

HACCP is essential for food safety that identifies, evaluates, and prevents biological, chemical, and physical risks rather than relying solely on end-product testing. The HACCP system involves seven key steps: conducting a hazard analysis, identifying critical control points (CCPs), setting critical limits, monitoring procedures, establishing corrective actions, verification procedures, and maintaining documentation. By implementing HACCP, manufacturers can ensure the safety and quality of VMP, comply with regulatory requirements, and protect consumer health through proactive risk management (Bereda 2021).

Process monitoring and preventive controls are essential practices in ensuring the consistent quality, maintaining product integrity, reducing variability, and ensuring regulatory compliance of VMP production. Process monitoring involves the continuous observation and recording of key parameters such as mixing time, temperature, humidity, and equipment performance during manufacturing. Preventive controls are proactive measures implemented to reduce or eliminate potential risks, such as microbial contamination, nutrient degradation, or cross-contamination. These controls may include equipment calibration, hygiene protocols, the use of controlled environments, and adherence to SOPs.

Documentation and traceability are critical components of quality assurance in the production of VMP. Documentation involves maintaining accurate and complete records of each stage of production, including raw material sourcing, batch formulation, processing parameters, in-process checks, and final product testing. Traceability ensures that every ingredient and finished product can be tracked backward (to its source) and forward (to its distribution). This system helps to identify and isolate any issues quickly in case of a defect, contamination, or recall. Effective documentation and traceability not only support regulatory compliance but also enhance transparency, accountability, and consumer safety (Tola 2025).

5.5.3 Quality Control

Raw Material Testing

Raw material testing is the crucial step, which involves verifying the identity, purity, potency, particle size, moisture content, and absence of contaminants (such as heavy metals, pathogens, or adulterants) in all incoming ingredients. This testing ensures that only high-quality, compliant raw materials enter the production process, which directly influences the stability, efficacy, and safety of the final premix. Additionally, microbial analysis is used to detect and quantify microorganisms such as bacteria, yeasts, moulds, and pathogens in food products. Common tests include total plate count, coliform count, yeast and mould count, and specific pathogen detection (e.g., *Salmonella*, *E. coli*, and *Listeria*). The results guide producers in evaluating hygiene practices, validating shelf life, and meeting regulatory standards. Regular microbial testing is essential to maintain food safety, quality assurance, and consumer protection (Godswill, Somtochukwu, and Kate 2020).

5.6 ISSUES AND CHALLENGES IN VMP MANUFACTURING

The production of VMP involves several key challenges that can affect the quality, stability, and effectiveness of the final product:

a. Ensuring uniform mixing and distribution of micronutrients, especially when ingredients vary widely in particle size, density, and flow properties.
b. Maintaining the stability of sensitive vitamins and minerals, which can degrade due to heat, moisture, light, or interaction with other components.
c. Preventing cross-contamination, meeting strict regulatory standards, and achieving high bioavailability are also significant hurdles.
d. Bioavailability of the micronutrient is another issue as not all micronutrients are equally bioavailable.
e. Raw ingredient handling, as it directly affects product quality, safety, and consistency. Factors such as moisture control, protection from light and heat, and segregation of incompatible materials are important to preserve ingredient integrity. Effective raw ingredient handling lays the foundation for a high-quality and compliant premix formulation.
f. Complying with distinct regulatory requirements demands careful formulation of each ingredient.

5.7 TRENDS AND INNOVATIONS IN VMP PRODUCTION

The demand for VMP is flourishing worldwide due to emphasis on health, wellness, and preventive care. The VMP is highly sensitive to physical contaminations, reacts with other ingredients, and results in a decline in efficacy. The latest research on microencapsulation, nanotechnology, advanced mixing and weighing by automation technologies, and more focuses on the quality control during the production to storage of VMP. A novel biopolymeric coacervate utilizing Arabic Gum and lesser mealworm protein isolate (US-LMPI) encased Vitamin B_{12} was effectively generated by spray drying, leading to microparticles with semi-spherical morphologies and achieving 87.69% encapsulation efficiency (Gharibzahedi et al. 2025).

5.8 CONCLUSIONS

The production of VMP for food fortification, particularly in staple foods like rice, plays a crucial role for addressing micronutrient deficiencies in vulnerable populations. The essential prospect of VMP production includes the selection of raw material forms, maintaining physicochemical properties, and premix storage stability during subsequent processing. Furthermore, uniform dispersion, moisture resistance, and degradation of premix powder are essential for functional integration into fortified food matrices. However, several challenges, including the degradation of heat and light-sensitive vitamins during extrusion, interactions between minerals and matrix components, and scalability issues in low-resource settings, exist. Best practices should emphasize compliance with BIS and Codex standards through the application of GMP, HACCP protocols, and real-time monitoring of uniform mixing and nutrient retention analysis. It is also recommended to conduct quantitative nutrient analysis. Additionally, selection of good packaging material for VMP helps to maintain the stability and protects the product from external environmental conditions. VMP production can significantly contribute to sustainable nutrition security and improve the hidden hunger globally.

REFERENCES

Bereda, G. 2021. "Quality assurance and quality control. current furtherances and hereafter point of view." *Journal of Analytical and Pharmaceutical Research* 10: 212–215. https://medcraveonline.com/JAPLR/JAPLR-10-00386.php

Bett-Garber, K. L., E. T. Champagne, D. A. Ingram, and C. C. Grimm. 2004. "Impact of iron source and concentration on rice flavor using a simulated rice kernel micronutrient delivery system." *Cereal Chemistry* 81 (3): 384–388. https://doi.org/10.1094/CCHEM.2004.81.3.384.

BIS. 2021. Vitamin and Mineral Premix for Manufacturing Fortified Rice Kernels—Specification (IS 17781:2021). New Delhi: Bureau of Indian Standards, Government of India. https://standardsbis.bsbedge.com/BIS_searchstandard.aspx?Standard_Number=IS+17781&id=34818

Brennan, J. G. 2006. *Agglomeration in Industry Freeze-Drying Plastic Packaging Materials for Food and Pharmaceuticals Handbook of Flavourings Common Fragrance and Flavor Materials Coffee: Growing, Processing, Sustainable Production Agrochemicals.* In Traders.

Carrasco, E., A. Morales-Rueda, and R. M. García-Gimeno. 2012. "Cross-contamination and recontamination by *Salmonella* in foods: A review." *Food Research International* 45 (2): 545–556. https://doi.org/10.1016/j.foodres.2011.11.004.

Engisch, W. 2014. "Loss-in-weight feeding in continuous powder manufacturing." *Chemical and Biochemical Engineering*. New Brunswick: Rutgers, The State University of New Jersey.

Fischer, J. A. J., A. M. Cherian, J. N. Bone, and C. D. Karakochuk. 2023. "The effects of oral ferrous bisglycinate supplementation on hemoglobin and ferritin concentrations in adults and children: A systematic review and meta-analysis of randomized controlled trials." *Nutrition Reviews* 81 (8): 904–920. https://doi.org/10.1093/nutrit/nuac106.

FSSAI. 2021. "Food Safety and Standards (Fortification of Foods) Regulations, Version–I (02.03.2021)." New Delhi: FSSAI Regulatory Document. https://fssai.gov.in/upload/uploadfiles/files/Compendium_Food_Fortification_Regulations_04_03_2021.pdf

FSSAI. 2022. "Food Safety and Standards (Food Products Standards and Food Additives) Regulations, Version-XXVI (20.12.2022)." New Delhi: FSSAI Regulatory Document. https://fssai.gov.in/upload/uploadfiles/files/Compendium_Food_Additives_Regulations_20_12_2022.pdf.

Gharibzahedi, S. M. T., G. K. Hasabnis, E. Akin, and Z. Altintas. 2025. "Molecularly imprinted polymers-based electrochemical sensors for tracking vitamin B_{12} released from spray-dried microcapsules during in vitro simulated gastrointestinal digestion." *Sensing and Bio-Sensing Research* 47: 100759. https://doi.org/10.1016/j.sbsr.2025.100759.

Godswill, A. G., I. V. Ikechukwu Somtochukwu, A. O., and E. C. Kate. 2020. "Health benefits of micronutrients (vitamins and minerals) and their associated deficiency diseases: A systematic review." *International Journal of Food Sciences* 3 (1): 1–32. https://doi.org/10.47604/ijf.1024.

Hurrell, R. F., M. B. Reddy, J. Burri, and J. D. Cook. 2000. "An evaluation of EDTA compounds for iron fortification of cereal-based foods." *British Journal of Nutrition* 84 (6): 903–910. https://doi.org/10.1017/S0007114500002531.

Kilcast, D., and P. Subramaniam. 2000. *The Stability and Shelf-Life of Food*. Boca Raton, Washington DC: Woodhead Publishing Limited and CRC Press LLC.

Pendergast, D. 2021. *Food Safety and Hygiene*. Springer eBooks, 145–165. https://doi.org/10.1007/978-981-16-5415-2_6

Roe, M. A., R. Collings, J. Hoogewerff, and S. J. Fairweather-Tait. 2009. "Relative bioavailability of micronized, dispersible ferric pyrophosphate added to an apple juice drink." *European Journal of Nutrition* 48 (2): 115–119. https://doi.org/10.1007/s00394-008-0770-3.

Sachdeva, B., R. Kaushik, S. Arora, and S. Kapila. 2015. "Bioavailability of iron in multiple fortified milk." *Journal of Food Science and Technology* 52 (9): 6017–6023. https://doi.org/10.1007/s13197-015-1711-9.

Sampers, I., H. Toyofuku, P. A. Luning, M. Uyttendaele, and L. Jacxsens. 2012. "Semi-quantitative study to evaluate the performance of a HACCP-based food safety management system in Japanese milk processing plants." *Food Control* 23 (1): 227–233. https://doi.org/10.1016/j.foodcont.2011.07.018.

Scheers, N., L. Rossander-Hulthen, I. Torsdottir, and A.-S. Sandberg. 2016. "Increased iron bioavailability from lactic-fermented vegetables is likely an effect of promoting the formation of ferric iron (Fe^{3+})." *European Journal of Nutrition* 55 (1): 373–382. https://doi.org/10.1007/s00394-015-0857-6.

Steiger, G., N. Müller-Fischer, H. Cori, and B. Conde-Petit. 2014. "Fortification of rice: technologies and nutrients." *Annals of the New York Academy of Sciences* 1324: 29–39. https://doi.org/10.1111/nyas.12418.

Tola, G. B. 2025. "Food Contaminants: A Scoping Review of Sources, Toxicity, Pathophysiological Insights, and Mitigation Strategies." *Preprints.org* https://doi.org/10.20944/preprints202502.0030.v1.

Wine, R. N., L. H. Li, L. H. Barnes, D. K. Gulati, and R. E. Chapin. 1997. "Reproductive toxicity of di-n-butylphthalate in a continuous breeding protocol in Sprague-Dawley rats." *Environmental Health Perspectives* 105 (1): 102–107.

Wusigale, and L. Liang. 2020. "Folates: Stability and interaction with biological molecules." *Journal of Agriculture and Food Research* 2: 100039. https://doi.org/10.1016/j.jafr.2020.100039.

Yusuf, D. N., D. M. M. Khatun, D. N. Nahar, D. R. Akhter, D. M. Z. Rahman, D. M. N. Parvej, and D. A. Ahmed. 2022. "Efficacy of ferrous fumarate in comparison to carbonyl iron in the treatment of iron deficiency anaemia during pregnancy in Rajshahi, Bangladesh." *Scholars International Journal of Obstetrics and Gynecology* 5 (5): 216–222. https://doi.org/10.36348/sijog.2022.v05i05.001.

Ziegler, E. E., S. J. Fomon, S. E. Nelson, J. M. Jeter, and R. C. Theuer. 2011. "Dry cereals fortified with electrolytic iron or ferrous fumarate are equally effective in breast-fed infants." *The Journal of Nutrition* 141 (2): 243–248. https://doi.org/10.3945/jn.110.127266.

Zimmermann, M. B., P. Winichagoon, S. Gowachirapant, S. Y. Hess, M. Harrington, V. Chavasit, S. R. Lynch, and R. F. Hurrell. 2005. "Comparison of the efficacy of wheat-based snacks fortified with ferrous sulfate, electrolytic iron, or hydrogen-reduced elemental iron: randomized, double-blind, controlled trial in Thai women." *The American Journal of Clinical Nutrition* 82 (6): 1276–1282. https://doi.org/10.1093/ajcn/82.6.1276.

6 Pulverization of Rice

Shubham Mandliya, Siddhesh Kumar Dubey, Arpan Dubey, Sourav Misra, Siddharth Vishwakarma, and Hari Niwas Mishra

6.1 INTRODUCTION

Rice is a major food mainly used globally for more than fifty percent of the population in several forms. In food processing, rice is used as flour, which is produced through several milling methods. Milling has the leading role in making rice flour, and it varies in particle size distribution and degrees of damaged starch. These variations in rice flour properties have an important role on dough quality and may affect the final product. Consequently, the method of pulverization is crucial when exploring new food processing applications for rice flour in making new products (Kamata and Nei 2024). Pulverization plays a vital role in converting whole rice grains into flour, which is used in several food production units such as gluten-free baking, instant food formulations, and snack production (Sapna et al. 2019). Pulverization increases the digestibility of food items, allows for more efficient and homogeneous mixing, and improves the solubility and suspension in liquid systems features, which are difficult to achieve with other milling methods because of the coarser particle size of rice flour. Additionally, the higher surface area of rice flour after pulverization accelerates biochemical processes like enzymatic activity and fermentation. In essence, pulverization provides higher functional and processing benefits than other milling methods. It also contributes to better texture, extended shelf life by reducing moisture retention, and easier packaging and handling due to better powder flow properties (Tian et al. 2023).

Therefore, optimizing the pulverizer parameter is crucial for rice flour to have better properties for specific industrial and culinary applications. The pulverization of rice is a major preliminary step in manufacturing fortified rice kernels. These kernels are designed to face nutritional deficiencies by adding essential micronutrients such as iron, folic acid, and vitamin B_{12}. Rice flour with coarser particle size is difficult to mix with micronutrients. Therefore, pulverization of rice is crucial to ensure the homogeneous mixing of rice flour with nutrients, allowing for uniform micronutrients distribution throughout each kernel. Furthermore, pulverized rice provides better flowability for extrusion processing, the most common process for manufacturing FRK and providing a better shape. The fine particle size of rice flour with micronutrients provides better binding, shaping, and textural integrity of the extruded kernels, making them similar to raw/parboiled rice grain visually and physically. This similarity of kernels with normal rice grains is necessary for public

DOI: 10.1201/9781003661245-6

acceptance and blending operation. Additionally, extruded kernels manufactured from finely pulverized rice flour provide better mechanical strength and cooking stability than normal rice, ensuring that they retain their shape and do not leach micronutrients or disintegrate during washing or cooking.

6.2 SIZE REDUCTION VS PULVERIZATION AND ITS MECHANISM

Size reduction is a mechanical process used to reduce the size of solid materials into smaller particles, ranging from large fragments to fine particles. It includes several methods, such as crushing, grinding, shearing, and cutting, depending on the material type and the desired final particle size. In size reduction, impact, attrition, compression, and shear are the main mechanisms involved. Major size reduction equipment such as hammer mill, cutter mill, jaw crusher, and roller mill are commonly used. It works on the principle of applying mechanical force to overcome the internal binding forces of particles, therefore breaking them into smaller sizes. The desirable finer particle size is achieved after selecting proper grinding equipment and optimizing process parameters such as speed, temperature, feed rate, and feed moisture content. These parameters affect the final particle size of the product and energy consumption. These parameters are required to adjust to achieve the desired particle size at minimum energy consumption. Size reduction is primarily used in the food processing industries, improved mixing, increased reaction rate, and proper material handling.

Conversely, pulverization is a more specific form of size reduction that aims to produce ultra-fine powders, typically below 100 microns in size. It is primarily achieved through high-energy impact and attrition forces using specialized equipment such as jet mills, ball mills, and ultra-fine grinders. The principle behind pulverization centers on imparting high-velocity impact and frictional forces to grind the material into a dust-like consistency. In industrial applications, pulverization is typically used when a uniform and very fine particle size is essential, such as in producing fine powders for pharmaceuticals, pigments, or explosives. The working mechanism is characterized by rapid repeated impacts, with materials colliding against hard surfaces or other particles to fracture into even smaller pieces.

6.3 MECHANISMS AND ENGINEERING PRINCIPLES

6.3.1 Mechanisms

Compression/Crushing

- Material is compressed by an external force beyond its strength.
- Slow application of force causes rupturing in many directions, resulting in irregularly shaped particles.
- Preferred for crystalline, hard, friable materials with abrasive or non-sticky surfaces.
- Used for grains, sugars, salt, mineral stones, and oilseed crushing to increase surface area.

Impact

- Sudden force applied beyond strength of the material leads to fracture through weak regions.
- Creates new surfaces and utilizes energy primarily for fracturing.
- Used to create fine powder from a wide range of feed, e.g., nut cracking and size reduction of fibrous food.

Cutting

- Involves a sharp knife forced through material at high speed to create smaller pieces.
- Forces include compression, friction, deformation, and separation.
- The cutting process is divided into start-up, deformation, separation, and detaching phases.
- Dominating force depends on food material's nature, with friction dominant in homogeneous solids and separation dominant in porous solids and vegetable tissues.

Shearing

- Combines cutting and crushing actions using a knife and a bar.
- Thickness of knife edges determines whether cutting or crushing occurs.
- Ideal shearing unit has sharp, thin-edged knife with small clearance between bar and knife.
- Preferred for soft to medium-hard materials producing relatively coarse particles.

Attrition

- Involves shear and compression causing rubbing or wearing action due to friction.
- Material placed between two rotating metal discs moving in opposite directions.

6.3.2 Engineering Principles

Kick's Law

Kick reckoned that the energy required for a given size reduction was proportional to the size reduction ratio, and took the value of the power n as 1. In such a way, the following relation (Eq. 6.1), known as Kick's law, is obtained.

$$E = K \ln \frac{x_1}{x_2} \tag{6.1}$$

Where x_1/x_2 is the size reduction ratio and K is Kick's law constant. Kick's law has been found to hold more accurately for coarser crushing, where most of the energy is used in causing fracture along existing cracks. Kick's law is more suitable where coarser grinding takes place.

Rittinger's Law

Rittinger's assumption was based on the fact that particles do not deform before breaking, hence being infinitely brittle. They considered that for the grinding of solids, the energy required should be proportional to the new surface produced, and gave to the power n the value of 2, thus obtaining the so-called Rittinger's law (Eq. 6.2).

$$E = K\left[\frac{1}{x_1} - \frac{1}{x_2}\right] \tag{6.2}$$

Where E is the energy per unit mass required for the production of a new surface by reduction, K is called Rittinger's constant and is determined for a particular equipment and material, x_1 is the average initial feed size, and x_2 is the average final product size. Rittinger's law has been found to hold better for fine grinding, which results in a large increase in surface.

Bond's Law

According to this law, the work necessary for reduction was inversely proportional to the square root of the size produced. In Bond's consideration, n takes the value of 3/2, giving the following version of Bond's law (Eq. 6.3).

$$E = 2K\left[\frac{1}{\sqrt{x_p}} - \frac{1}{\sqrt{x_f}}\right] \tag{6.3}$$

Where x_p and x_f are measured in μm and E in kWh/ton, $K = 5Ei$, with Ei being the Bond Work Index, defined as the energy required to reduce a unit mass of material from an infinite particle size to a size such that 80% passes a 100 μm sieve. This law is suitable for a variety of materials undergoing coarse, medium, and fine size reduction.

6.4 METHODS AND EQUIPMENT

6.4.1 Methods

Traditional Methods

The traditional method of rice pulverization is mainly used in Asian countries such as India and China. This method, widely used to produce fine rice flour for culinary and ceremony purposes, predates modern milling methods. Manual or semi-manual tools are needed to maintain the nutritional integrity of grain. Traditionally, this happened using stone grinders composed of two circular stones, one stationary and the other rotating manually or with the help of animals. Although this traditional process is time-consuming and labor-intensive, it is highly valued for retaining the rice's nutritional qualities and authentic flavor. It also reduced the heat generated during the grinding process, decreasing the nutritional loss (Yu et al. 2024a).

Novel Methods

A novel method of rice pulverization uses newer technological advances to increase efficiency, consistency, and nutritional retention. These modern technologies provide a cleaner, faster, and more scalable alternative than the traditional method. They also enable better control over particle size distribution and texture, making the resulting rice flour highly suitable for specialized industrial and health-focused food products. With the advancement of food processing technologies, novel methods like cryogenic grinding have revolutionized rice pulverization. Unlike traditional dry or wet grinding, cryogenic grinding uses liquid nitrogen to supercool rice grains, making them brittle and easier to pulverize. A novel approach to rice pulverization involves using extremely low temperatures to make the grains brittle for easy grinding without heat damage (Boland and Bridget 2014). After getting cooled, the grains are ground using jet or impact mills, which work at high speed without producing heat. This low temperature helps to preserve the functional properties of rice such as gelatinization and water absorption, which are generally degraded in conventional methods due to heat generation. As a result, the final rice flour is ultra-fine, uniform in particle size, and ideal for use in gluten-free baking, baby food, and other high-performance applications such as fortification.

Another method applies high-frequency sound waves to create micro-vibrations that enhance grinding efficiency (Liu et al. 2025). Additionally, high-speed steam jets or air break down the grains through particle collisions and uniformly produce flour of fine particle size (Angelidis et al. 2016). These modern techniques of rice pulverization reduce the processing time, maintain nutritional and functional properties, and provide better flour texture, making it perfect for gluten-free and infant food applications (Gao et al. 2020).

6.4.2 Equipment

Conventional Equipment

Roller Mill

Roller mill consists of two or more heavy steel cylinders rotating towards each other, effectively compressing and pulling particles of feed through the gap between them (Figure 6.1). As a result, the particles experience compressive forces, leading to a reduction in size. Some designs also maintain differential speeds to apply shearing forces on the particles. The surface of the rollers can be smooth or feature corrugations, breaker bars, or teeth, which enhance friction and aid in trapping particles between the rolls. Toothed-roll crushers may be arranged in pairs similar to smooth-roll crushers or with a single roll working against a stationary curved breaker plate. Toothed-roll crushers offer greater versatility compared to smooth-roll crushers, although they may struggle with very hard solids. They operate through a combination of compression, impact, and shear forces, unlike smooth-roll crushers, which rely solely on compression. Roller mills are primarily used in commercial wet milling processes for rice flour production.

Roller mills have substantial advantages, particularly their capacity to produce uniformly sized particles with minimum fines or big fragments, hence improving

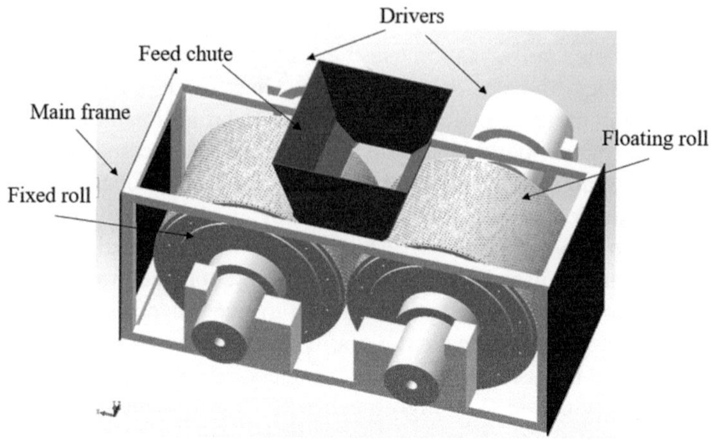

FIGURE 6.1 Schematic of roller mill. (Adapted from Wang, Du, and Guosheng 2023.)

the quality of the finished product. They function with diminished heat production and moisture loss, aiding in the preservation of the material's nutritional integrity. These devices operate at comparatively low noise levels, fostering a safer and more agreeable work environment. Roller mills exhibit superior efficiency by consuming less energy, rendering them economically beneficial for large-scale milling operations. However, they are unsuitable for processing fibrous materials or objects with flat, two-dimensional configurations. Optimal performance is observed with readily processed materials, such as wheat grains. Frequent manual changes are essential to sustain the appropriate particle size, requiring the operator to meticulously calibrate the roll settings.

Hammer Mill

A standard hammer mill primarily operates based on the principle of impact, with some attrition occurring as well. Illustrated in Figure 6.2, a hammer mill consists of a high-speed rotor rotating inside a cylindrical case. The rotor carries multiple hammers around its perimeter. As the rotor rotates, the hammers swing in a circular path within the casing, which contains a sturdy breaker plate. The material feed enters the action zone where the hammers propel it against the breaker plate, compelling it to pass through a bottom-mounted screen due to gravity once it reaches the appropriate size. Reduction primarily occurs due to impact forces, although attrition forces can also contribute to reduction under certain feeding conditions.

The hammers can be substituted with knives or other tools to enable the mill to handle tough, ductile, or fibrous materials. The hammer mill is a versatile piece of equipment capable of achieving high reduction ratios and accommodating a wide range of materials, including hard and abrasive as well as fibrous and adhesive substances.

Although hammer mills are known for their high efficiency and performance, they come with certain limitations. The rapid rotation of internal components can produce considerable heat, which may be detrimental to heat-sensitive materials and

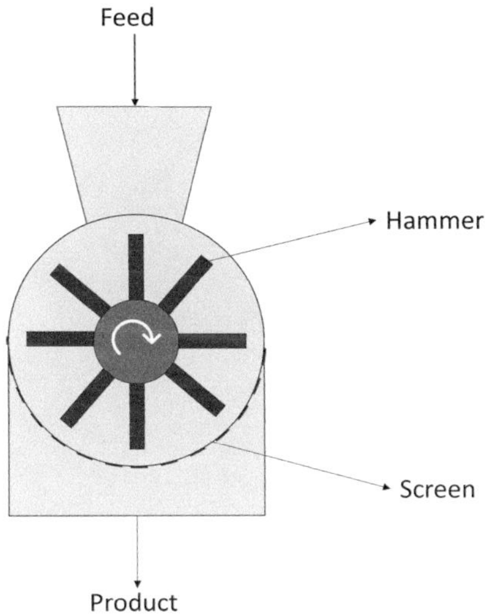

FIGURE 6.2 Schematic of hammer mill (inner view).

could compromise the nutritional integrity of grains and other food items. In addition, the intense friction from high-speed hammer movement leads to mechanical wear, necessitating frequent upkeep. Nevertheless, their ability to operate continuously, process materials quickly, and produce a range of particle sizes—from coarse to moderately fine—makes them a widely utilized milling solution.

Ball Mill

A ball mill is a widely used grinding device that operates on the principle of impact and attrition. It consists of a hollow cylindrical container mounted on a metallic frame, which rotates along its longitudinal axis (Figure 6.3). The cylinder is partially filled with balls, typically made of hard materials such as steel or ceramic, occupying about 30%–50% of its volume. These balls are responsible for crushing and grinding the material placed inside the mill. The process begins by loading the material and the balls into the container, which is then sealed. As the cylinder rotates, the balls tumble and collide with each other and the walls of the cylinder, causing the material to break down into smaller particles.

Ball mills are effective for both wet and dry grinding processes, making them highly versatile in various industries, including food processing. The speed of rotation is critical in determining the efficiency of the size reduction. The balls merely slide over one another at low speeds, producing minimal impact. At higher speeds, however, the centrifugal force can cause the balls to be thrown against the walls, enhancing the grinding process through both impact and abrasion. Once the grinding is complete, the mill is stopped, and the finely ground material is discharged.

Ball mills provide multiple benefits, such as the capability to generate finely ground materials and the option for continuous operation. They can also be paired with sieving systems or classifiers to reach the desired particle size. Despite these advantages, the grinding process in a ball mill tends to be relatively slow, and the

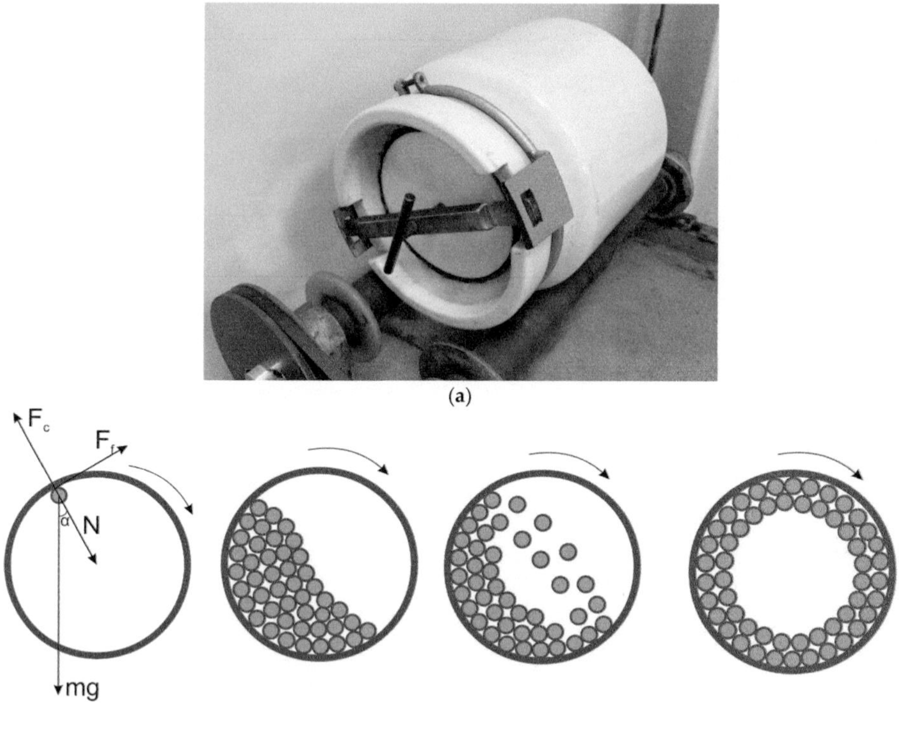

(a) (b) (c) (d) (e)

FIGURE 6.3 Ball mill (a), the effect of forces (b), and the types of motion of grinding bodies in a ball mill: (c) rolling over, (d) falling, and (e) rolling. (Adapted from Lomovskiy et al. 2020.)

constant movement of the grinding media produces considerable noise. Nevertheless, due to their reliable performance and adaptability, ball mills remain a preferred choice for many grinding applications.

Disc Attrition Mill

These mills, depicted in Figure 6.4, utilize shear forces to reduce the size of particles, particularly in the fine size range. Various basic designs of attrition mills exist. The single disc mill (Figure 6.4) features a high-speed rotating grooved disc that maintains a narrow gap with its stationary casing, resulting in intense shearing action to comminute the feed. The gap can be adjusted according to the feed size and desired product specifications. In the double disc mill (Figure 6.4), the casing houses two rotating discs that counter-rotate, generating a higher degree of shear compared to the single disc mill. The Buhr mill (Figure 6.4), an older type originally used in flour milling, comprises two circular stones mounted on a vertical axis. The upper stone is typically fixed with a feed entry port, while the lower stone rotates, and the product is discharged over its edge. Attrition mills find extensive applications in the food industry.

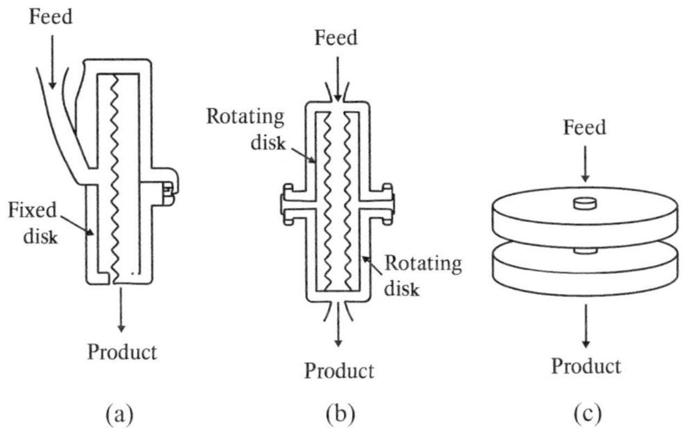

FIGURE 6.4 Disc attrition mill schematic: (a) single disc, (b) double disc, and (c) buhr mill.

A primary advantage of the disc attrition mill is its capacity to generate fine particle sizes while minimizing heat production. This trait is especially crucial when managing temperature-sensitive goods, such as grains, as excessive heat can undermine their nutritious value. Disc attrition mills are available in various designs, particularly single-disc and double-disc variants (Figure 6.4). In the single-disc variant, a grooved disc rotates swiftly against a stationary surface with minimal clearance, generating significant shear forces. The double-disc variation incorporates two discs revolving in opposing directions, hence augmenting the shearing effect and facilitating finer grinding.

Modern Equipment

Jet Milling

Jet milling operates through a collision-based mechanism wherein compressed air or inert gas (such as nitrogen) is injected through nozzles into a grinding chamber at high velocity. This creates a turbulent airflow that accelerates rice particles to speeds exceeding 500 m/s (Yu et al. 2024b). High-speed collisions between particles result in size reduction via impact and attrition, producing ultrafine powders without needing mechanical grinding components. This working principle allows precise particle size control while avoiding heat generation and material contamination. It is an advanced, non-mechanical pulverization method that is particularly effective for making ultrafine rice flour with particle sizes ranging from 1 to 10 µm. Jet milling reduces starch damage and thermal degradation and preserves the nutritional value and functional properties of rice flour. These characteristics are primarily important for the process involving heat-sensitive and good-quality products with high purity, such as gluten-free foods, infant formulas, and pharmaceutical excipients (Zhang et al., 2021). The absence of a grinding surface reduces the risk of metal contamination and ensures a pure and consistent product with a uniform particle size distribution. Its ability to make fine flour of good quality and contaminant-free makes it highly valuable in food and pharmaceutical industries.

- **Advantages**
  i. **Ultra-fine particles**: Can achieve sizes <10 μm.
  ii. **No heat generation**: Maintains rice's nutritional and functional properties.
  iii. **No contamination**: No metal contact with the product.
  iv. Uniform particle size distribution.
- **Considerations**
  i. **Cost**: Jet milling equipment is expensive to purchase and operate.
  ii. **Energy consumption**: It is relatively high due to the need for compressed air.
  iii. **Moisture sensitivity:** Input rice must be low-moisture (typically <6%) to prevent clogging.

Cryogenic Grinding

It is the most effective method for producing good quality, ultra-fine rice flour with good texture, particularly in areas where product purity, functionality, and better thermal stability are required. Cryogenic grinding involves the process of very low temperatures, typically using liquid nitrogen, to brittle the material prior to and during milling. Cryogenic grinding significantly reduces the elasticity and plasticity of rice by decreasing the temperature below the glass transition point of material, making it easier to fracture (Saxena et al. 2018). This process is particularly effective for rice which may be soft, elastic, or heat-sensitive under normal grinding conditions.

This method first cooled by using nitrogen to reduce its temperature below its glass transition point. This cooling makes the rice brittle and reduces its plasticity significantly. After getting brittle, rice is subjected to mechanical force using conventional milling equipment such as hammer mills or pin mills, which allows rice to be easily fractured into fine particles with minimum resistance. The minimum temperature also suppressed heat outcome during grinding, which is essential for preserving heat-sensitive components like starches and volatile compounds.

- **Advantages**
  i. Prevents thermal degradation, microbial growth, and oxidation and preserves native starch structure, antioxidants, vitamins, and phenolics.
  ii. Minimizes starch gelatinization and agglomeration.
  iii. Enhances powder flowability and reduces clumping.
  iv. Maintains flavor, color, and nutritional quality.
  v. Produces fine, uniform particle size distribution.
  vi. Reduces overall grinding energy due to material embrittlement.
- **Limitations**
  i. Higher initial and operational costs due to the use of liquid nitrogen.
  ii. Requires specialized equipment and safety protocols.
  iii. Not always economically viable for large-scale, low-cost applications.

Wet Milling

It is a mostly used and effective method for manufacturing rice flour, particularly valued for its advantageous features to produce good quality, fine-textured flour with minimum starch damage. The process involves soaking, water milling, and cleaning, which softens the grains and facilitates particle breakdown while preserving starch integrity and reducing heat generation. Wet-milled rice flour typically exhibits lower protein and ash content, higher crystallinity, and fewer starch granule aggregates, resulting in improved swelling power, peak viscosity, and gelatinization enthalpy in certain rice varieties (Leewatchararongjaroen and Anuntagool 2016). The resulting flour has a bright color and narrow particle size distribution, making it suitable for high-quality food applications such as rice noodles, batters, and baby food.

In this method, rice is soaked in water for a specific period to gain moisture content and soften the endosperm. The hydrated rice grains are then ground using mills like stone mills, colloid mills, or wet grinders. In this milling process, the high moisture of rice acts as a lubricant and cooling agent, preventing heat generation and reducing nutritional loss and starch damage. The slurry or paste resulting from milling is then filtered and dried at the required moisture level to obtain a fine particle size of rice flour (Wu et al. 2019).

The wet milling method for rice pulverization follows a sequential step to produce good quality rice flour with fine particle size and preserved starch functionality. Firstly, the impurities like dust, stones, particle, and husks of raw rice are removed in the cleaning process by using a sieve and destoner. Then, the cleaned raw rice is then soaked for 4–24 h in water for easier milling. After steeping, the extra water is removed to achieve the essential moisture content for grinding. The steeped rice is then ground into a slurry using wet milling equipment like stone mills and colloid mills. This slurry is sieved through fine mesh sieves to remove coarser particles and collect uniform fine and smooth texture rice flour slurry. The obtained rice slurry is subjected to drying to reduce the moisture content at a suitable level by drying methods like tray drying, spray drying, and drum drying for easier handling and storage (Leewatchararongjaroen and Anuntagool 2016).

The obtained dried fine rice flour is packaged in moisture-proof containers and kept under controlled environmental conditions to maintain its nutritional value and functional properties. This milling method is primarily effective for manufacturing good quality fine rice flour in applications like batters, rice noodles, and infant foods.

- **Advantages**
  - i. Minimizes starch damage due to reduced mechanical and thermal stress.
  - ii. Improves flour functionality for food applications (e.g., gelatinization, viscosity).
  - iii. Enhances the fineness and smooth texture of the final product.
  - iv. Reduces dust formation, making the process safer and cleaner.
  - v. Improves separation of bran and endosperm, enhancing purity.

- **Limitations**
  i. High water usage and longer processing time.
  ii. Requires drying step post milling, which increases energy consumption.
  iii. Potential for microbial contamination if not properly managed.
  iv. Not suitable for all types of rice products (e.g., quick-drying formulations).

6.5 IMPACT OF PULVERIZATION (PARTICLE SIZE) ON RICE FLOUR

The milling method particularly affects rice flour's qualities, physicochemical and functional properties, including starch content, gelatinization, particle size distribution, and thermal properties. These factors generally affect final product quality (Yu et al. 2024a). Rice flour with finer particle size results in a smoother texture, higher hydration, and faster gelatinization, making it ideal for food applications in gluten-free baking and infant food. Flour with fine particle size enhances the surface area to volume ratio, improving interaction with water and other ingredients during food applications. This improves viscosity, digestibility, and pasting properties. However, exceeding pulverization may damage starch granules, change viscosity, minimize the gel strength, and decrease nutrient retention. Therefore, optimizing the rice flour particle size through pulverization parameters is important for achieving final product performance in fortified and functional foods.

6.5.1 Effect of Grinding Parameters on Particle Size

The particle size of rice flour is generally controlled by several grinding parameters, which play an essential role in characterizing the final product's quality, functionality, and suitability. The main key parameters are grinding method, milling temperature, operational time, speed, moisture content, and feed rate of grain. These factors affect the particle size distribution, surface quality, and functional properties of flour (Protonotariou, Ritzoulis, and Mandala 2021).

Grinding method: Types of grinding significantly affect the particle size of rice flour. Modern rice milling methods like cryogenic and jet milling generally produce ultra-fine particles of flour.

Temperature: Heat developed during the grinding process may cause partial gelatinization of starch, leading to agglomeration and coarser particle size (Barnwal et al. 2014). Grinding processes that work under low temperature prevent heat generation, resulting in uniform finer particle size and preserving starch functionality and nutritional value.

Grinding time and speed: Maximum grinding time and maximum speeds generally lead to fine particle sizes due to high mechanical force (Hanumanthappa et al. 2020). However, over-grinding can cause particle agglomeration or damage starch due to heat generation.

Moisture content: The moisture content of rice prior to milling affects brittleness and milling efficiency. Higher grain moisture may lead to clumping or smearing and produce non-uniform particle size. Therefore, feed grains of lower moisture are required for higher milling efficiency, lower energy consumption, and high quality of fine particle size.

Feed rate: A slow feed rate provides consistent flour but may increase the milling time, and a high feed rate may lead to uneven particle size distribution.

6.5.2 Starch Modification and Functional Properties

Pulverization changes the starch structure, significantly affecting its functional characteristics like water absorption, gelatinization temperature, paste viscosity, and gel strength. These changes depend on the grinding process and the final particle size of the flour.

Starch Modification

Pulverization decreases the particle size of rice and may lead to physical changes in starch granules, especially at maximum mechanical force or higher temperature. Fine grinding process particularly uses high energy or prolonged methods that can disrupt starch crystallinity, enhancing amorphous regions (Dome et al. 2020). This modification increases the ability of starch to swell, gelatinize, and interact with water. In cryogenic method of grinding, heat is minimized, and starch granules remain highly intact, preserving native properties while achieving a finer particle size flour. Moreover, starch modification through pulverization may increase its emulsifying and film-forming characteristics.

Water Absorption and Gelatinization

Fine particles show maximum water absorption capacity due to their higher surface area. This provides rapid and uniform gelatinization when heated, which is crucial for foods like sauces, kid foods, baking foods, and soups. Higher water absorption also provides good texture and moisture retention in gluten-free and fermented products. Small particles usually result in more damaged starch content. Damaged starch is highly reactive with water and enzymes, leading to rapid gelatinization and maximum water absorption capacity.

Pasting and Rheological Behavior

Finer pulverized rice flour tends to show better pasting viscosity and gel strength, which are required for structural stability in food products (Sharma et al. 2023). However, excessive pulverization can maximize the enzyme susceptibility or decrease the retrogradation stability, which affects the shelf life or texture of the product. Finer particles show a higher viscosity peaks and shorter pasting times, which is better for bakery batters or extruded products.

6.5.3 Effect on Nutrient Stability

The pulverization operation can affect the stability and retention of micronutrients in rice flour, especially when subjected to maximum speed or maximum milling temperature. Major micronutrients are affected, including vitamins, proteins, minerals, and antioxidants. Heat-sensitive micronutrients such as vitamin B_{12} and folic acid may degrade because of high heat generation during dry milling at high speed. In contrast, wet or cryogenic milling preserves the micronutrients more effectively due to the minimum processing temperature. Pulverization also exposes nutrients to oxidative stress, especially when the flour has a high surface area (as in superfine milling). This can degrade lipophilic vitamins (e.g., vitamin E) and promote lipid peroxidation, affecting both nutritional value and shelf stability. Packaging in low-oxygen or vacuum-sealed containers can help mitigate these losses.

Proteins and amino acids generally withstand pulverization but may denature under high heat. Pulverization can also improve protein digestibility, especially when partial gelatinization of starch makes the protein matrix more accessible to digestive enzymes. Furthermore, finer particle size promotes uniform distribution of added micronutrients during fortification. However, if the flour is overly refined, some naturally occurring nutrients in the rice bran or germ may be lost.

To maximize nutrient retention:

- Use low-temperature grinding methods (e.g., cryogenic or wet milling).
- Minimize exposure to air and light during processing.
- Combine milling with protective microencapsulation of sensitive micronutrients.

6.5.4 Microstructural Changes in Pulverized Rice Flour

Pulverization causes significant microstructural changes in rice flour, directly influencing its functionality, processing behavior, and end-product quality. These changes are mainly observed in starch granules, protein matrix, and the overall flour morphology. Under moderate grinding, starch granules largely retain their polyhedral or spherical shape, with slight surface roughening. However, as grinding intensity increases, particularly in jet or cryogenic milling, granules may rupture, collapse, or show signs of fissuring and micro-cracks. These deformations increase surface area and water absorption capacity, which is beneficial for applications requiring rapid hydration.

Micrographs obtained via scanning electron microscopy (SEM) reveal that fine pulverization leads to:

- Loss of granular integrity.
- Formation of amorphous zones.
- Disruption of the native starch crystalline structure.

These alterations can enhance enzyme accessibility, thus improving digestibility, but may also result in higher damaged starch content, which affects pasting behavior and shelf life. In addition to starch, pulverization also disrupts the protein-starch-lipid matrix. The loosening of this matrix facilitates better nutrient release and mixing during fortification (Krishnan et al. 2022). However, excessive mechanical stress can denature proteins, affecting their solubility and emulsification properties.

The degree of microstructural damage depends on:

- Grinding method (dry, wet, cryogenic).
- Speed and duration.
- Initial rice moisture content.

These microstructural changes have downstream effects on flow properties, compressibility, and rehydration time critical for industrial food processing.

6.6 PARTICLE SIZE DISTRIBUTION AND ITS ANALYSIS

In a sample containing particles with uniform size and fixed density (ρ_p), the number of particles and total surface area (A_w) in the total mass (m) are estimated as follows:

$$A_w = Ns_p = \frac{6m}{\rho_p \ \phi_p \ D_p} \tag{6.4}$$

Here, ϕ_p represents the particle's sphericity, and D_p is the particle diameter (mm). The bulk solid is divided into several fractions by sieving, each containing particles of about equivalent size. It then calculated the specific surface area (A_w) of particles (per unit mass) in the mixture as follows:

$$A_w = \frac{6}{\rho_p \ \phi_p} \sum_{i=1}^{n} \frac{x_i}{\overline{D_{pi}}} \tag{6.5}$$

where x_i is the mass fraction in the ith fraction, n is the total number of fractions, and $\overline{D_{pi}}$ is the average particle diameter in the ith fraction. The particle size distribution can be expressed in either differential or cumulative particle size distribution (PSD) curves.

6.6.1 Differential Analysis

The mass fraction (the ratio of mass collected in a particular fraction to the total mass taken for size fractionation) or the number of particles in each size fraction is represented as a function of average particle size or size range. These discrete data for each of the individual fraction over the entire range are called the differential PSD (Figure 6.5a). The trend line is called the size frequency curve. The most frequently occurring particle size (maximum value in the curve) is called the 'mode particle size'.

6.6.2 Cumulative Analysis

A second way of representation for size distribution is arrived with cumulative analysis. The mass fraction or the number of particles is added consecutively starting from fractions containing the smallest size particles until the mass fraction reaches 1.0. A continuous plot between the cumulative mass fraction (y-axis) and the average particle size (x-axis) is the cumulative distribution curve (Figure 6.5b).

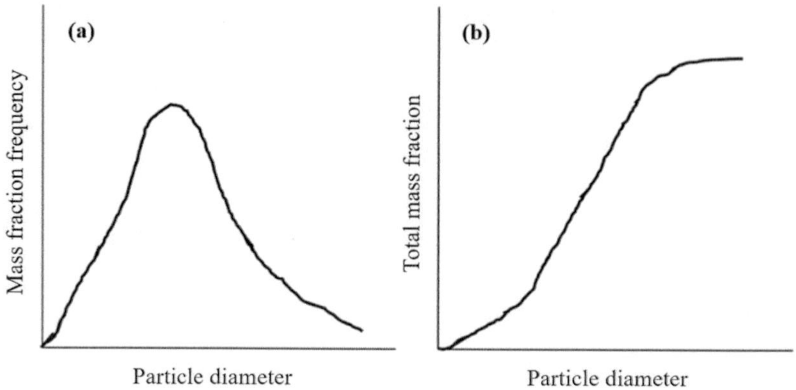

FIGURE 6.5 (a) Differential and (b) cumulative particle size distribution.

Estimation of average particle size, specific surface area, or particle population in the mass may be carried out with either of the two analyses. In principle, cumulative analysis is considered to be more precise as it does not depend upon the assumption of equal size of all the particles in any single fraction.

6.6.3 Laser Diffraction Analysis

Laser diffraction analysis is a widely used technique for determining particle size distribution, covering a broad range from nanometers to millimeters. It is based on the principle that larger particles scatter light at smaller angles, while smaller particles scatter at larger angles. The particle size distribution can be determined accurately by analyzing the scattering pattern and intensity. A sample (either as a dry powder or a liquid suspension) is prepared and properly dispersed to prevent aggregation. A laser beam passes through the sample, and scattered light is detected at various angles. The scattering pattern reflects the particle sizes present.

The interpretation of scattering data relies on two main theories:

- **Fraunhofer diffraction theory**: Assumes large, opaque particles and is suitable for sizes above 10 μm.
- **Mie scattering theory**: More accurate for a wide size range, considering particle transparency and refractive index.

Specialized software processes the data using these models to calculate the particle size distribution, typically presented as a volume-based curve or size percentiles (e.g., D10, D50, D90).

6.7 ENVIRONMENTAL AND ECONOMIC CONSIDERATIONS

Rice pulverization, while essential for producing fine rice flour used in various food products, presents several environmental and economic challenges. From an environmental perspective, advanced techniques such as jet milling and cryogenic grinding require substantial energy input. Jet mills operate using high-speed compressed air, while cryogenic systems rely on maintaining ultra-low temperatures with liquid nitrogen, both of which demand significant electricity, especially in continuous, large-scale operations. This maximum energy consumption allows higher greenhouse gas emissions when powered by non-renewable energy sources, resulting in a considerable carbon footprint. On the other hand, wet milling is less energy intensive but totally water dependable, creating challenges in scarce water areas. It produces excessive amounts of wastewater containing starch and organic residues that can pollute surrounding areas or water bodies.

Economically, the investment cost of milling equipment and installation for advanced milling is relatively high. Cryogenic grinders and jet mills require customized infrastructure and skilled labor to operate, increasing setup and operational cost. Maintenance charges, energy utilization, and specialized inputs like liquid nitrogen increase the financial burden. However, these modern technologies allow maximum efficiency and minimum product waste, providing consistent, good quality finer particle flour that meets the requirement for food processing.

6.7.1 Sustainability of Pulverization

Sustainability in rice pulverization refers to minimizing environmental impacts while maintaining product quality and economic viability. As global food systems strive to reduce their carbon and water footprints, the processing stage, particularly energy-intensive steps like pulverization, becomes a focal point for sustainable innovation. Energy consumption is a significant concern in dry and cryogenic pulverization. Cryogenic grinding, for instance, uses liquid nitrogen, which requires energy-intensive production and transportation. Jet milling consumes substantial electricity to maintain high-velocity airflows. In contrast, wet milling uses less electrical energy but has high water consumption, contributing to water scarcity concerns and wastewater discharge.

To address these, several sustainability strategies are being explored:

- **Energy efficiency improvements**: Variable frequency drives, energy-efficient motors, and heat recovery systems can significantly reduce power consumption.
- **Renewable energy integration**: Solar or biomass-powered systems for grinding can offset fossil fuel dependence.
- **Water recycling systems:** In wet milling, fresh water use and environmental discharge are reduced.

6.7.2 Economic Feasibility in Large-Scale Fortification

The economic feasibility of large-scale rice pulverization for fortification depends on several critical factors, including initial investment, operational efficiency, market demand, and long-term profitability.

Initial Capital Investment

Large-scale rice pulverization with advanced technologies like jet milling and cryogenic grinding requires high setup costs due to specialized equipment, infrastructure, and skilled labor. Fortification processes, such as micronutrient blending, add to the cost. However, these methods ensure precise control over particle size and micronutrient distribution, producing high-quality fortified rice flour.

Operational Efficiency

Several large-scale rice pulverization industries use AI and machine learning to optimize the milling time, adjusting parameters such as temperature, feed moisture, and energy consumption. These increase the pulverization efficiency by minimizing waste, lowering energy consumption, ensuring good quality, decreasing operational costs, and increasing the financial viability of large-scale fortification facilities.

Market Demand and Profitability

The demand for fortified rice is increasing globally, mostly in region affected with micronutrient deficiencies, which provides profitability for large-scale pulverization facilities.

6.8 FUTURE TRENDS AND INNOVATIONS IN RICE PULVERIZATION FOR FORTIFICATION

The demand for fortified, nutrient-rich, good-quality, sustainable food products is increasing globally. Rice pulverization is a technique used to produce fine rice flour for manufacturing fortified rice kernels and several new product developments.

6.8.1 Advanced Pulverization Techniques for Micronutrient Carriers

The future rice pulverization process is being engineered with a particular target of manufacturing ultrafine, stable rice flour that acts as an efficient micronutrient carrier.

Nano-pulverization and superfine grinding: These processes decrease the rice flour to sub-10-µm particle size, providing better micronutrients such as iron, vitamin B_{12} and folic acid distribution. This improves the surface area to volume ratio, enhancing water absorption and bioavailability in the digestive tract.

6.8.2 AI-ML in Optimizing Pulverization Parameters

Artificial intelligence and machine learning are set to revolutionize the rice pulverization process by enabling grinding time optimization, quality assurance, and predictive maintenance.

- **Parameter optimization:** AI algorithms may analyze large amounts of data from pulverization equipment to determine the efficient speed, temperature, feed rate for rice varieties, and required particle size. This allows better flour quality and prevents excessive grinding or nutrient loss.

- **Defect detection and prediction:** Machine learning models may identify anomalies like non-uniform particle size distribution or excessive heating in operation time, allowing for automatic adjustment or turning off the process, minimizing waste and machine wear.
- **Energy efficiency modeling:** AI and machine learning may be used to simulate many energy inputs and predict the most efficient grinding conditions, particularly reducing operational costs.

6.9 CONCLUSIONS

Rice pulverization is a crucial method to produce high-quality rice flour, especially for producing fortified rice kernels that focus on reducing malnutrition problem. By selecting the proper method and equipment and optimizing the equipment parameter, the textural, functional, nutritional, and overall quality of rice flour can be improved. Using modern grinding equipment and smart tools like Artificial Intelligence and Machine learning can make the pulverization process better and highly efficient. It produces flour with fine particles size, which improved the functional, rheological, and nutritional stability of flour that is beneficial in making fortified rice kernels.

REFERENCES

Angelidis, Georgios, Styliani Protonotariou, Ioanna Mandala, and Cristina M. Rosell. 2016. "Jet Milling Effect on Wheat Flour Characteristics and Starch Hydrolysis." *Journal of Food Science and Technology* 53: 784–91.

Barnwal, Pradyuman, Ashish Mohite, K. K. Singh, P. Kumar, T. JOHN Zachariah, and S. N. Saxena. 2014. "Effect of Cryogenic and Ambient Grinding on Grinding Characteristics of Cinnamon and Turmeric." *International Journal of Seed Spices* 4 (2): 26–31.

Boland, Tim, and Bridget Nyland. 2014. "Practical Low-Temperature Size Reduction." *Air Products and Chemicals*. https://www.airproducts.com/-/media/files/en/312/312-14-021-us-practical-low-temperature-size-reduction.pdf

Dome, Karina, Ekaterina Podgorbunskikh, Aleksey Bychkov, and Oleg Lomovsky. 2020. "Changes in the Crystallinity Degree of Starch Having Different Types of Crystal Structure after Mechanical Pretreatment." *Polymers* 12 (3): 641.

Gao, Wenjie, Feng Chen, Xi Wang, and Qingran Meng. 2020. "Recent Advances in Processing Food Powders by Using Superfine Grinding Techniques: A Review." *Comprehensive Reviews in Food Science and Food Safety* 19 (4): 2222–55.

Hanumanthappa, Harish, Harsha Vardhan, Govinda Raj Mandela, Marutiram Kaza, Rameshwar Sah, and Bharath Kumar Shanmugam. 2020. "Estimation of Grinding Time for Desired Particle Size Distribution and for Hematite Liberation Based on Ore Retention Time in the Mill." *Mining, Metallurgy & Exploration* 37 (2): 481–92.

Kamata, Tatsuki, and Daisuke Nei. 2024. "Effect of Different Rice Pulverization Methods on the Printability of a 3D Food Printer." *Engineering in Agriculture, Environment and Food* 17 (3): 104–14.

Krishnan, Veda, Maharishi Tomar, Lovemore N. Malunga, and Sijo Joseph Thandapilly. 2022. "Food Matrix: Implications for Nutritional Quality." In *Conceptualizing Plant-Based Nutrition: Bioresources, Nutrients Repertoire and Bioavailability*, edited by Ramesh S. V., Shelly Praveen, 43–60. Singapore: Springer Nature Singapore.

Leewatchararongjaroen, Jitranut, and Jirarat Anuntagool. 2016. "Effects of Dry-Milling and Wet-Milling on Chemical, Physical and Gelatinization Properties of Rice Flour." *Rice Science* 23 (5): 274–81.

Liu, Can, Yong Zhang, Lida Zhu, Qiang Li, Xin Shu, Shaoqing Qin, Dazhong Wang, and Wentian Shi. 2025. "A Review of Ultrasonic Vibration-Assisted Grinding for Advanced Materials." *Intelligent and Sustainable Manufacturing* 2 (1): 10001.

Lomovskiy, Igor, Aleksey Bychkov, Oleg Lomovsky, and Tatiana Skripkina. 2020. "Mechanochemical and Size Reduction Machines for Biorefining." *Molecules* 25 (22): 5345.

Protonotariou, S., C. Ritzoulis, and I. Mandala. 2021. "Jet Milling Conditions Impact on Wheat Flour Particle Size." *Journal of Food Engineering* 294:110418.

Sapna, Indrakumar, Moirangthem Kamaljit, Ramakrishna Priya, and Padmanabhan Appukuttan Jayadeep. 2019. "Milling and Thermal Treatment Induced Changes on Phenolic Components and Antioxidant Activities of Pigmented Rice Flours." *Journal of Food Science and Technology* 56: 273–80.

Saxena, S. N., P. Barnwal, S. Balasubramanian, D. N. Yadav, G. Lal, and K. K. Singh. 2018. "Cryogenic Grinding for Better Aroma Retention and Improved Quality of Indian Spices and Herbs: A Review." *Journal of Food Process Engineering* 41 (6): e12826.

Sharma, Rajiv, Nachiket Kotwaliwale, Navdeep Jindal, and Dharmesh C. Saxena. 2023. "Principal Component Analysis for Particle Size-Based Classification of Gorgon Nut Flour and Their Morphological, Structural, Thermal and Pasting Characteristics." *International Journal of Food Science and Technology* 58 (6): 3144–53.

Tian, Yaning, Lan Ding, Yonghui Liu, Li Shi, Tong Wang, Xueqing Wang, Bin Dang, Linglei Li, Guoyuan Gou, and Guiyun Wu. 2023. "The Effect of Different Milling Methods on the Physicochemical and in Vitro Digestibility of Rice Flour." *Foods* 12 (16): 3099.

Wang, Fulin, Jin Du, and Guosheng Su. 2023. "Investigation of Roller Press Surface and Stud Based on FEM Simulation." *Applied Sciences* 13 (19): 11032.

Wu, Tong, Lili Wang, Yan Li, Haifeng Qian, Liya Liu, Litao Tong, Xianrong Zhou, Li Wang, and Sumei Zhou. 2019. "Effect of Milling Methods on the Properties of Rice Flour and Gluten-Free Rice Bread." *Lwt* 108:137–44.

Yu, Tian, Sun Jing, Li Jiaxin, Wang Aixia, Nie Mengzi, Gong Xue, Wang Lili, Liu Liya, Wang Fengzhong, and Tong Litao. 2024a. "Effects of Milling Methods on Rice Flour Properties and Rice Product Quality: A Review." *Rice Science* 31 (1): 33–46.

Yu, Y., K. Chen, D. Wang, Y. Qin, and J. Liu. 2024b. "Large Eddy Simulation of Swirl Topology Evolution in Strong Oxidant Jet Milling Force Field." *Journal of Applied Fluid Mechanics* 18 (2): 374–88.

Zhang, Yiqin, Ruijia Hu, Michael Tilley, Kaliramesh Siliveru, and Yonghui Li. 2021. "Effect of Pulse Type and Substitution Level on Dough Rheology and Bread Quality of Whole Wheat-Based Composite Flours." *Processes* 9 (9): 1687.

7 Mixing of Vitamin-Mineral Premix in Pulverized Rice Flour and Conditioning

Shubham Mandliya, Indrajit Dalai, Sourav Misra, Siddharth Vishwakarma, Chandrakant Genu Dalbhagat, and Hari Niwas Mishra

7.1 INTRODUCTION

Mixing is crucial in manufacturing rice flour, especially when producing Fortified Rice Kernels (FRK), where uniform nutrient distribution is essential. Vitamin-mineral premixes (VMP) are mixed with rice flour in order to ensure that every kernel provides a balanced nutritional profile. The enhanced rice fulfills the nutritional requirements in every grain since proper mixing ensures that micronutrients are dispersed equally throughout the flour. Factors like mixing time, intensity, and adding ingredient speed all affect the mix's uniformity. For example, a more uniform distribution of moisture and nutrients in the rice flour is made possible by a slow incorporation of premix and water. In providing complete integration, longer mixing times lower the possibility of clumping or uneven distribution. Optimum texture and quality in the FRK product depended equally on uniform moisture content. The procedure also helps the starch granules hydrate, which affects the dough's structure and elasticity. Consequently, effective mixing is essential for creating high-quality FRK, providing both uniformity in the nutritional profile and the final product's desirable sensory qualities like texture and color.

Conditioning is the process of carefully adjusting the temperature and moisture content of rice flour throughout processing. This is essential for increasing the texture, consistency, and overall quality of the dough when producing FRK. This method promotes the appropriate hydration of the starch granules by continually adding moisture to the rice flour–VMP mixture. Water is added slowly and gradually to help ensure that the dough is evenly moistened throughout, avoiding the development of dry patches or areas that are too wet, which could compromise the dough's structural integrity.

Conditioning also entails changing other elements, like temperature, to further control the hydration process, affecting how proteins interact and how starches

DOI: 10.1201/9781003661245-7

expand their operations. A homogeneous, more elastic dough that is easier to mold and extrude into FRK results from proper conditioning. By producing a stable internal structure, the conditioning process also improves the final product's texture, appearance, and shelf life. By managing the distribution of moisture during conditioning, manufacturers may guarantee the production of quality fortified rice that maintains its original attributes throughout the following processing steps.

7.2 MECHANISMS AND ENGINEERING PRINCIPLES OF MIXING

7.2.1 Mechanisms

A mixture consists of two or more components, which may exist in solid, liquid, or gaseous state. These components blend or dissolve to form a single homogeneous phase. The mixing process changes these key attributes (texture, color, and appearance) that are critical to consumer perception and product quality. According to Pachón et al. (2015), a wide variety of food powders, such as flour, salt, sugar, dried milk, fruits and vegetables (dried), dry additives, cereal grains, bakery mixes, spice blends, and soup bases, are processed using ribbon mixers/blenders either individually or in combination.

Mixing rice flour and VMP is an essential step in producing FRK through extrusion. The process ensures the uniform distribution of micronutrients present in VMP throughout the rice flour matrix, thus achieving consistent fortification in each kernel, thereby achieving homogeneity. This homogeneity minimizes variability in nutrient content between kernels, thereby guaranteeing reliable and standardized dose of vitamins and minerals in each kernel. Various mixing mechanisms for solid–liquid systems are illustrated in Figure 7.1.

7.2.2 Engineering Principle of Mixing

The engineering principle of mixing is that they are grounded in solid dynamics, particle mechanics, and transport phenomena, aiming to achieve the uniform distribution of components within a system through controlled mechanical agitation. Fundamentally, mixing involves the interplay of convective transport, shear deformation, and diffusive motion, each contributing at different scales to blend homogeneity.

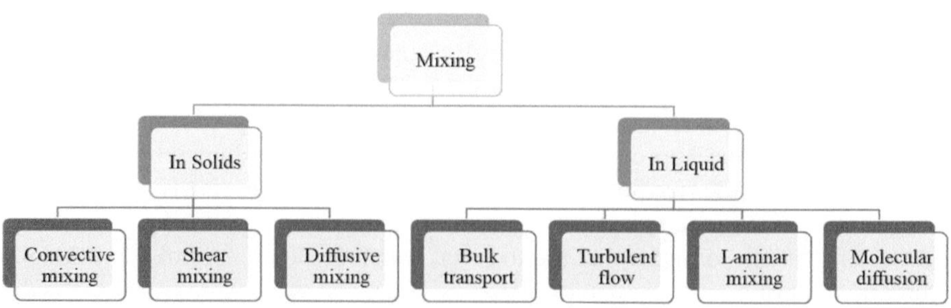

FIGURE 7.1 Types of mixing.

In particulate systems like fortified rice flour blends, shear forces promote intimate contact between particles. The diffusive mechanisms have a fine-scale uniformity. Mixing efficiency depends on factors such as mixing element geometry, vessel configuration, operational parameters (e.g., speed), and material properties (particle size, density, and moisture content). Dry powder mixing has low segregation and electrostatic effects, which is crucial.

Convective Mixing

Convective mixing is the bulk transport and macroscopic distribution of solids or fluid particles within a mixing vessel, primarily driven by mechanical agitation mechanisms such as impellers, ribbons, paddles, or gas-induced circulation (Tsugeno et al. 2021). This mixing significantly increased the uniformity of large-scale particle mobility, reducing concentration gradients and improving homogeneity. Several variables, including mixture geometry and particle characteristics such as size, density, and rotational speed, affect convective mixing. The advantage of batch processes is their ability to homogenize materials quickly; convective mixing is less efficient than continuous mixing processes. Optimized convective mixing is critical for maintaining product quality, such as pharmaceuticals, food processing, and the grain manufacturing industry.

Diffusive Mixing

Diffusive mixing is a mixing process in which particles are randomly moved, i.e., it is an entirely random particle motion. This motion causes adjacent particle layers or molecular species to interpenetrate, resulting in homogenization spontaneously. In contrast to convective mixing, which uses bulk flow to function on a macroscopic scale, diffusive mixing takes place at the particle or molecular level and is controlled by concentration gradients. The process is particularly relevant in systems with low Reynolds numbers and minimal external agitation or when mixing involves fine powders or solutes in liquids. Fick's laws of diffusion describe the underlying kinetics, with the diffusion coefficient being a key parameter influenced by particle size, temperature, and medium viscosity. It was slower than convective or shear mixing; diffusive mixing is crucial for achieving final homogeneity in mixtures, especially after macro-scale mixing is completed. It plays a vital role in dissolution, polymer blending, and nanomaterial synthesis processes.

Shear Mixing

Shear mixing is a mechanical technique that generates relative motion between adjacent layers by utilizing velocity differences inside a material. As a result, it improved the proper homogenization (Spicer and Gilchrist 2015). It works exceptionally well for mixing high-viscosity materials and dispersing cohesive powders. The material is exposed to limited shear pressures during the process, which can break up droplets, align polymer chains, and disturb agglomerates, all of which aid in fine-scale mixing. Compared to convective mixing, which moves large sections of material, and diffusive mixing, which relies on random molecular motion, shear mixing excels in achieving microscale blending under high-intensity mechanical stress. This mechanism is widely employed in high-shear mixers, twin-screw extruders, rotor-stator devices, and colloid mills, where narrow gaps or rotating elements generate intense shear fields. The efficiency of shear mixing is governed by factors such as shear rate, temperature, and equipment geometry. In food processing, pharmaceuticals, and polymer manufacturing, shear mixing plays an important role in texture modification, dispersion uniformity, and phase stability (Vanarase, Osorio, and Muzzio 2013).

Hybrid Mixing Strategies

Hybrid mixing strategies integrate two or more fundamental mixing mechanisms, such as convective, diffusive, and shear mixing, to achieve enhanced homogeneity, particularly in complex or multiphase systems. These strategies are designed to overcome the limitations inherent in individual mixing modes by combining their strengths. For example, convective mixing is a bulk material movement, while shear mixing fine-scale dispersion and structural disintegration improves macroscopic and microscopic uniformity. Hybrid mixers often feature engineered geometries such as twin-screw extruders, ribbon-blade agitators, or impeller-stator configurations that impart axial transport, radial dispersion, and local deformation. These designs are particularly advantageous in food processing, polymer compounding, pharmaceuticals, and additive manufacturing, where the processed materials exhibit complex rheological behaviors or contain immiscible components. The hybrid strategies are essential for innovative manufacturing approaches such as 3D food printing and responsive material design.

7.3 MIXING INDICES

Mixing index reflects the degree of homogeneity achieved during the mixing process. The purpose of mixer is to facilitate the random intermingling of materials. A mixture is considered homogeneous when one substance is uniformly dispersed throughout another. This uniformity is evaluated by analyzing the concentration of a specific target component (tracer) across multiple spot samples that deviates least from its mean value. In an unmixed condition, i.e., at the beginning of mixing, one component dominates the second component, and with time and mixing, the composition of each sample becomes consistent and tends toward average composition (Das and Das 2019; Premi and Sharma 2022; Smith 2011). One common method for tracking this progression involves calculating the standard deviation of each component across subsequent samples, as shown in the Eq. 7.1.

$$\sigma_m = \sqrt{\frac{\left(\sum_{i=1}^{N} (x_i)^2 - \overline{x} \cdot \sum_{i=1}^{N} x_i\right)}{N-1}} \tag{7.1}$$

where σ_m, N, x_i, and x are the standard deviation, number of samples, concentration of individual sample, and mean concentration of samples, respectively. With time, mixing becomes consistent and thus the standard deviation decreases. Different mixing indices (Eq. 7.2–7.4) for the measurement of mixing level are shown below:

$$M_1 = \frac{\sigma_m - \sigma_\infty}{\sigma_0 - \sigma_\infty} \tag{7.2}$$

$$M_2 = \frac{\log \sigma_m - \log \sigma_\infty}{\log \sigma_0 - \log \sigma_\infty} \tag{7.3}$$

$$M_3 = \frac{\sigma_{\mathrm{m}}^2 - \sigma_\infty^2}{\sigma_0^2 - \sigma_\infty^2} \tag{7.4}$$

where σ_m, σ_0, and σ_∞ are the standard deviation of individual sample at a particular time, the standard deviation initially, and the standard deviation when mixed completely, respectively. Depending upon the rate of mixing or quantity of compositions, Eqs. (7.2–7.4) were utilized. When components are mixed in equal quantity at lower rate of mixing, M_1 was calculated; when component A (low quantity) is mixed with component B (large quantity) at higher rate of mixing, M_2 was evaluated; when instead of solid–solid mixing, solid–liquid was used, M_3 was evaluated. The mixing index is correlated with mixing time and is shown in the Eq. 7.5.

$$\ln M = -K \times t_m \tag{7.5}$$

where K represents the mixing rate constant, and t_m denotes the mixing duration (s).

7.3.1 Coefficient of Variance (CV)

The coefficient of variance (CV) is a critical statistical metric used to evaluate the degree of homogeneity in nutrient blending, particularly in dry mixing operations for fortified food products such as rice flour and VMP. CV is defined as the ratio of the standard deviation to the mean ($CV = \sigma/\mu$) and is typically expressed in percentage. The CV quantitatively reflects the relative dispersion of micronutrient concentrations across sample replicates. A lower CV indicates better uniformity, implying that micronutrients or moisture are consistently distributed throughout the bulk material. CV values below 10% are generally considered acceptable for most micronutrient blends in fortified rice production, although thresholds may vary depending on regulatory requirements and nutrient sensitivity. The CV is susceptible to mixing efficiency, particle size differences, and sampling methods, making it a valuable index for process validation and quality control. However, its accuracy is contingent on representative sampling and sufficient replicate measurements. In large-scale operations, elevated CVs may signal inadequate mixing, ingredient segregation, or poor equipment design. As part of modern process control, real-time monitoring systems and AI-driven predictive models can now estimate or correct CV trends during production, enabling adaptive control strategies to maintain product consistency and minimize batch-to-batch variation.

7.3.2 Mixing Time and Efficiency

Mixing time and efficiency are fundamental parameters that dictate the quality and consistency of fortified food blends, such as rice flour combined with VMP. Mixing time refers to the duration required to achieve a specified degree of homogeneity within a mixture, which is typically determined by reaching a stable, acceptable CV across representative samples. Shorter mixing times are desirable for process efficiency but must not compromise blend uniformity. Conversely, excessively long mixing times can lead to overmixing, resulting in particle attrition, segregation upon discharge, electrostatic buildup, or nutrient degradation, particularly in sensitive components like vitamins A and C. Mixing efficiency is a measure of how effectively a mixer achieves the desired homogeneity relative to energy input, time, and material properties. It is influenced by factors such as mixer design (e.g., ribbon, paddle, and plowshare), fill level, rotational speed, particle characteristics (size, shape, and density), and preconditioning treatments. In fortified rice production, achieving high mixing efficiency ensures uniform micronutrient distribution, stable moisture content, and optimized processing conditions for extrusion or drying.

7.4 METHODS AND EQUIPMENT

The various types of mixing equipment used in the fortified rice industries are discussed below.

7.4.1 Tumbler Mixer

Tumbler mixers are available in various configurations, including horizontal cylinders, V-type cones, Y-type cones, and double cone designs (Figure 7.2). These mixers are primarily used in batch operations for blending free-flowing, non-segregating powders. Their fundamental operating principle involves the rotation of a vessel containing the solid materials, producing a gentle tumbling action that facilitates uniform mixing. This design offers specific advantages based on the nature of the materials and mixing requirements. A brief overview of different types of tumbler mixers is provided below.

Horizontal Cylinder Mixer

The horizontal cylinder mixer consists of a cylinder mounted on trunnions with internal baffles or lifter bars (Figure 7.2a). As the vessel rotates, the baffles gently lift and aerate the material, effectively preventing it from merely sliding along the bottom surface. This lifting action promotes better mixing and minimizes the risk of lump formation.

Double Cone Mixer

The double cone mixer consists of two conical sections joined by a central cylindrical portion, which rotates around a horizontal axis (Figure 7.2b). The symmetrical conical design ensures uniform mixing and facilitates easy discharge of the product. Typically, the mixer is filled up to two-thirds of its total volume to achieve optimal

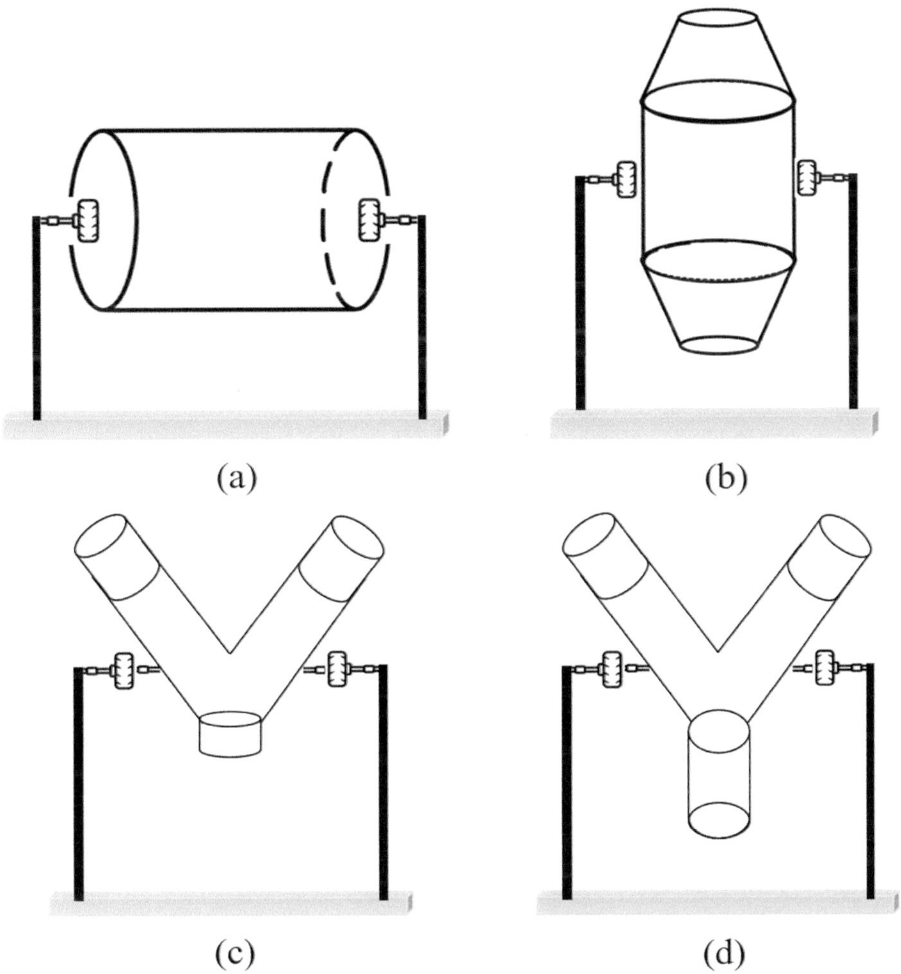

FIGURE 7.2 Schematic for (a) horizontal cylinder mixer, (b) double cone mixer, (c) V-cone mixer, and (d) Y-cone mixer.

mixing efficiency. As the vessel rotates at speeds ranging from 30 to 100 rpm, the tumbling motion causes the material to spread and fold, particularly at the transition zone between the cones. This results in gentle mixing with minimal shear, making it ideal for sensitive or cohesive powders. The product is discharged through an outlet at the bottom, and the tank can be tilted freely between 0° and 360° for easy cleaning and emptying. Advanced models are equipped with automated drum systems and PLC programming, enabling complete product transfer between the mixer and shaker with each full revolution. These mixers offer high safety, long service life, and significantly reduced mixing cycles, often completing a batch in just 10 min with minimal risk of spillage.

V-Cone and Y-Cone Mixer

The V-cone and Y-cone mixers, also known as twin-shell mixers, consist of two hollow cylindrical shells joined at an angle between 75° and 90° (Figure 7.2c and d). These mixers are especially suited for applications requiring precise formulation and uniform blending of powders and granules, including the addition of lubricants. The unique "V" or "Y" geometry creates a continuous pattern of splitting and recombining the material, promoting efficient mixing with minimal particle attrition (Sharma and Kumar 2022). While the tumbling motion resembles that of the double-cone blender, the greater surface contact between the material and the vessel walls generates slightly higher friction, resulting in less gentle but more thorough mixing. Widely used in the pharmaceutical, food, cosmetic, nutraceutical, plastic, and related industries, these mixers offer ease of cleaning, optional intensifier bars for specialized needs, and a wear-resistant, contamination-free design. Their low maintenance requirements and hygienic construction make them ideal for high-purity applications (You et al. 2021).

7.4.2 Ribbon Mixer

Ribbon mixers are widely used for blending dry solids, particularly powders and granules. These mixers feature a horizontally oriented agitator in the form of a helical ribbon, which rotates within a U-shaped or W-shaped trough (Figure 7.3). The ribbon agitator consists of inner and outer helical blades that move materials in opposite directions, creating a thorough and uniform mixing action. Due to the large open spaces between the ribbons, materials can slip through with minimal resistance, resulting in lower energy consumption while maintaining efficient mixing. Ribbon mixers offer several advantages, including wear and corrosion resistance, easy material discharge, belt-driven power transmission, and optional cooling jackets for temperature-sensitive applications. Their gentle mixing mechanism minimizes product degradation, making them ideal for dry blending tasks across various industries.

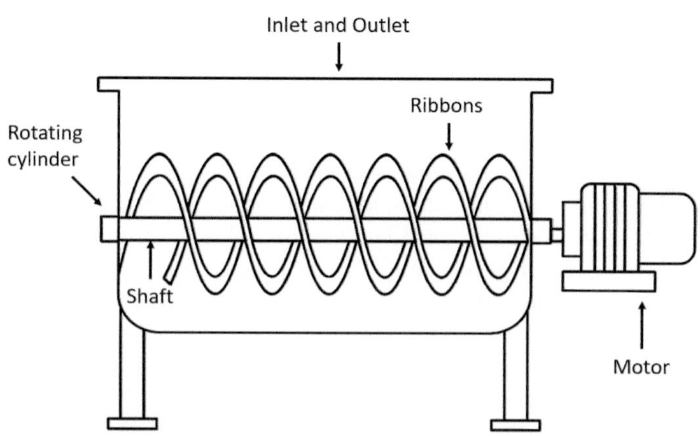

FIGURE 7.3 Schematic of ribbon mixer (inner view).

7.4.3 Paddle Mixer

Paddle mixers consist of a horizontal trough equipped with paddle-shaped agitators that rotate to move materials within the vessel (Figure 7.4). These mixers are particularly effective for handling cohesive or sticky materials, making them well-suited for various applications in the food industry. These are operating primarily on the principle of convective mixing (Ebrahimi et al. 2018). This mixing method operates by continuously repositioning and blending materials within a horizontal trough, ensuring thorough and uniform distribution of particles.

The paddle mixer consists of paddles, which are attached on a shaft. As the shaft rotates, the paddles continuously lift and fold the material from the bottom of the trough to the top and back down, generating a cascading motion throughout the batch. This tumbling effect ensures a steady and uniform flow of ingredients, allowing each particle to be evenly distributed within the mixture. The strategic design of the paddles enables the application of both convective and shear forces, effectively breaking down lumps and agglomerates. This mixing principle is crucial for achieving consistent and homogeneous blends, particularly in food processing industries where uniformity in product quality is essential.

7.4.4 Innovative Mixing Technologies

Innovative mixing technologies have revolutionized the blending of sensitive and low-dose ingredients such as VMP in food fortification systems, especially in products like FRK. Traditional mixers often fall short in achieving uniform distribution and protecting nutrient stability, prompting the adoption of advanced mixers like high-intensity paddle mixers (plowshare and ribbon hybrids), twin-screw continuous mixers, fluidized bed mixers, and resonant acoustic mixers. These technologies

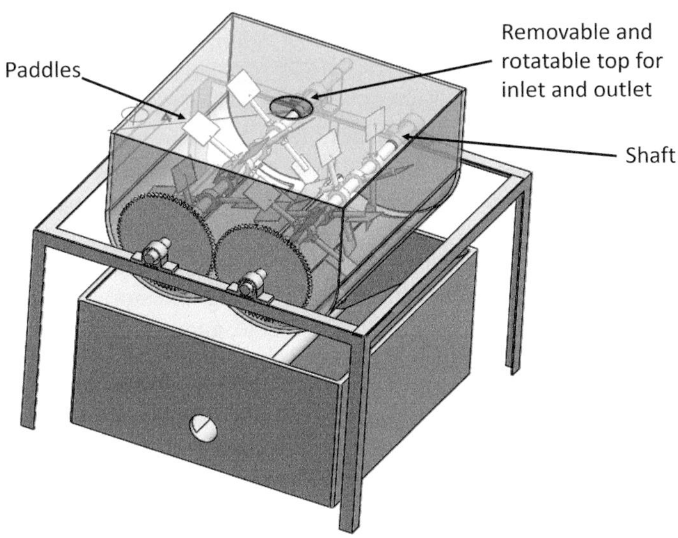

FIGURE 7.4 Double shaft paddle mixer.

offer superior control over mixing dynamics, allowing the effective handling of ingredients with varying particle sizes, densities, and flow properties. For example, high-shear mixers enable rapid and uniform dispersion of micronutrients, while fluidized bed mixers combine mixing with drying and coating, which is helpful for encapsulated or hygroscopic components. Smart mixers with IoT integration monitor mixing homogeneity, temperature, and batch consistency, ensuring compliance with GMP and HACCP standards. The AI-assisted systems optimize mixer settings based on predictive models, reducing energy use, mixing time, and batch-to-batch variability. These innovations improve nutrient stability and product safety and enhance process scalability, making them indispensable in modern, high-throughput fortification plants.

7.5 IMPACT OF MIXING OF VMP–RICE FLOUR

The dry mixing of VMP with rice flour is an essential step in the production of FRK, directly influencing nutrient uniformity, product quality, and extrusion performance. A homogeneous dry blend ensures that micronutrients such as iron, folic acid, and vitamins are evenly distributed throughout the rice flour matrix. This is essential for the accurate distribution of micronutrients in the final fortified product. The physical properties of the dry blend, such as flowability, particle size uniformity, and bulk density, play a vital role in ensuring smooth feeding into the extruder and consistent FRK formation. Thus, optimizing the dry blending process ensures nutrient stability and bioavailability and enhances process efficiency and consumer safety in large-scale fortification programs.

7.5.1 FRK Manufacturing Using Extrusion

The manufacturing of FRK using extrusion is a key technology for addressing micronutrient deficiencies through staple food fortification. In this process, a well-mixed blend of rice flour and VMP is fed into an extruder, which undergoes mixing, cooking, and shaping under controlled temperature, pressure, and shear conditions. Extrusion not only gelatinizes the starch in rice flour to form a cohesive kernel structure matrix, thereby protecting sensitive vitamins like folic acid and vitamin A from degradation. Extrusion parameters such as moisture content, screw speed, and barrel temperature critically influence the texture, appearance, and nutritional retention of the FRK. Proper control ensures the production of uniform, durable kernels with good cooking behavior and minimal nutrient loss during processing and storage. Extrusion-based FRK manufacturing enables cost-effective, scalable, and nutritionally effective fortification, especially suitable for distribution through public health programs.

7.5.2 Nutrient Bioavailability

The bioavailability of nutrients in FRK is an essential determinant of the effectiveness of rice fortification. Bioavailability refers to the proportion of ingested nutrients absorbed and utilized by the body. In FRK production, factors such as the choice of vitamin-mineral compounds (e.g., NaFeEDTA, ferrous fumarate), particle size, encapsulation techniques, and interaction with the rice flour matrix significantly influence nutrient release and absorption. Proper mixing and extrusion processing can enhance bioavailability by improving the dispersion of micronutrients and reducing inhibitory interactions with food components. Additionally, protective strategies such as coating or encapsulation help maintain nutrient stability during storage and cooking, ensuring that nutrients remain intact until consumed. Optimizing nutrient bioavailability in FRK ensures that the intended health benefits, such as improved iron status, are met.

7.5.3 Sensory and Textural Attributes

The sensory and textural attributes of FRK are critical for consumer acceptance and successful integration into regular diets. Given that FRK are typically blended with raw/parboiled rice at a ratio of 1:100. The appearance, texture, color, aroma, and taste of FRK and raw rice are the same. The extrusion process plays a central role in defining these attributes by controlling starch gelatinization, moisture content, and die design. Improper processing can result in FRK that are too hard, rubbery, discolored, or have an off-flavor, leading to consumer rejection. After cooking, texture parameters such as hardness, cohesiveness, and stickiness should align with local rice preferences. Sensory trials evaluate cooked FRK blends for mouthfeel, taste, color uniformity, and overall acceptability. Additionally, the presence of off-odors from certain iron compounds or the visual detection of VMP particles must be minimized through proper formulation and coating.

7.5.4 Regulatory Compliance

Regulatory compliance in FRK production is essential to ensure food safety, nutritional adequacy, and public health effectiveness. It involves adherence to national standards that govern the type and levels of micronutrients added, the quality of raw materials, the fortification process, and the labeling requirements. In India, FRK production must comply with regulations followed by the Food Safety and Standards Authority of India (FSSAI), which specify permissible forms and concentrations of iron, folic acid, and vitamin B_{12} in fortified rice. Manufacturers must also follow Bureau of Indian Standards (BIS) specifications (e.g., IS 17782:2021 for FRK), including physical parameters, storage stability, and shelf life. In addition, Good Manufacturing Practices (GMP) and Hazard Analysis and Critical Control Points (HACCP) systems must be implemented to prevent contamination, ensure process hygiene, and maintain product integrity. Regular quality checks for nutrient homogeneity, shelf life, cooking behavior, and sensory attributes, along with batch-wise documentation and traceability, are mandatory. The regulatory requirements are critical for legal and market authorization and ensure that fortified rice reliably delivers its intended health benefits to target populations.

7.6 CONSIDERATIONS FOR MIXING VMP WITH PULVERIZED FLOUR

Mixing VMP with pulverized rice flour is a critical step that directly influences the homogeneity, stability, and nutritional efficacy of FRK. Proper consideration must be given to the particle size compatibility between the VMP and the flour to prevent segregation during mixing or extrusion. Uniform particle size distribution helps to ensure consistent nutrient dispersion and stable flow behavior. Additionally, the hygroscopic nature and sensitivity of certain micronutrients, such as iron or folic acid, necessitate careful control of humidity and temperature during mixing to prevent clumping or degradation. The mixing time, equipment type (e.g., high-intensity paddle mixer or ribbon blender), and sequence of ingredient addition must be optimized to achieve thorough blending without inducing nutrient loss or powder layering. Mixing too long may result in the degradation of sensitive vitamins or undesirable changes in the texture of the rice flour. Pre-blending VMP with a small quantity of flour before full-scale mixing can enhance uniformity, especially when VMP concentration is very low. Maintain a consistent particle size for both the rice flour and the VMP. Uniform particle size facilitates even blending and prevents segregation during mixing, ensuring that each rice kernel receives adequate fortification. Use high-quality rice flour and VMP to ensure the nutritional integrity of the final product. Check the specifications and quality of raw materials to meet industry standards and regulatory requirements. Furthermore, maintaining closed, dust-controlled environments is essential to avoid nutrient loss and ensure worker safety. Maintain strict hygiene and sanitation practices throughout the mixing process to prevent contamination. Contaminants can compromise the safety and quality of the fortified rice product. Use mixing equipment designed for food-grade materials to prevent contamination and ensure compliance with food safety standards. The design should facilitate easy cleaning to prevent cross-contamination between batches. Implement regular testing procedures to verify the homogeneity and nutritional content of the mixture. Quality assurance measures should be in place to monitor and control the entire mixing process. Maintain detailed records of the mixing process, including ingredient quantities, mixing times, and any deviations from the standard operating procedures. This documentation is crucial for quality control and traceability. These considerations collectively help achieve a homogeneous, flowable, and extrusion-ready blend that meets regulatory and performance standards in fortified rice manufacturing.

Example 7.1

In multi-stage mixing, how much quantity (g) of rice flour must be added in 25 g of VMP to complete 3 stages of dry mixing?

Solution

For 25 g of VMP to be mixed in 3 stages.

First stage = 25 g VMP in 25 g Rice flour = 50 g blend 1
Second stage = 50 g blend 1 + 50 g rice flour = 100 g blend 2
Third stage = 100 g blend 2 + 100 g rice flour = 200 g final blend
So, the final blend is 200 g, from which 25 g is VMP and **175 g is rice flour.**

7.7 CONDITIONING OF RICE FLOUR–VMP BLEND

Rice flour is primarily composed of starch as a carrier matrix. At the same time, the VMP typically includes sensitive and hygroscopic micronutrients such as iron, zinc, folic acid, and vitamins A, B-complex, and D. These micronutrients segregate, degrade, or interact with the matrix under uncontrolled environmental conditions. Conditioning involves equilibrating the moisture content and temperature of the blend between the rice starch particles and the micronutrient powder, thereby reducing segregation during handling and processing. Moisture facilitates partial gelatinization or surface swelling of starch granules, which enhances the binding capacity and encapsulation potential for micronutrients. This is particularly important to preserve sensitive vitamins' bioavailability and thermal stability during high-temperature processes such as extrusion. The conditioning enables the VMP particles to integrate more homogeneously with the rice. A mixture of rice flour and VMP was conditioned for uniform distribution of moisture throughout the mixture sample, which helps in producing uniform FRK when extruded. During conditioning, the moisture was adjusted to influence the texture and structure of FRK. For conditioning, the initial moisture of the mixture was measured and the desired amount of water was added.

7.7.1 Vitamin-Mineral Premix (VMP)

VMP formulations are strategically designed based on nutritional targets, processing methods, and stability requirements. A typical formulation includes iron, folic acid, and vitamin B_{12} blends for combating anemia (details can be seen in Chapter 5). In contrast, more comprehensive blends may incorporate zinc, vitamin A, vitamin D, thiamine, riboflavin, niacin, and vitamin B_6. These premixes are available in various physical forms, such as encapsulated powders, spray-dried blends, or microencapsulated granules, often combined with carriers like maltodextrin or rice flour to aid dispersion and improve handling. The choice of nutrient compounds such as ferrous sulfate, ferrous fumarate, or NaFeEDTA depends on their bioavailability, sensory compatibility, and regulatory acceptance. However, during processing, especially extrusion-based FRK production, nutrient stability is challenged by factors such as high temperature, moisture, oxygen exposure, mechanical shear, and pH changes. Heat-sensitive vitamins like vitamin A, folic acid, and vitamin B_{12} are easily degraded, while moisture can lead to hydrolysis or microbial growth if not adequately controlled. Oxygen promotes oxidative losses, particularly in iron and fat-soluble vitamins. Physical stress from mixing or extrusion can further reduce the efficacy of micronutrients. Therefore, successful rice fortification depends on the precise formulation of the VMP and the careful optimization of processing conditions to retain nutrient functionality, ensure bioavailability, and maintain product shelf stability.

The VMP should be made with the same carrier material in which it is being used, so as to get the better uniformity and similar physical properties. VMP should be purchased from the vendors with a valid FSSAI license in 99.5 category, and the chemical salts must meet the required FSSAI guidelines (FSSAI, 2018). The VMP premix composition should meet IS 17781 specifications. The details regarding the formulation and guidelines of VMP can be seen in Chapter 5. For preparing FRK as per the guidelines, the required amount of VMP that needs to be added is calculated using the Eq. 7.6.

$$X = \frac{A}{B} \tag{7.6}$$

where X is the amount of VMP (g) required in 100 g rice flour, and A and B are the concentration of vitamins and minerals in 100 g of FRK (mg/100 g) and 1 g of VMP (mg/g).

Example 7.2

A FRK manufacturer has procured a VMP with iron concentration of 168 mg/g. Calculate the amount of VMP to be added to 20 kg rice flour to get final Fe concentration of FRK 345 mg/100 g.

Solution

Given, iron concentration in VMP = 168 mg/g

Iron required in FRK = 345 mg/100 g

Amount of VMP required in FRK = 345/168 = 2.05 g VMP is required in 100 g mixture

In 100 g mixture = 100 – 2.05 = 97.95 g rice flour

So for 20 kg rice flour = mixture will be 20,000/0.9795 = 20418.58 g mixture

So, VMP required = 20,418.58–20,000 = **418.58 g**

7.7.2 Hydration and Conditioning Steps

Moisture adjustment is a critical factor in optimizing the extrusion performance of FRK, as it directly influences starch gelatinization, nutrient retention, and the physical integrity of the extrudate. Proper moisture preconditioning of the rice flour VMP blend, typically to a range of 30%–34%, ensures sufficient plasticization of starch, leading to improved flow, expansion control, and uniform shaping during extrusion. If the moisture is too low, the blend becomes excessively viscous and can cause poor die flow, rough surface texture, or nutrient scorching. Conversely, excessive moisture may lead to incomplete cooking, soft structure, or sticky material buildup in the barrel. Thus, calculating the appropriate amount of water is essential in developing FRK and is shown in the Eq. 7.7.

$$Q = \frac{W(b-a)}{100-b} \tag{7.7}$$

where a and b are the initial and final moisture content in wet basis, respectively, and Q is the amount of water required (mL) to raise the moisture from a to b (Meghwal and Goswami 2017).

Strategies for moisture adjustment include premixing water or steam injection during feeding, using moisture-retaining carriers, such as pregelatinized starch or maltodextrin, and real-time moisture monitoring using inline sensors. Differential moisture feeding, where a small portion of water is added at multiple barrel zones, also helps to manage localized shear and heat profiles. Controlled drying post extrusion is also important to stabilize final moisture content to ~12% for storage stability. Altogether, precise moisture management enhances process efficiency, ensures consistent FRK quality, and protects sensitive micronutrients during thermal and mechanical stress.

Controlled hydration of the rice flour–VMP mixture ensures optimal flow properties and uniformity during FRK extrusion. Adequate hydration allows for the even distribution of moisture throughout the powder matrix, promoting uniform starch gelatinization and softening of particles, which enhances flowability and prevents clogging or inconsistent feeding into the extruder. When hydration is appropriately managed, the blend exhibits improved bulk density, reduced powder segregation, and cohesive flow behavior, minimizing variability in kernel shape, size, and texture. On the other hand, poor hydration can lead to the formation of dry pockets or overly wet clumps, causing inconsistent torque, poor nutrient dispersion, and defective extrudates. Controlled hydration stabilizes the thermal and mechanical energy transfer during extrusion, allowing for better cooking and shaping performance. Techniques such as preconditioning in a closed mixer, staged water addition, or inline moisture control systems help to consistently achieve the desired hydration level. Maintaining controlled hydration is a key to achieving high-quality, uniform, and nutrient-stable FRK with reliable processing performance.

Conditioning, the pre-extrusion step involving moisture and heat adjustment of the rice flour VMP blend, is crucial for optimizing extrusion performance. Still, it poses a risk of nutrient degradation, especially for heat and moisture-sensitive vitamins like folic acid, vitamin A, and vitamin B_{12}. To minimize nutrient loss during conditioning, carefully controlling temperature, moisture level, and residence time is essential. Conditioning temperatures should typically not exceed 60°C–70°C, and exposure time should be minimized to prevent thermal breakdown. Using indirect steam or warm water instead of direct high-pressure steam can help achieve gradual and uniform moisture penetration without overheating. By implementing these strategies, producers can maintain the nutritional integrity and bioavailability of micronutrients while still achieving the desired material properties for efficient and stable FRK extrusion. For conditioning, the hydrated sample is kept in a ziplock pouch at 5°C temperature for the whole night for uniform moisture distribution, if mixed manually. The conditioned sample is brought to room temperature before extrusion.

Example 7.3

Calculate the amount of water to be added to 1 kg flour if the initial moisture of mixture is 12% and the final moisture needs to be adjusted to 30%.

Solution

$$\text{Water to be added } (Q) = \text{Weight } (\text{g}) \times \frac{(\text{final moisture} - \text{initial moisture})}{(100 - \text{final moisture})}$$

$$\text{Water to be added } (Q) = 1{,}000 \times \frac{(30-12)}{(100-30)}$$

$$\text{Water to be added } (Q) = 1{,}000 \times \frac{18}{70}$$

$$\text{Water to be added } (Q) = \mathbf{257.143\ g}$$

7.8 CHALLENGES AND FUTURE TRENDS IN MIXING FOR RICE FORTIFICATION

Mixing in rice fortification presents several technological and operational challenges that can critically influence added micronutrients' uniformity, stability, and bioavailability. One major challenge is the heterogeneous nature of rice flour and VMP components, which vary widely in particle size, density, hygroscopicity, and electrostatic behavior, leading to segregation during processing or storage and achieving consistent micronutrient dispersion without overexposing sensitive vitamins to oxygen, moisture, or high temperature. There are nutrient losses during high-shear or thermal processing (e.g., extrusion), off-flavor development from mineral interactions with rice proteins and lipids, and sensory and stability concerns.

7.8.1 Effect of Uneven Mixing (Micronutrient and Moisture)

Uneven mixing of micronutrients and moisture in fortified rice formulations can lead to significant quality, nutritional, and safety issues. Inadequate dispersion of vitamin and mineral premix results in spatial variability in nutrient concentration, often leading to over-fortification in certain portions and under-fortification in other portions. Another reason is that high concentrations of minerals like iron or zinc may induce undesirable interactions with rice proteins or lipids, leading to off-flavors, discoloration, and reduced consumer acceptance.

Similarly, non-uniform moisture distribution can cause challenges during thermal processing. Regions with excess moisture may undergo premature starch gelatinization or nutrient leaching, whereas drier zones may remain underprocessed, affecting texture and structural integrity. Moreover, uneven moisture facilitates microbial contamination, reducing shelf life and food safety. The synergistic imbalance of micronutrients and moisture can also distort extrusion behavior, leading to inconsistent expansion, density, and surface finish of FRK.

7.8.2 Scalability Issues in Large-Scale Mixing

Scaling up mixing processes from laboratory or pilot-scale to industrial-scale operations introduces various technical challenges that can compromise blend uniformity, product quality, and process efficiency. In rice flour fortified with micronutrient premix, scalability issues often stem from the nonlinear behavior of mixing dynamics, where factors such as particle size disparity, density differences, and flowability are amplified in larger volumes. As mixer size increases, achieving consistent micronutrient dispersion becomes more difficult, with uneven shear distribution and longer residence times.

7.8.3 AI-Driven Predictive Modeling for Optimized Nutrient Blending

AI-driven predictive modeling is emerging as a powerful tool to optimize nutrient blending in VMP formulations, enhancing both processing efficiency and nutritional consistency in FRK production. By leveraging machine learning algorithms such as artificial neural networks (ANN), support vector machines (SVM), or ensemble methods, predictive models can learn from historical blending data, ingredient characteristics, and processing parameters to forecast optimal mixing conditions, ingredient ratios, and potential stability issues. These models account for complex, nonlinear interactions among variables such as particle size, hygroscopicity, flowability, and shear sensitivity, which are difficult to optimize manually. AI tools can also simulate different formulation scenarios and predict homogeneity levels, nutrient loss during processing, and final product quality before actual production. The integration of AI with sensor-based IoT systems allows real-time data feedback during blending and conditioning, enabling dynamic adjustments to maintain target specifications. This predictive capability reduces material waste and quality rejections and ensures compliance with regulatory standards for nutrient delivery. The AI-enabled nutrient blending represents a significant step toward precision fortification, delivering scalable, consistent, and cost-effective nutrition solutions in public health and commercial food applications.

7.8.4 Real-Time Quality Monitoring

Real-time quality monitoring is a critical component in the production of FRK, ensuring consistency, regulatory compliance, and nutrient integrity throughout the processing line. By integrating sensor-based technologies, such as near-infrared (NIR) spectroscopy, moisture sensors, torque and temperature sensors, and machine vision systems, manufacturers can continuously track key quality parameters like moisture content, nutrient homogeneity, blend uniformity, temperature fluctuations, and physical characteristics of the extrudate. These systems provide immediate feedback, allowing for automated adjustments to feed rate, mixing time, hydration levels, and extrusion conditions, thereby minimizing product variability and nutrient loss. Advanced systems may also employ AI algorithms and digital twins to analyze trends and predict deviations before they occur, enhancing preventive control. Real-time monitoring not only reduces the reliance on manual sampling and laboratory testing but also improves process traceability, reduces rework and waste, and supports GMP and HACCP compliance. Ultimately, this approach ensures that the fortified rice consistently meets nutritional targets and sensory expectations, making it more effective for large-scale public health interventions and consumer markets.

7.9 CONCLUSIONS

The effective mixing of VMP with pulverized rice flour is a main process for the production of nutritionally consistent and high quality of FRK. The particle size of VMP and pulverized rice flour play an important role during the mixing. Proper understanding of mixing mechanism, time, equipment, and conditioning techniques ensures proper distribution of moisture and VMP for the extrusion process. Future advancements like AI and real-time monitoring promise greater precision and scalability.

REFERENCES

BIS. 2021. Fortified Rice Kernels — Specification (IS 17782:2021). New Delhi: Bureau of Indian Standards, Government of India. https://standardsbis.bsbedge.com/BIS_search-standard.aspx?Standard_Number=IS+17782&id=34819

Das, S. K., and M. Das. 2019. *Fundamentals and Operations in Food Process Engineering*. 1st ed. Boca Raton, FL: CRC Press. https://doi.org/10.1201/9780429058769.

Ebrahimi, M., A. Yaraghi, F. Ein-Mozaffari, and A. Lohi. 2018. "The Effect of Impeller Configurations on Particle Mixing in an Agitated Paddle Mixer." *Powder Technology* 332 (June):158–70. https://doi.org/10.1016/j.powtec.2018.03.061.

Meghwal, M., and T. K. Goswami. 2017. "Flow Characterization of Ambiently and Cryogenically Ground Black Pepper (*Piper nigrum*) Powder as a Function of Varying Moisture Content." *Journal of Food Process Engineering* 40 (1): e12304. https://doi.org/10.1111/jfpe.12304.

Pachón, H., R. Spohrer, Z. Mei, and M. K. Serdula. 2015. "Evidence of the Effectiveness of Flour Fortification Programs on Iron Status and Anemia: A Systematic Review." *Nutrition Reviews* 73 (11): 780–95. https://doi.org/10.1093/nutrit/nuv037.

Premi, M., and V. Sharma. 2022. "Mixing and Forming." In *Agro-Processing and Food Engineering: Operational and Application Aspects*, edited by Harish Kumar Sharma, and Navneet Kumar, 253–305. https://doi.org/10.1007/978-981-16-7289-7_7.

Sharma, H. K., and N. Kumar, eds. 2022. *Agro-Processing and Food Engineering: Operational and Application Aspects. Agro-Processing and Food Engineering*. Singapore: Springer Singapore. https://doi.org/10.1007/978-981-16-7289-7.

Spicer, P. T., and J. F. Gilchrist. 2015. "Microstructure, Rheology, and Processing of Complex Fluids." *Advances in Industrial Mixing: A Companion to the Handbook of Industrial Mixing*, no. 3b, 87.

Smith, P. G. 2011. *Introduction to Food Process Engineering*. https://doi.org/10.1007/978-1-4419-7662-8.

Tsugeno, Y., M. Sakai, S. Yamazaki, and T. Nishinomiya. 2021. "DEM Simulation for Optimal Design of Powder Mixing in a Ribbon Mixer." *Advanced Powder Technology* 32 (5): 1735–49. https://doi.org/10.1016/j.apt.2021.03.026.

Vanarase, A. U., J. G. Osorio, and F. J. Muzzio. 2013. "Effects of Powder Flow Properties and Shear Environment on the Performance of Continuous Mixing of Pharmaceutical Powders." *Powder Technology* 246 (September): 63–72. https://doi.org/10.1016/j.powtec.2013.05.002.

You, Y., J. Guo, X. Lv, S. Wu, Y. Li, K. Tang, and Y. Yu. 2021. "Numerical Simulation of Particle Mixing Behavior in High Speed Shear Mixer and Cylinder Mixer." *ISIJ International* 61 (7): ISIJINT-2020–768. https://doi.org/10.2355/isijinternational.ISIJINT-2020-768.

8 Manufacture of Fortified Rice Kernels Using Extrusion Technology

Chandrakant Genu Dalbhagat, Siddharth Vishwakarma, Shubham Mandliya, and Hari Niwas Mishra

8.1 INTRODUCTION

Fortified foods are considered one of the best approaches or strategic interventions to address micronutrient deficiencies, often called 'hidden hunger'. A significant portion of the global population is affected by micronutrient malnutrition, particularly in developing countries like India. Fortification refers to adding the essential micronutrients, i.e., vitamins and minerals, to the food that might have a low/nil quantity. To tackle iron-deficiency anaemia, the micronutrients iron, folic acid, and vitamin B_{12} are effective; therefore, the regulatory bodies of different countries suggest the fortification of these micronutrients. A suitable vehicle for these vitamins and minerals is required to deliver the required amount to the targeted population. In India, the Food Safety and Standards Authority of India (FSSAI) provided a gazette on fortified foods that suggests the micronutrient dose and suitable vehicle. The identified vehicles are the staple foods, including rice, atta (flour), oil, milk, and salt. According to FSSAI guidelines, rice is the best vehicle for supplying iron, folic acid, and vitamin B_{12} to the anaemic population. Therefore, rice fortification strategy as a food-based approach is a steady and sustainable way to address the issue of iron-deficiency anaemia.

Rice fortification is achieved by fortifying the essential micronutrients in/on the rice analogues/rice, which is called as fortified rice kernels (FRK) (Dalbhagat and Mishra 2019). The quantity of fortificants depends on the method used to manufacture the FRK. The FRK can be prepared using (a) dusting, (b) coating, and (c) extrusion technology (Steiger et al. 2014). The dusting and coating methods involve the application of layers of micronutrients on the surface of the milled rice using electrostatic force and water-resistant edible coating, respectively. However, during washing and/or cooking, water-soluble micronutrients are lost, causing a reduction in fortification efficiency. Moreover, the surface appearance, colour, and taste change due to micronutrient layers affecting consumer acceptability. In comparison, the extrusion process infuses micronutrients into the kernels, making micronutrients less vulnerable to washing and cooking losses (Steiger et al. 2014).

DOI: 10.1201/9781003661245-8

Extrusion is a versatile technology in which different unit operations such as conveying, mixing, kneading, shearing, heating, cooking, shaping, and forming are combined to produce novel products (Dalbhagat et al. 2024; Dalbhagat, Mahato, and Mishra 2019). The operating parameters include barrel and die temperatures, screw speed, feed rate, cutter speed, and raw ingredient properties such as moisture content, particle size, composition, and nutritional profile (Singh Gujral and Singh 2002; Xu et al. 2016). The flexibility of operational parameters combined with ingredients parameters allows the development of well-balanced, nutritionally dense functional foods. In other words, it can be said that extrusion is a high-temperature, short-time shearing process that modifies the starch and protein material under different operating conditions (Liu et al. 2011). The extrusion processing conditions promote the starch gelatinisation, protein degradation, and different complex formations such as amylose-lipid, starch-protein, and starch-fibre, causing changes in the micro- and macrostructure, functional properties, and nutritional characteristics (Hagenimana et al. 2006). Nowadays, extrusion has become a mostly adopted technique for manufacturing breakfast cereals, savoury snacks, pre-cooked flours, cereal-based baby foods, and textural proteins. The knowledge of the extrusion process makes it easy to develop food formulations or recipes by choosing any cereal, protein, or fibre-containing ingredients.

Different studies (Nithya, Dalbhagat, and Mishra 2022, 2023; Dalbhagat and Mishra 2019, 2021) have proven that extrusion is a process for manufacturing the FRK using rice flour and vitamin-mineral premix (VMP). The recent technological advancement has enabled control of the extrusion process to achieve the desired characteristics in the FRK. Therefore, this chapter comprehensively discusses the framework of extrusion-based FRK manufacture, including types of extruders, raw material selection, optimisation techniques, product evaluation, operational strategies, automation, and future scope.

8.2 EXTRUSION PROCESS

The extrusion process involves shaping food materials (starchy or proteinaceous) by forcing them through the specially designed die opening after a previous heating of the food material. In the extrusion process, the food material is heated either by an external heat source or heat produced through the mechanical friction to cook or expand and extrude in the desired shape.

8.2.1 Types of Extrusion Process

The extrusion process can be categorised based on temperature profile, product moisture content, and extruder design.

Cold, Warm, and Hot Extrusion

In the cold extrusion process, the food or feed material is processed in the extruder barrel and die head at a lower temperature, generally in the range of 30°C–60°C, i.e., above the ambient temperature (Mandliya et al. 2025). Under this condition, the dough is only shaped to the desired product output. No starch gelatinisation or protein denaturation occurs while the high-moisture products need to be dried. The appearance of the product remains opaque. For example, pasta, noodles, and macaroni.

Warm extrusion is the intermediate process that works between the temperature range of 60°C and 90°C, generally below 100°C. The starchy feed material undergoes partial gelatinisation, while proteins are also partially denatured. The product seems to be translucent or opaque depending on the food formulation. For example, if only rice flour is processed to make extruded rice kernels, the kernels produced may become translucent, while adding VMP makes the kernels opaque.

Hot extrusion refers to a process when the feed material is processed above 100°C. The product is translucent and has a shiny appearance. Most of the expanded products are manufactured using the hot extrusion process. For example, Kurkure and cheese balls.

Low and High Moisture Extrusion

Low moisture extrusion is also called dry extrusion as the feed material is extruded at a lower moisture content, i.e., 18% wet basis (w.b.) or lower. Cooking is achieved at low moisture through friction and external heating. For example, expanded products such as Kurkure. On the other hand, when the moisture content of feed materials exceeds 18% w.b.–40% w.b., the process is called high moisture extrusion. Nowadays, the moisture content for high moisture meat analogue is set higher than 40% w.b. Ready-to-cook (RTC) extruded products are made using this process. For example, pasta and noodles.

Single- and Twin-Screw Extrusion

When the extruder contains a smooth, deep-flighted Archimedes screw within a grooved barrel and is used for making extruded product, it is called as single-screw extrusion. On the contrary, when the barrel is housed, two screws rotating in either the same or opposite direction to process the food material is called the twin-screw extrusion process.

8.2.2 Extruder

Extruder is a pump that transports, mixes, shears, stretches and shapes material under elevated pressure and temperature. It may be considered a screw reactor for physical, chemical, and biochemical transformations. In general, extruders (Figure 8.1) can be classified as single-screw extruder (SSE) and twin-screw extruder (TSE). The type, design, and operating power significantly influence the process efficiency, product quality, nutrient retention, and operational flexibility. Each type of extruder has its benefits and limitations, depending on the application scale, formulation complexity, and required product characteristics.

Types of Extruders

SSE consists of a precisely designed rotating single screw inside a barrel. These extruders are easier to operate, more economical, and most suitable for continuously extruding the simple formulation. However, their mixing and shearing capabilities are limited, making it inconvenient for complex formulations requiring precise nutrient dispersions.

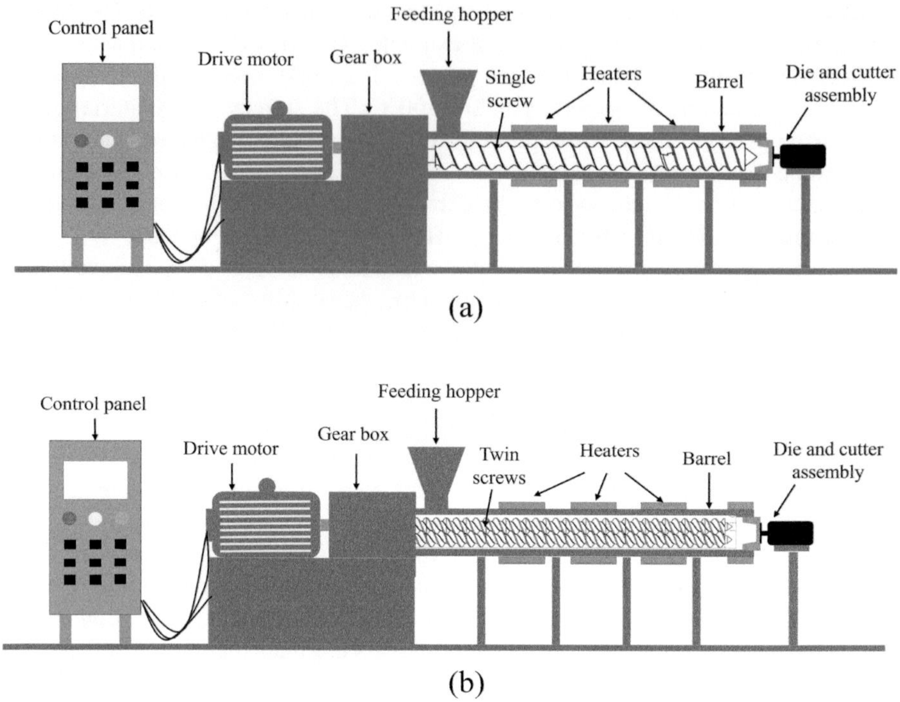

FIGURE 8.1 Types of extruders: (a) single-screw extruder and (b) twin-screw extruder.

On the contrary, the TSE consists of twin rotating screw housed inside a barrel. These screws may be intermeshing or non-intermeshing, either rotating in the same direction, i.e., corotating or in the opposite direction, i.e., counterrotating. The TSE offers superior mixing, flexibility, and control over the process parameters compared to SSE.

8.2.3 Parts of Extruder

The extruder has several key parts that function together to transform the food material into the desired extruded product. Understanding these parts and their role in operating the extruders is essential for maintaining performance, addressing troubleshooting issues, and achieving a good quality product (Figure 8.2). The parts of extruder are explained below.

Feeding System

The feeding system ensures a consistent and accurate flow rate of moistened and conditioned feed material in the extruder barrel to attain the desired production capacity. In most of the FRK manufacturing line, the feeding system involves an inclined screw conveyor (Figure 8.2), which has a hopper to receive the conditioned material from the mixer-cum-conditioner. It operates smoothly without interruption to provide a consistent feed rate. However, the feed rate may vary due to moisture content

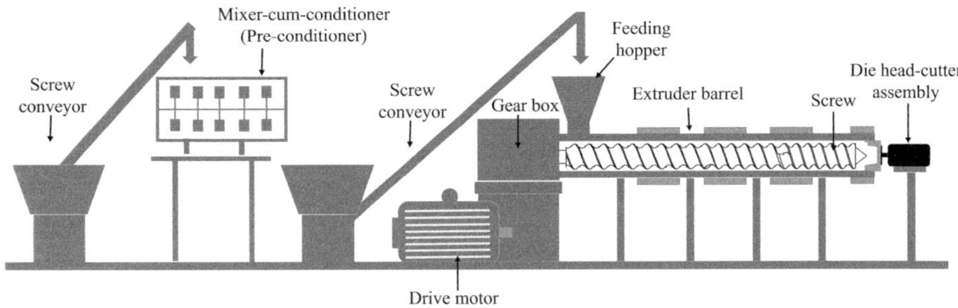

FIGURE 8.2 General overview of the commercial extruder system.

and density of feed material, screw design, and screw speed. Sometimes, instead of inclined screw feeders, the belt conveyors may be used to deliver the weighed quantity of feed material using a load cell to the hopper of the extruder barrel.

Preconditioner

A preconditioner is an optional unit between the feeding system and the extruder barrel. It is a large cylindrical or tubular chamber in which the raw ingredients are fed, and moisture is added to condition the raw ingredient mix. It is a mixer-cum-conditioner, having the mixing chamber (U or W-shaped) with one or twin shafts on which the paddles are mounted to mix the dry material. It is also equipped with a spray nozzle or pipe through which water can be sprayed on the feed material to moisten it. In some preconditioners, the facility to introduce steam is also provided, with proper sealing of the lid cover. Mixing material depends on the paddle design, retention time, and moisture adjustment. Longer residence time allows proper conditioning and better moisture absorption in the feed material.

Barrel

A barrel is a cylindrical housing that encloses the screws. It is the main part of the extruder, which is heated using electric heaters or other heating systems (e.g., hot oil circulation over the barrel) to melt the feed material during the extrusion process. It can be a single unit or divided into multiple heating zones, each equipped with a heating element. The material is conveyed, compacted, mixed, sheared, and cooked as it moves through the barrel. The barrel unit is fabricated using high-strength hardened stainless steel material, which is wear and corrosion resistant and withstands higher temperature and pressure during the extrusion. The temperature and pressure sensors are installed in the barrel to record the data and analyse the thermal and pressure profile during the operation.

Screw

The screw is also called an auger, which is a rotating shaft inside the extruder barrel that transports, compresses, and melts the material. The screw is designed in such a way that it pushes the material forward, causing frictional heat generation. The external heat supply and pressure generated during extrusion melt and soften the material. The screw can be divided into three significant sections, viz., feed, compression (transition), and metering zones (Figure 8.3).

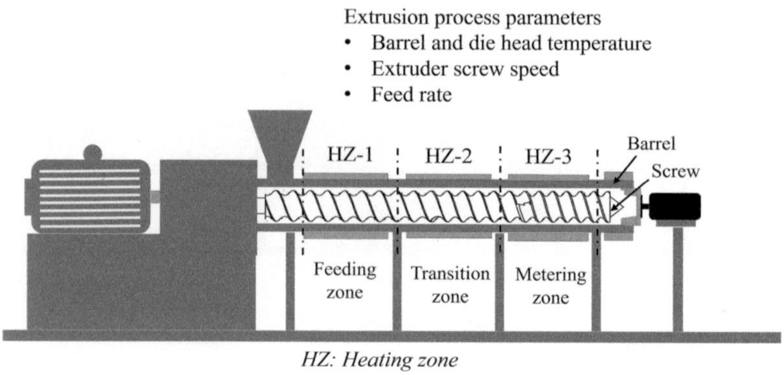

FIGURE 8.3 Different zones of the screw.

The feeding zone is the area where the preconditioned, low-density, moistened ingredient mix is introduced into the barrel. In this section, the material is conveyed and compressed slightly. In the transition zone, the material is compressed, causing a high degree of fill as the screw channel volume and pitch decrease. The shearing of materials begins because of compression and high fill of material in the screw channel. The increase in temperature, pressure, and extrudate density can be observed. In the metering zone, temperature and pressure increase rapidly, and shear rate is higher due to the screw design and configuration. Under this condition, the feed material melts and softens to extrude in a desired shape.

Die and Cutter Assembly

The die decides the final shape of the extrudate, which is installed at the exit of the extruder. The dimensions and geometry of the opening at the die surface determine the shape of the extruded product. For FRK, the rice shape is cut at the die surface to get rice-shaped kernels. The cutter assembly is essential to cut the extrudate precisely with the die. In FRK manufacturing, the cutter must be matched face to face with the die without any gap to produce FRK separately; otherwise, the FRK may stick together while cutting at the die.

Heating and Cooling System

Each barrel section has an independent heater wrapped around it to provide the heat and melt the material inside. The electric heaters may be ceramic band, probe type, or aluminium casted heaters. In some extruders, a cooling system maintains the barrel and die head temperature uniform throughout production. The temperature profile is maintained using the heating and cooling system.

Drive System

The drive system includes an electric motor and a gearbox, which provide the power to rotate the screw and move the material in the extruder. The electric motor is used to drive the gearbox, while the gearbox transmits the power from the motor to the screw. The gearbox provides the torque and speed reduction to operate the extruder at the desired speed. The variable frequency drive (VFD) is used to control the speed of an electric motor, which in turn controls the screw speed and specific mechanical energy.

Control Panel

The control panel consists of various sensors such as temperature, pressure, torque sensors, controllers (temperature and torque), and feedback mechanisms to monitor and adjust the process parameters such as temperature, screw speed, and feed rate during the extrusion process. It ensures that the extruder operates within the desired parameters.

8.3 SELECTION OF RAW MATERIALS

The selection of raw ingredients is critical for the quality of FRK. The primary materials are rice flour and VMP. In addition to improving the textural and cooking properties, the permitted additives are added.

8.3.1 Rice Flour

Rice flour is the main ingredient, and it is often prepared using the broken rice obtained from the rice mill. It is a bulk material in which other ingredients are mixed. Therefore, it should be compatible with VMP and other additives so that they can be uniformly distributed throughout the rice flour. The broken rice should be cleaned for foreign matters (organic and inorganic) with a moisture content of 11%–12% w.b. before grinding. The broken rice should be ground in a suitable grinder (e.g., hammer mill) to reduce the particle size to less than 250 μm. The particle size is generally preferred below 150 μm. The lower particle size makes it compatible with the VMP for mixing (BIS 2021a). The starch composition, i.e., the ratio of amylose to amylopectin, plays a significant role in determining the texture, expansion behaviour, and retrogradation properties of FRK; therefore, an amylose content of 20%–25% provides better textural and cooking characteristics to the FRK (Nithya, Dalbhagat, and Mishra 2022; Nithya et al. 2024).

8.3.2 Vitamin-Mineral Premix

VMP is an essential ingredient in FRK manufacturing. It is mixed with rice flour and should have good stability, bioavailability, and compatibility with the rice flour. The VMP should be free flowing without any lumps, white to off-white in colour, without any off-flavour. The micronutrient source should be food-grade micronised ferric pyrophosphate, folic acid, and vitamin B_{12} (BIS 2021b). The base or carrier material (e.g., rice starch) in VMP should be white in colour so that during extrusion, there would not be a significant colour change of FRK.

8.3.3 Additives

The additives include hydrocolloids, gums, binding agents, acid regulators, antioxidants, and emulsifiers to improve the textural properties of FRK (BIS 2021a).

The portion of permitted food-grade additives should be such that there would not be any adverse changes in FRK during extrusion.

8.3.4 Water

The potable water or reverse osmosis purified water should be used in the preparation of FRK (BIS 2021a).

8.4 PROCESS OPTIMISATION

Process optimisation is important to identify the set of operating process parameters for the TSE and raw ingredient composition to produce good quality FRK.

8.4.1 Raw Material Characteristics

Raw material significantly affects the quality of FRK as the material transformation in the extruder is dependent on the composition, particle size, and moisture content of the raw ingredient mix. These parameters affect the specific mechanical energy, degree of gelatinisation, textural and structural properties, cooking characteristics, etc. Fine particle size of rice flour promotes better hydration and distributes moisture evenly throughout the rice flour mix. Larger particles may cause uneven heating and cooking in the extruder and subsequently jamming of the extruder. Undercooking rice flour–VMP mix in the extruder might lead to reduced binding, affecting the cooking properties of FRK. Amylose and amylopectin ratios also influence the degree of gelatinisation, binding, and cooking characteristics. High amylose content in rice flour produces good FRK quality with improved textural and cooking characteristics (Nithya et al. 2024). The ideal moisture content of the feed material mix ready for extrusion should be 28%–32% w.b., or, as per extruder design, the moisture content can be optimised (Yadav, Dalbhagat, and Mishra 2022).

8.4.2 Extruder Process Parameters

Extruder process parameters include the die and barrel temperature, extruder screw speed, feeder speed or feed rate, and cutter speed. The combination of these parameters governs the degree of starch gelatinisation, micronutrient retention, texture and structure formation, and FRK production capacity. The extruder barrel is divided into three to four zones, each equipped with a heater. The temperature of each zone is set in increasing order from the feeding section to the die head section (e.g., 30°C–50°C–70°C–90°C). The lower temperature at the die head section may cause jamming because of undercooking, while a higher temperature leads to expansion in the FRK. Higher temperature improves the starch gelatinisation; however, the expansion and nutrient degradation may occur in FRK.

The extruder screw speed controls the residence time, shear rate, and mixing efficiency and heat distribution in the melt inside the extruder. Typical values range from 50 to 350 rpm, which must be synchronised with the feed rate and cutter speed. A high screw speed reduces residence time but increases shear, which can enhance mixing and expansion and increase the thickness of FRK, while a lower speed produces thinner FRK. The feed rate determines the production capacity of FRK and directly affects the back pressure and shear in the extruder. Consistent feed rate using a gravimetric or volumetric feeding system leads to uniform FRK formation. The unstable feed rate may cause variation in the size and shape of FRK and a non-uniform degree of gelatinisation. Die pressure affects the density and shape of FRK. Die pressure is dependent upon the set extruder process parameters. Optimum higher pressure causes the compaction of FRK if the moisture in the rice flour–VMP mix is around 28%–32% w.b. The lower moisture and higher pressure may cause expansion in FRK. The higher torque can be experienced at lower moisture and temperature.

8.4.3 Product Parameters

The quality of FRK is evaluated based on physical dimensions, bulk and true densities, colour, textural properties, and cooking characteristics, and the extruder process parameters significantly affect these properties. The extruder parameters, such as high temperature and low moisture, significantly affect the size and shape of FRK and sometimes may cause expansion in FRK. The expansion causes a decrease in bulk density, which leads to the floating of FRK during cooking and segregation of FRK during blending. The higher true density in FRK can be obtained at lower die head temperature and higher moisture content (Dalbhagat and Mishra 2019). Cooking characteristics such as cooking time, solid losses, water absorption ratio, textural properties, and functional properties are also significantly affected by the extruder process parameters. Optimum process parameters should be obtained for the production of FRK (Yadav, Dalbhagat, and Mishra 2022; Dalbhagat, Mahato, and Mishra 2019). The effect of extrusion process parameters (Dalbhagat and Mishra 2019), rice varieties (Nithya et al. 2022), and amylose content (Nithya et al. 2024) on the quality of FRK is well discussed in the literature.

8.4.4 Design of Experiments

Design of experiments (DoE) is a statistical methodology used to identify, quantify, and optimise the influence of multiple extruder operational parameters on desired properties of FRK during the extrusion process. It allows systematic planning and analysis of trials to understand parameter interactions and build predictive models with fewer experimental runs. The experimental designs include a full factorial design, a central composite rotatable design, and a Box Behnken design with response surface methodology (RSM) to find the best polynomial equation to fit the experimental data and predict the responses (Dalbhagat and Mishra 2019). The dependent variables include the physicochemical attributes, cooking characteristics, and textural properties, which are evaluated using ANOVA to determine the statistical significance. The experimental values are used to develop the predictive model using RSM. The validation of the model is done by comparing the predicted values with the experimental values using goodness-of-fit measures such as R^2, adjusted R^2, and RMSE. DoE helps to develop a data-driven, systematic approach for FRK process optimisation.

8.4.5 Optimisation Approach

The FRK manufacturing process is optimised to fine-tune formulation and extrusion process parameters to achieve maximum performance using the FRK quality parameters. It shows the relationship between independent and dependent parameters to explore their interaction. The optimisation techniques include RSM, desirability function analysis, and machine learning-based regression models such as artificial neural network (ANN)-genetic algorithm (GA) approach. ANN can model complex nonlinear relationships between inputs and responses. Coupling ANN with GA allows global optimisation even when datasets are noisy or contain nonlinear dependencies. Once the optimal conditions are obtained, the validation experiments are carried out to confirm the predicted results using real experimental data. Optimisation ensures not only technical feasibility but also economic efficiency.

8.5 EXTRUDER OPERATION

The successful operation of the extruder for FRK production requires a well-defined standard operating procedure that includes startup, steady state processing, shutdown, and cleaning.

8.5.1 Operational Procedure

The operational procedure for TSE is relatively easy and simple; however, the operator needs proper training.

- The TSE, including all its accessories, should be cleaned thoroughly before startup. It should be ensured that there is no waste or burnt material inside the barrel and in the feeding system. The TSE barrel should be flushed with compressed air to remove residual material. Simultaneously, the nuts and bolts should be tightened.
- The heating and cooling system should be inspected so that the TSE stabilises at a uniform temperature. The electric connections of the heater element should be checked. The cooling liquid level should be checked to ensure regular flow of cooling liquid through the barrel jacket. The temperature and pressure sensors should be appropriately checked and calibrated if required. The gearbox also needed the lubricating oil; the level should be checked before the start of TSE.
- The breaker plate should be appropriately placed while mounting the die at the die head section. The cutter assembly should be set on the die, ensuring a face-to-face match between the blade and die surface.
- The control panel should be switched on once the TSE is cleaned and set. The temperature profile for TSE should be set with a gradual increase from the feeding section to the die head section. Wait until the temperature at each barrel section stabilises. The set temperatures in each zone should be as per the optimum conditions. To maintain the uniform temperature, the cooling system should also be activated. Monitor the temperature of each

barrel and die head sections through the control panel and record the data for further analysis.

- The TSE operating conditions, such as temperature, screw speed, and feed rate, may differ according to rice varieties. However, there might be a slight change in the operating conditions to make FRK from any rice variety.
- Once the TSE stabilises, start the feeder screw at the specific speed to deliver the desired feed rate of feeding material, followed by the extruder screw speed at the optimum speed. Keep checking the pressure at the die head. When the pressure starts increasing, switch on the cutter to cut the FRK.
- The TSE operating process parameters include die head section temperature, barrel temperature (Zones 1, 2, and 3), extruder screw speed, feed rate/feeder screw speed, and material moisture content.
- The TSE can run continuously to produce FRK until an interruption or an electric shutdown. Once TSE stops, the viscoelastic material in the die hardens, and it is difficult to restart the TSE under the same conditions. The TSE must clean again to restart production.

8.5.2 Cleaning Protocol

Once the production of FRK is over, the TSE should be switched off to cool down the die head section. The die should be dismantled, and the main screw should be run with dry flour to remove the viscoelastic melt from the barrel. The dry flour removes maximum residual material from the barrel. Still, there might be some material stuck to the screw and inner wall of the barrel. The barrel may be heated to a higher temperature, such as 120–150°C for 20 min, to burn the residual material, followed by flushing with high-pressure compressed air. The die and cutter are cleaned manually using hot water (80°C–85°C) and normal water and disinfected.

8.5.3 Troubleshooting

While operating the extruder, technical issues might arise that must be addressed promptly to avoid quality issues in FRK (Table 8.1). Proper extruder maintenance

TABLE 8.1
Troubleshooting and Probable Solutions

S. No	Troubleshoot	Probable Solution
1	Overheating	Control barrel and die head temperature
2	Blocking/jamming	Maintain feed material properties and feed rate
3	Improper gelatinisation	Check for temperature, residence time, and moisture
4	Poor FRK colour	Set a proper temperature profile and avoid overheating
5	Inconsistent FRK size	Check the die design and die–cutter setting
6	Excessive fines	Maintain uniform pressure and the die–cutter assembly
7	Micronutrient losses	Check the temperature profile and extruder screw speed

ensures smooth operation and consistent product quality during the manufacturing of FRK. The TSE should be inspected for visible wear, loose parts, or unusual vibrations. It should also ensure no leaks in the barrel, seals, or feed system. The screw alignment should also be verified to prevent damage to the barrel or other components.

8.6 EXTRUDER MAINTENANCE

Maintenance of the extruder is critical to ensure consistent production without interruption, prolonged.

8.6.1 Screw and Barrel

Regular cleaning of the screw and barrel is essential to prevent the buildup of material inside the barrel, which may cause blockage during the fresh production. The wear and tear and corrosion of the screw and barrel should be checked to avoid any physical hazard. The damaged screw and barrel should be replaced to maintain production and avoid damage to the whole extruder system.

8.6.2 Die and Cutter

The die and cutter play a key role in producing a consistent size and shape of FRK. While cutting the FRK at the die face, the blade may lose sharpness, break, or be damaged. The damaged blade should be replaced immediately. Heavy friction between the die and cutter may cause damage to both, which requires properly setting the cutter on the die. Therefore, the cutter blade should be appropriately aligned with the die face.

8.6.3 Heating and Cooling System

For consistent FRK production, the heating and cooling system maintenance must be done periodically to achieve a uniform temperature profile. If required, temperature sensors should be calibrated, and their correct working and functioning should be ensured. The heating elements and cooling liquid should be replaced at regular intervals.

8.6.4 Gearbox and Drive System

Wear and tear of the gears in a gearbox should be checked periodically. If any unusual sound, vibration, or noise persists, the maintenance should be done, and the damaged part may be replaced. The drive motor, belt, and pulley should also be checked to ensure TSE works efficiently. The lubricating oil level should be checked; the oil may be refilled or replaced.

8.6.5 Electrical and Control Systems

A control system involves the sensors and electrical components such as controllers, variable frequency drives, and relays. Temperature, pressure, and torque sensors and encoders should be checked for proper working to ensure an effective feedback control system. The electric wiring should be inspected periodically; if required, it may be changed wherever necessary to avoid a short circuit or damage.

8.6.6 Safety Checks

Safety precautions and alarms must be associated with TSE to ensure safety and prevent accidents. TSE should have an emergency stop system to shut it down quickly under unavoidable or damaging conditions. A fire safety system should also be checked. A detailed preventive maintenance schedule must be followed for regular system inspection, damaged part replacement, and maintaining the system in working condition. The record of all activities related to repair and maintenance should be kept to track the performance of TSE and prevent any issues that cause damage to it.

8.7 CHALLENGES AND FUTURE SCOPE

8.7.1 Challenges

- Inconsistent moisture distribution in the rice flour–VMP mix results in uneven heating in the barrel and die head, sometimes leading to jamming the extruder or blocking the openings in the die, causing size and shape distortion in FRK. Expansion of FRK also occurs if the temperature and pressure are high in the extruder.
- The changes in rice varieties and VMP can also lead to changes in FRK quality and the extruder behaviour, causing variations in quality from batch to batch of FRK.
- Replicating the FRK quality with exact size and shape, density, colour, cooking characteristics, and textural properties is always challenging due to the non-precise control of extruder process parameters and variation in the raw material characteristics.
- The extreme extruder process parameters (e.g., higher temperature and screw speed) significantly degrade the micronutrients, especially the folic acid and vitamin B_{12}.
- High initial investment in TSE and control systems restricts its adoption by small-scale manufacturers.
- Operating and maintaining the extruder system requires trained personnel, which is always a bottleneck in producing good quality FRK.
- Frequent cleaning is essential to prevent contamination, and safety checks are required to maintain the extruder system; however, disassembling extruders and ensuring complete sanitisation is time-intensive.

8.7.2 Future Scope

- Development of heat-stable, encapsulated VMP enhances the retention of micronutrients during the extreme extrusion conditions.
- Innovating extruders that operate at lower temperatures, providing good quality FRK, and maintaining the micronutrients and consistent quality of FRK.
- Use of real-time AI models for predictive process control, process parameters adjustments, and failure prevention through sensor data analysis.
- Creating a digital twin for the extrusion system or whole process line to simulate and optimise the FRK quality under varying raw material and operating conditions.
- Designing the compact and affordable TSE with a robust control system for precise control of operating parameters.

8.8 CONCLUSIONS

Rice fortification is a strategic intervention to address the issue of micronutrient malnutrition. It can be done by mixing the FRK in normal rice in a specific ratio (e.g., 1:100). The technology available for FRK manufacturing includes dusting, coating, and extrusion; however, the extrusion technique provides better nutrient retentions in FRK as well as offers easy process and handling. This method infuses the micronutrients in the FRK, thereby enhancing their stability during washing and cooking. In extrusion, particularly, the TSE offers substantial advantages as it ensures controlled heating profile in barrel zones and extruder speed, which provides a proper degree of starch gelatinisation and produces a FRK with desirable size and shape, cooking characteristics, and textural properties. The raw material selection (rice flour and VMP) and its properties (composition, particle size) and extruder process parameters (barrel and die temperature, extruder screw speed, and feed rate) are important to optimise and control the FRK manufacturing process. The operational procedure and control highlight its relevance in improving the quality and efficiency of the extrusion process. Proper synchronisation of the operating parameters is required for maximum micronutrient retention and achieving the desired texture and cooking quality. The optimisation techniques and experimental designs help improve the quality of FRK and find the optimal operating conditions for the extruder. In conclusion, extrusion-based FRK production is a proven and validated technology that is a cornerstone for producing fortified rice and fighting against the hidden hunger. The scalability, consistency, and operational flexibility of extrusion make it a method of choice for producing FRK.

REFERENCES

BIS. 2021a. *Fortified Rice Kernels — Specification (IS 17782:2021)*. New Delhi: Bureau of Indian Standards, Government of India. https://standardsbis.bsbedge.com/BIS_searchstandard.aspx?Standard_Number=IS+17782&id=34819

BIS. 2021b. *Vitamin and Mineral Premix for Manufacturing Fortified Rice Kernels—Specification (IS 17781:2021)*. New Delhi: Bureau of Indian Standards, Government of India. https://standardsbis.bsbedge.com/BIS_searchstandard.aspx?Standard_Number=IS+17781&id=34818

Dalbhagat, Chandrakant Genu, Dipendra Kumar Mahato, and Hari Niwas Mishra. 2019. "Effect of Extrusion Processing on Physicochemical, Functional and Nutritional Characteristics of Rice and Rice-Based Products: A Review." *Trends in Food Science and Technology* 85: 226–40. https://doi.org/10.1016/j.tifs.2019.01.001.

Dalbhagat, Chandrakant Genu, and Hari Niwas Mishra. 2019. "Effects of Extrusion Process Conditions on System Parameters; Physicochemical Properties and Cooking Characteristics of Extruded Fortified Rice Kernels." *Journal of Cereal Science* 89: 102782. https://doi.org/10.1016/j.jcs.2019.05.016.

Dalbhagat, Chandrakant Genu, and Hari Niwas Mishra. 2021. "Effect of the Drying Process on the Color Change, Fissure Development, and Morphology of Fortified Rice Kernels." *Journal of Food Process Engineering* 44 (7). https://doi.org/10.1111/jfpe.13719.

Dalbhagat, Chandrakant Genu, Ajay Kumar Swarnakar, Rakesh Kumar Raigar, and Hari Niwas Mishra. 2024. "Extrusion Technology." In *Unit Operations in Food Processing*, edited by Deepak Mudgil and Shweta Mudgil, 1st ed., 299–322. New Delhi: Scientific Publisher.

Hagenimana, Anastase, Xiaolin Ding, and Tao Fang. 2006. "Evaluation of Rice Flour Modified by Extrusion Cooking." *Journal of Cereal Science* 43 (1): 38–46. https://doi.org/10.1016/j.jcs.2005.09.003.

Liu, Chengmei, Yanjun Zhang, Wei Liu, Jie Wan, Weihua Wang, Li Wu, Naibei Zuo, Yiran Zhou, and Zhonglin Yin. 2011. "Preparation, Physicochemical and Texture Properties of Texturized Rice Produce by Improved Extrusion Cooking Technology." *Journal of Cereal Science* 54 (3): 473–80. https://doi.org/10.1016/j.jcs.2011.09.001.

Mandliya, Shubham, Siddharth Vishwakarma, Mahima Bora, Subrata Kumar Bag, Chandrakant Genu Dalbhagat, and Hari Niwas Mishra. 2025. "Ready-to-Eat Foods Development through Composite Extrusion Technological Development." In *Recent Advances in Ready-to-Eat Food Technology*, edited by Hari Niwas Mishra, Pradyuman Kumar, and Ajay Singh, 35–51. CRC Press.

Nithya, A., Chandrakant Genu Dalbhagat, and Hari Niwas Mishra. 2022. "A Comparative Study on the Physicochemical, Cooking and Textural Properties of Fortified Rice Kernels Prepared from Raw and Parboiled Rice." *International Journal of Food Science and Technology* 57 (2): 1325–32. https://doi.org/10.1111/ijfs.15529.

Nithya, A., Chandrakant Genu Dalbhagat, and Hari Niwas Mishra. 2023. "Development of a Simple Method for Extraction and Analysis of Folic Acid in Fortified Rice Kernels by HPLC and Its Application in Vitamin Retention Studies." *Journal of Food Measurement and Characterization* 17 (6): 6440–51. https://doi.org/10.1007/s11694-023-02107-z.

Nithya, A., Siddharth Vishwakarma, Chandrakant Genu Dalbhagat, and Hari Niwas Mishra. 2024. "Apparent Amylose Content Positively Influences the Quality of Extruded Fortified Rice Kernels." *Carbohydrate Polymers* 338: 122213. https://doi.org/10.1016/j.carbpol.2024.122213.

Singh Gujral, Hardeep, and Narpinder Singh. 2002. "Extrusion Behaviour and Product Characteristics of Brown and Milled Rice Grits." *International Journal of Food Properties* 5 (2): 307–16. https://doi.org/10.1081/JFP-120005787.

Steiger, Georg, Nadina Müller-Fischer, Hector Cori, and Béatrice Conde-Petit. 2014. "Fortification of Rice: Technologies and Nutrients." *Annals of the New York Academy of Sciences* 1324 (1): 29–39. https://doi.org/10.1111/nyas.12418.

Xu, Enbo, Xiaowei Pan, Zhengzong Wu, Jie Long, Jingpeng Li, Xueming Xu, Zhengyu Jin, and Aiquan Jiao. 2016. "Response Surface Methodology for Evaluation and Optimization of Process Parameter and Antioxidant Capacity of Rice Flour Modified by Enzymatic Extrusion." *Food Chemistry* 212: 146–54. https://doi.org/10.1016/j.foodchem.2016.05.171.

Yadav, Gorenand Prasad, Chandrakant Genu Dalbhagat, and Hari Niwas Mishra. 2022. "Effects of Extrusion Process Parameters on Cooking Characteristics and Physicochemical, Textural, Thermal, Pasting, Microstructure, and Nutritional Properties of Millet-based Extruded Products: A Review." *Journal of Food Process Engineering* 45 (9): 1–25. https://doi.org/10.1111/jfpe.14106.

9 Drying and Polishing of Fortified Rice Kernels

Madhuresh Dwivedi, Ganga Kishore S, K. Kamaleeswari, Chandrakant Genu Dalbhagat, and Hari Niwas Mishra

9.1 INTRODUCTION

Rice (*Oryza sativa*) is a staple food for nearly half the global population, contributing significantly to daily caloric intake. Conventional milling depletes essential nutrients, impacting glucose metabolism (Kishore et al. 2024). Iron deficiency remains a widespread global health concern, particularly among vulnerable populations. Targeted fortification strategies address micronutrient deficiencies, particularly IDA, by enriching staple foods like rice with essential nutrients, including iron, folic acid, and vitamin B_{12}, which are crucial for haemoglobin synthesis (Dalbhagat and Mishra 2019). Fortified rice kernels (FRK) produced via extrusion, coating, and dusting enhance nutritional intake while leveraging rice's widespread consumption and physicochemical adaptability. Utilizing rice broken in FRK production improves resource efficiency and adds value to rice processing.

Drying is a cost-efficient preservation process essential for FRK, reducing moisture content below 12% w.b. to inhibit microbial metabolism and enzymatic activity, ensuring shelf stability while retaining structural and nutritional integrity. Drying methods significantly influence physicochemical properties, with improper drying causing colour changes and fissure formation, compromising cooking quality (Dalbhagat et al. 2024). Advances in drying technologies aim to optimize efficiency, reduce energy consumption, and enhance product characteristics, driving innovation in food processing. Tray drying utilizes continuous hot air circulation to dehydrate rice kernels, ensuring simplicity and reliability. Microwave drying accelerates moisture removal via electromagnetic radiation, enhancing efficiency and product quality (Pande et al. 2025). Superheated steam drying is a sustainable, energy-efficient innovation with minimal environmental impact.

Polishing is a crucial post-processing step in extruded rice production, enhancing surface characteristics, visual appeal, and consumer acceptance. It refines texture and improves uniformity while optimizing cooking properties. Polishing effectiveness depends on technique, speed, duration, pressure, and physicochemical properties (Xu et al. 2021). Advanced polishing technologies focus on minimizing nutrient losses while maintaining fortified rice quality. However, research on its specific impact on FRK remains limited. Addressing challenges in drying and polishing

DOI: 10.1201/9781003661245-9

through technological advancements will support improved nutrient retention and processing efficiency.

9.2 DRYING OF FRK

Drying is a simultaneous heat and mass transfer process crucial for extending food shelf life by reducing moisture content to safe levels. Lower moisture prevents microbial growth, enzymatic activity, and structural degradation while enhancing storability. FRK require controlled drying to maintain physicochemical, microbial, and structural integrity (Dalbhagat et al. 2021a). The mode of heat transfer on drying is shown in Figure 9.1. Fortification methods are categorized as wet (soaking, extrusion, parboiling) and dry (dusting, coating), with wet-method rice requiring post-fortification drying (Saha and Roy 2020). Drying techniques include sun drying, hot air drying, microwave drying, vacuum drying, fluidized bed drying (FBD), and superheated steam drying, each influencing nutrient retention and product quality differently (Dalbhagat et al. 2021a; Kyritsi et al. 2011; Mandliya and Mishra 2021; Mandliya et al. 2022; Kumari et al. 2025; Saniso et al. 2022; Ojediran et al. 2021). Understanding these drying methods optimizes milling recovery, head rice yield, and product enhancement.

9.2.1 Sun Drying

Sun drying is a traditional, eco-friendly method used globally to reduce paddy moisture post harvest, processing, and parboiling. It enhances nutritional retention and does not require electricity. FRK containing calcium, iodine, zinc, folic acid, iron,

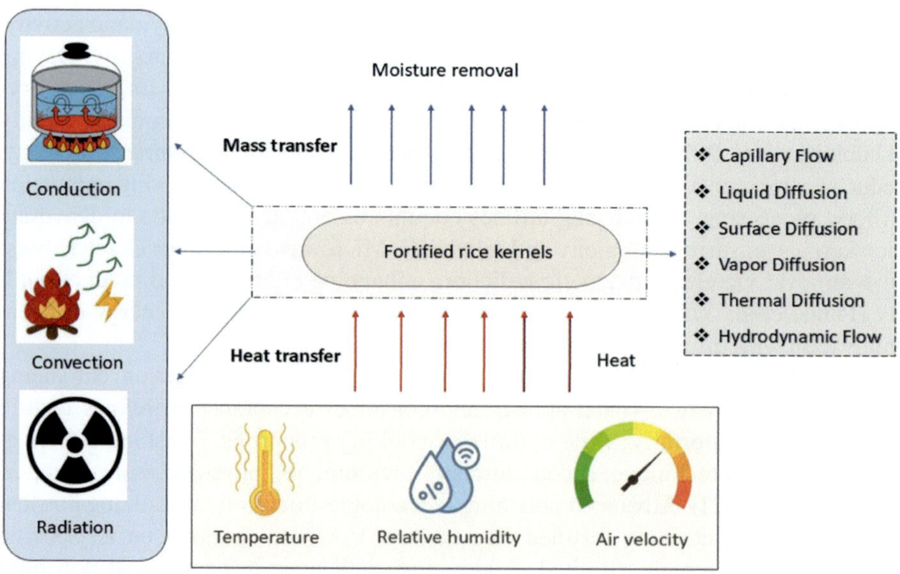

FIGURE 9.1 Mode of heat transfer on drying mechanism of FRK.

and vitamin A undergo soaking in micronutrient-rich solutions, allowing diffusion via concentration gradients. Parboiling gelatinizes starch and improves micronutrient penetration. Studies show that sun drying optimizes bioavailability, with zinc and iron-fortified rice increasing absorption rates by 40%–45% (Prom-u-thai et al. 2011). Folic acid, β-carotene, and iron retention post drying were observed (Thiruselvam et al. 2014). While sustainable, sun drying demands high labour, large space, and results in post-harvest losses of ~30% (Hii et al. 2019).

9.2.2 Hot Air Drying

The traditional hot air drying method, which is widely recognized as being inexpensive, offers superior dried food materials, hygienic qualities, uniformity, simplicity, ease of handling, and affordability. To increase shelf life and produce higher-quality dried goods, it is crucial to identify the ideal drying process parameters for hot air drying. Dalbhagat and Mishra (2021b) found the solid loss (28.56%) and optimal water absorption (2.66) in FRK dried at 40°C for 1 m/s, while temperatures above 40°C led to disintegration. FRK drying on hot air drying is illustrated in Figure 9.2. Dalbhagat and Mishra (2021a) found that FRK dried at the temperature of 50°C–70°C caused the formation fissure inside the FRK as seen in X-ray CT scan images while dried at 40°C for 120 min showed no evidence of fissures. Palamanit et al. (2020) developed turmeric-fortified instant parboiled rice using hot air drying, which required 95°C and 102 min to reach a moisture content of 16% d.b. affecting the colour value of the fortified rice.

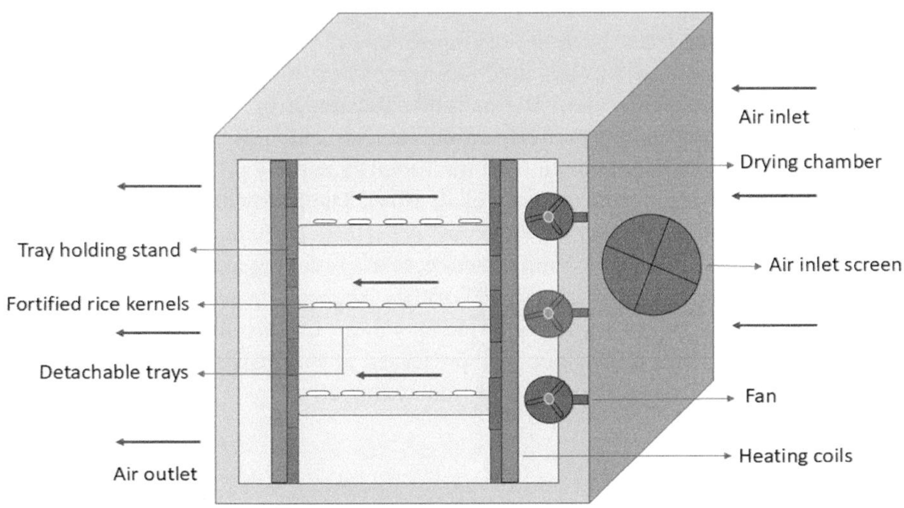

FIGURE 9.2 Hot air drying of FRK.

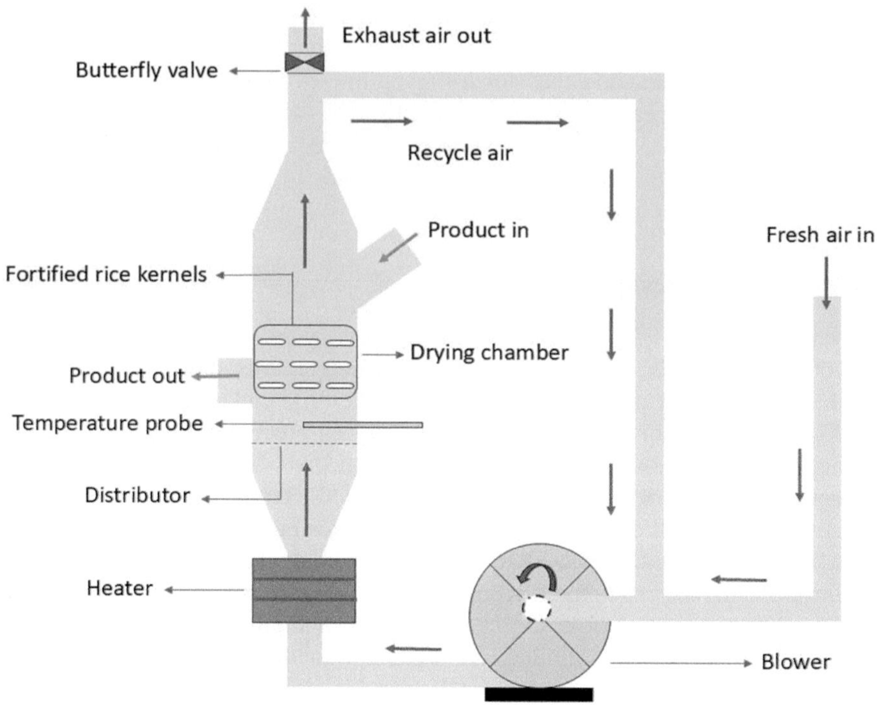

FIGURE 9.3 Fluidized bed drying of FRK.

9.2.3 Fluidized Bed Drying (FBD)

In FBD, heated air is pushed onto the bed through a distributor at an adequate speed to counteract the items' gravitational forces. In drying performance, it provides the highest heat transfer rates, minimal drying time, and uniform drying rates (Saniso et al. 2022). FBD of FRK was depicted in Figure 9.3. In some cases, the germinated fortified rice is dried using the FBD method (Taechapairoj and Kaewyot 2020). It is a reason to improve the GABA (γ-aminobutyric acid) level in germinated rice, increase the head rice count, and inhibit microbial growth by providing great mixing and a better mass transfer rate (Srisang et al. 2011). The paddy dried at 110°C–150°C using a fluidized-bed dryer provided the positive effects of increasing head rice yield, decreasing solid loss, retaining total phenolic, ferric reducing antioxidant power, and the 2,2-diphenyl-1-picrylhydrazyl (DPPH) radical scavenging activity with the drawbacks of increasing cooking time and reducing volume expansion. The amylose–lipid complex in germinated brown rice was preserved at the FBD temperature ranging from 110°C to 130°C (Taechapairoj and Kaewyot 2020).

9.2.4 Microwave Drying

Microwave drying utilizes dielectric heating, which causes polar molecules like water to oscillate, ensuring rapid and uniform drying (Mandliya et al. 2023).

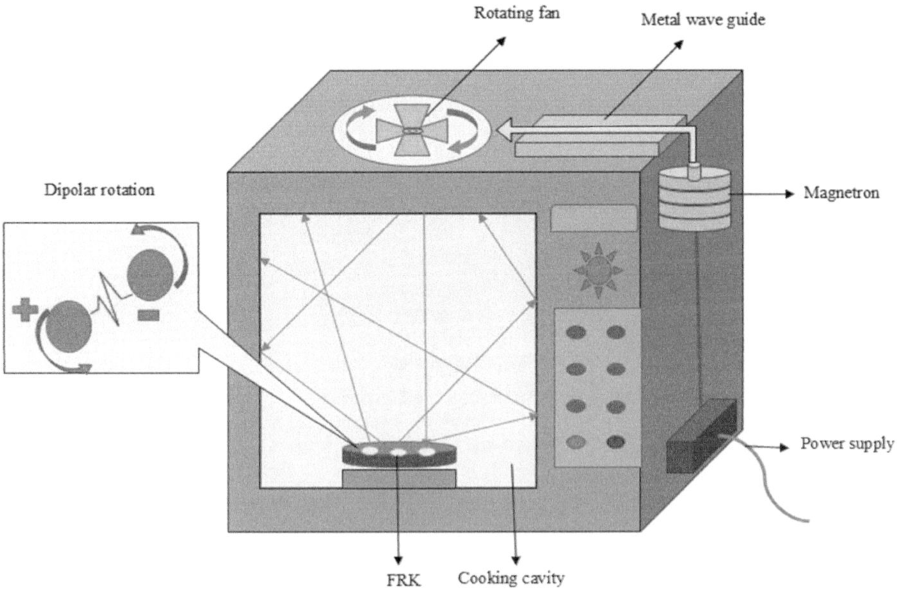

FIGURE 9.4 Microwave drying of FRK.

Dalbhagat et al. (2024) dried FRK at power (180, 360, and 540 W) and tempering time (1, 2, and 3 min) and studied physical properties (colour, fissure development, and cooking properties). Water absorption ratio (WAR) increased in the tempering with microwave power from 180 to 540 W. FRK drying in microwave dryer is illustrated in Figure 9.4. Kumari et al. (2025) optimized iron-fortified rice drying, achieving 97.12% micronutrient uptake and 58.87% gelatinization via microwave-assisted spraying. Palamanit et al. (2020) compared microwave and hot air drying for turmeric-enhanced FRK, finding that microwave-assisted hot air drying reduced drying time by 50% while maintaining phenolic (16.70–17.79 mg GAE/100 g) and antioxidant content (9.21–10.06 mg BHAE/100 g) with a soft texture with a hardness range of 31.03–33.16 N.

9.2.5 Freeze Drying

Freeze drying (lyophilization) is a method that works under the principle of sublimation (Figure 9.5). The freeze-drying process can enhance the rehydration process by evaporating solid ice crystals into vapour while preserving quality. It is responsible for the higher porous structure of the freeze-dried products and can lead to the highest rehydration ratio of the freeze-dried products. The capacity for rice to rehydrate depends on the integrity of the starch network when it is heated or rehydrated. Fu et al. (2021) boosted antioxidants in instant rice using tea powders (instant green tea (IGT), matcha (Mat), and instant black tea (IBT)). Rice was soaked in a 3% tea solution, cooked, frozen (–80°C), and dried (45°C, 25 Pa). Tea powder fortification significantly improved the rehydration ratio and WAI, doubling these values in all

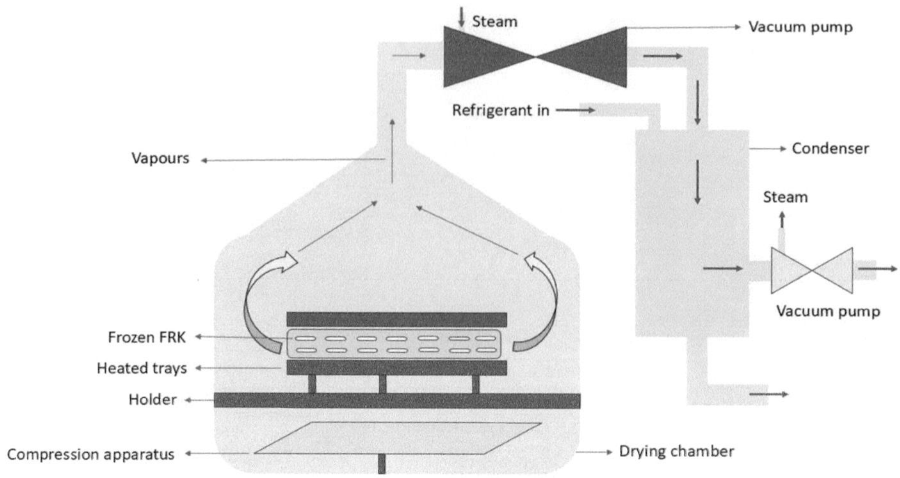

FIGURE 9.5 Freeze drying of FRK.

freeze-dried instant rice variants. This effect was attributed to structural changes in the rice and the presence of hydrophilic groups in the tea powders. Additionally, pre-treatment of cooked freeze-dried rice with α-amylase significantly improved its rehydration ratio, volume expansion, and swelling rate (Yussof et al. 2025).

9.2.6 Other Advanced Drying Techniques

Combined Drying

The combined drying method integrates multiple dehydration techniques to enhance drying efficiency and product quality. Various approaches include fluidized bed-assisted infrared (IR)-microwave drying (Nanvakenari et al. 2022a), hot air fluidized-bed drying with halogen lamp (Srisang et al. 2021), hybrid microwave-FBD (Nanvakenari et al. 2022b), and microwave-vacuum-assisted heat pump drying (Zielinska et al. 2013). Other methods such as osmotic dehydrated microwave-vacuum freeze drying (Duan et al. 2008) and microwave-vacuum-assisted atmospheric freeze drying (Eikevik et al. 2012) further optimize dehydration. Hybrid drying approaches reduce energy consumption while improving material properties. Osmotic dehydration enhances mass transfer, reduces processing time and costs, and lowers water activity by accelerating water loss (WL) and solids gain (SG) (Abrahão and Corrêa 2023). Pre-treated rice was dried using hot air (90°C, 80 min, moisture content: 40% w.b.), microwave (2.5 W/g for 10 min, moisture content: 20% w.b.), and osmotic dehydration (4% glucose, 25% NaCl, room temperature, 10 min mixing + 30 min standing). The final moisture content was 15% to 20% w.b. with a water activity of 0.5 after an additional hot-air drying step (90°C for 20 min). Instant rice produced via this combined method showed no significant difference from cooked rice, but exhibited increased hardness due to solute diffusion during osmotic dehydration (Kamaleeswari et al. 2025).

Infrared (IR) Drying

The principle behind IR drying is that the penetration of electromagnetic IR radiation induces intermolecular vibrations without needing a medium (Figure 9.6). During the process, intermolecular friction generates heat as a result of molecular vibrations (Sakare et al. 2020). The IR drying was used to remove the moisture content of grains during drying with the advantages of reduced drying time, highest energy efficiency, and minimal requirement. Timm et al. (2020) studied dehusking, polishing, and parboiled rice drying at temperatures of 70°C, 100°C, and 130°C with bed depths of 5, 10, and 15 mm and observed that longer exposure times had a greater effect on the fissures formation and reduced head rice yield. The drying time was directly proportional to the grain bed depth and indirectly proportional to the head rice yield (Sakare et al. 2020). Likitrattanaporn and Noomhorm (2011) reported that when rice was dried for 3–23 min using radiation power of 2,100 W and a bed depth of 15 mm, the head rice yield and energy consumption of IR-dried parboiled rice decreased compared to conventional process. However, research on the IR drying of FRK remains unexplored, presenting a significant and intensive area for further investigation.

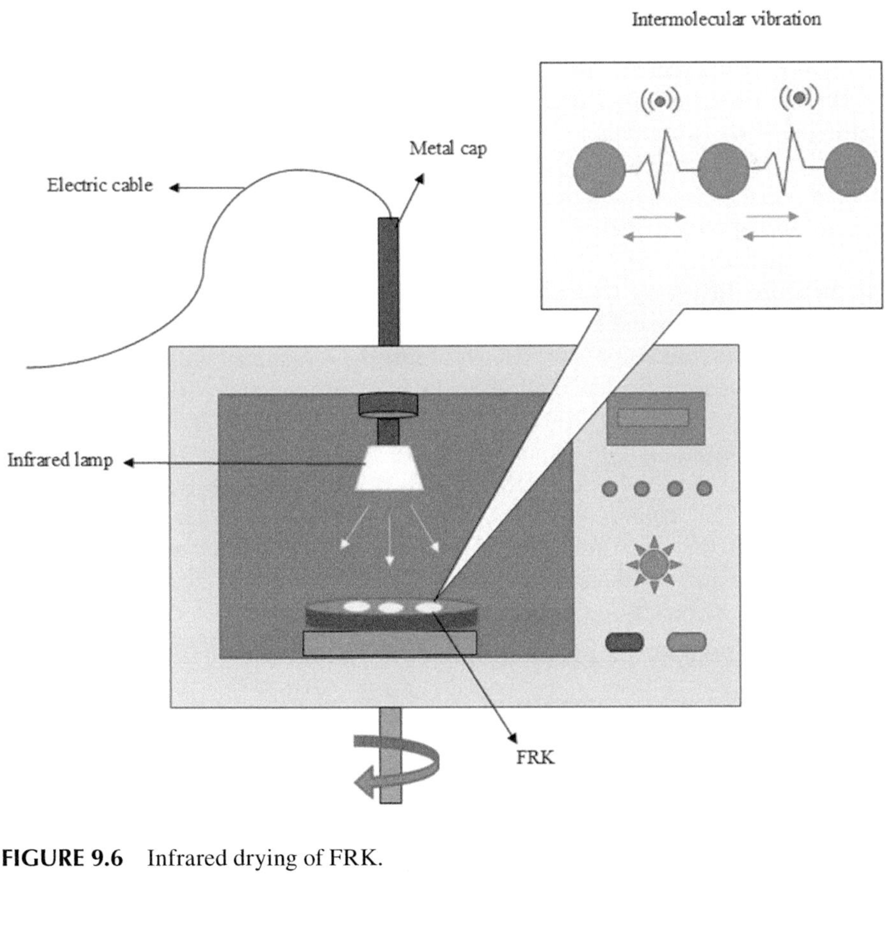

FIGURE 9.6 Infrared drying of FRK.

Ultrasound Pre-treated Hot Air Drying

In ultrasound treatment, acoustic waves cause compressions and decompressions in the medium particles as they travel through the food particles. As a result of increased mass transfer and turbulence, a significant amount of energy is subsequently produced. The ultrasound waves with low frequency and high intensity (frequency: 20–100 kHz and intensity: 10–1,000 W/cm^2) are considered disruptive; thus, it gives a considerable effect on the food characteristics such as physicochemical, mechanical, and biochemical properties during some unit operations like drying, freezing, tenderization, and thawing. Convective hot air drying (50°C–70°C, 1.5 m/s) accelerated drying due to ultrasound-induced microstructural changes, enhancing moisture diffusion and evaporation (Chemat et al. 2011).

9.3 EFFECT OF DRYING PARAMETERS ON NUTRITIONAL QUALITY OF FRK

According to WHO (2025), fortification enhances food micronutrient levels, improving nutritional availability and public health while minimizing risks. Micronutrient retention in FRK depends on processing methods, operational parameters, handling, and storage. Soaking followed by fortification improves retention, while spraying enhances vitamin stability, reducing deterioration caused by prolonged drying at 50°C–70°C (Kyritsi et al. 2011). Prom-u-thai et al. (2010) fortified rice with Zn by soaking, steaming, drying (70°C for 72 h), and husking. Hot air-dried unpolished rice had higher Zn concentration (23.5–78.8 mg/kg) than polished rice (13.2–44.1 mg/kg) due to phytic acid binding in bran. Both polished and unpolished fortified rice showed greater Zn retention than unfortified raw and parboiled rice. The impact of various drying on the quality characteristics of FRK is illustrated in Table 9.1. The drying efficiency of a food product is determined by drying rate, time, air velocity, and moisture diffusivity based on specific drying parameters. A tray drying study of FRK by Dalbhagat and Mishra (2021b) revealed that the effect of velocity on the effective moisture diffusivity of FRK was limited as compared to the temperature.

Methods like osmotic dehydration, soaking, blanching, and ultrasound treatment enhance drying efficiency (Yılmaz et al. 2025; Fu et al. 2021). Ultrasound treatment is particularly effective in reducing water diffusion barriers. Novel drying technologies, including superheated steam, fluidized bed, IR, freeze drying, microwave, and pulse electric field drying, optimize drying time, quality, and key parameters like temperature, airflow, and humidity. In irradiation drying, radiation frequency and wavelength variations influence moisture removal (Sakare et al. 2020).

9.4 CHALLENGES IN DRYING OF FORTIFIED RICE

Emerging drying technologies aim to overcome limitations in conventional dryers by optimizing drying time, rate, efficiency, cost, and heating medium availability while preserving nutrient stability. Sun drying requires extended exposure to achieve desired moisture levels, but its inconsistent temperature distribution can lead to nutritional loss, colour alteration, and sensory degradation. To overcome this issue, hot air

TABLE 9.1
Impact of Various Drying Methods on Nutritional, Physicochemical, Textural, and Microstructural Properties of FRK

Rice Variety	Method of Fortification	Drying Method	Drying Parameter	Major Findings	References
Long grain parboiled rice	Steaming and cooking **(Turmeric powder)**	MAHD	65°C, 80°C, 95°C and 110°C MPD: 0.588 W/g	• MAHD: Reduce the drying time by 50% than HA drying • Minimal variation in textural, rehydration, and microstructural properties • TPC and TAC improved	Palamanit et al. (2020)
Broken Swarna	**Extrusion (Vitamin and mineral)** Die head: 70–110°C Extruder screw speed: 40–80 rpm Feeder screw speed: 20–60 rpm feed moisture content: 26%–34% w.b.	HAD	40°C, 50°C, 60°C and 70°C 0–160 min 1.0 m/s	• No fissure formation in FRK: 40°C and 120 min • Increasing temperature: higher fissure formation and decrease L* and Whiteness index	Dalbhagat and Mishra (2021b)
Broken Swarna	**Extrusion (Vitamin and mineral)** Die head: 70°C–110°C Extruder screw speed: 40–80 rpm Feeder screw speed: 20–60 rpm Feed moisture content: 26%–34% w.b.	HAD	40°C, 50°C, 60°C, and 70°C 0.5 and 1.0 m/s	• Least solid loss (28.56%) and higher water absorption ratio (2.66): 40°C • Disintegration of FRK: 50°C–70°C	Dalbhagat and Mishra (2021a)
Long grain rice (Brown, white, and parboiled rice)	**B complex vitamins** Soaking: 90°C for 15 min Spraying: 20°C–25°C for 30-40 min	HAD	**Soaking:** 50°C–70°C for 2 h **Spraying:** 50°C for 30–60 min (60,000 Pa).	**Vitamin retention of dried rice** • Spraying: 54.3%–85.3% • Soaking: 53.5%–76.2% • >70% of vitamin content of cooked FRK retained • Drying at 50°C–70°C took around 4 h, compared to 1 h for spraying, leading to vitamin degradation	Kyritsi et al. (2011)

(Continued)

TABLE 9.1 (*Continued*)
Impact of Various Drying Methods on Nutritional, Physicochemical, Textural, and Microstructural Properties of FRK

Rice Variety	Method of Fortification	Drying Method	Drying Parameter	Major Findings	References
Paddy	**Germination (Antioxidants)**	FBD	110°C–150°C	• Final moisture content: 19%–21% • Elevated temperature: increasing cooking time (32 min) and decreasing volume expansion (426.67%) of FRK • Preserve antioxidant capacity	Taechapairoj and Kaewyot (2020)
Long grain white rice	**Antioxidant** (Tea- IGT, IBT, and Mat) Soaking: 3% w/w, 1 h	Freeze drying	Stage 1: frozen at 80°C for 12 h Stage 2: 45°C at 25 Pa	Rehydration ratio and WAI: doubled in FRK. TAC: 7.22–232.06 U/g TPC: 1.41–22.96 mg GAE/g DPPH: 89.04%	Fu et al. (2021)
Milled whole rice	**Spray soaking (Iron)** NaFeEDTA solution (2,125 mg/400 mL) MPD: 0.6 kW/kg for 20 min	MAHD	0.6 kW/kg for 65 min	Final moisture content: 13.21% (d.b.) **Micronutrient retention** Spray soaking: 97.12% Washing: 89.19% Cooking: 65.77%	Kumari et al. (2025)
Parboiled rice	**Blending (vitamin and mineral)** Extruded fortified rice analogue + parboiled rice	MW	MWP: 180, 360, and 540 W Tempering: 1, 2, and 3 min at 180 W	**Higher MWP** Drying time: 5 min (least) WI: 50.06 (least) L* value: 58.40 (least) **Tempering time:** 1 and 2 min: resist fissure formation in rice Appropriate MWP: < 180 W (better quality)	Dalbhagat et al. (2024)

TAC, total antioxidant capacity; TPC, total phenolic content; DPPH, 2,2-diphenyl-1-picrylhydrazyl; IGT, instant green tea; IBT, instant black tea; Mat, Matcha; MPD, microwave power density; MWP, microwave power; MAHD, microwave-assisted hot air drying; FBD, fluidized bed drying; MW, microwave drying; HAD, hot air drying; NaFeEDTA, sodium iron ethylenediaminetetraacetate.

drying was maintained at a constant temperature distribution. However, the drying parameters, viz., time–temperature combination, air velocity, and size of the rice (short, medium, or long rice) are the key influencing factors that create the impact on the drying of FRK (Xu et al. 2021). The optimization of the hot air drying process parameter results in high quality and retains the fortified nutrition in the FRK. In fluidized-bed drying, the solid rice particles are evenly distributed and blended efficiently (Inyang et al. 2017). However, there were certain changes (deterioration) in the FRK. It could be mitigated by irradiation drying instead of hot air drying, eliminating the need to heat the surrounding air or medium. The fundamental mechanism of this volumetric heating-drying process, which uses a microwave, makes it easier for heat to move from the FRK interior to its exterior. However, issues that could reduce FRK quality include inadequate heat management and mass transfer during the drying process. Combining the microwave with other drying techniques, like vacuum, heat pump, hot air, and FBD, resulted in a faster rate of dehydration, shorter drying times, and more accurate control. Combining convective, microwave, and ultrasound shortened drying time by approximately 88% (Szadzińska et al. 2017). On the other hand, freeze drying is highly effective in preserving volatile compounds, heat-sensitive nutrients, and flavours. Nevertheless, freeze drying has higher energy demands than other drying methods, leading to increased operational costs.

9.5 POLISHING OF FRK

Polishing is a key milling process applied to most commercially available rice varieties, removing 18%–20% of the oil-rich bran layer (Ojediran et al. 2020). In FRK production, polishing is the final step, following fortification processes such as soaking, coating, extrusion, spraying, germination, and drying, which enhance nutrient retention and stability. Based on processing and nutritional composition, rice is classified into three types: whole rice (minimally processed), brown rice (unpolished), and white rice (polished). Through polishing, brown rice is transformed into a smoother, visually enhanced grain with improved physical properties. The importance of polishing on nutritional, structural, functional, and safety aspects of FRK is shown in Table 9.2.

9.5.1 Purpose of Polishing

Polishing has a considerable impact on consumer acceptability on polished rice. Furthermore, polishing was used to improve the digestibility, water absorption capacity, foaming capacity, and apparent amylose content of the rice. Polishing enhances the optical properties of rice to achieve a white, smooth, shiny surface, making it more visually appealing to consumers. It is also used to remove the undesirable components that are present in rice bran, such as heavy metal contamination (lead and cadmium) and oil content (responsible for oxidation and rancidity) content present in the rice (Chen et al. 2022). Optimizing polishing techniques is essential for maintaining rice quality while balancing nutritional integrity, safety, and consumer preferences.

TABLE 9.2
Importance of Polishing on Nutritional, Structural, Functional, and Safety Aspects of FRK

Rice Variety	Polishing Method/ Machine	Method of Fortification	Major Findings	References
Polished rice (Boro rice)	Grainman tester milling machine	**Zinc and Iron fortification** Pre-steaming: 100°C for 2 min Soaking Parboiling: 121°C for 10 min	• Increasing zinc sulphate (soaking enhances rice zinc content) **Micronutrient retention** • Raw polished rice: 16.6–44.9 mg/kg • Parboiled polished rice: 12.6–32.9 mg/kg • Washing and cooking: 70%–81% • Polished rice: suboptimal for iron fortification	Hotz et al. (2015)
Processed rice	-	-	**Removal range (Pb)** Brown rice (40%), Lightly processed brown rice (62%) Germinated rice (74%) Polished rice (79%) Cadmium: Difficult to remove	Chen et al. (2022)
Paddy	Non-ferrous milling machine	**Soaking (Iron and zinc)** NaFeEDTA: $ZnSO_4$ solution	**Micronutrient retention (dried rice)** Fortified polished rice: Fe – 106.31% and Zn – 97.72 ppm Non-fortified polished rice: Fe – 7.44 ppm and Zn – 4.74 pm **Cooked rice** Fortified polished rice: Fe – 43.54 ppm and Zn – 36.7 ppm Non fortified rice: Fe – 4.24 ppm and Zn – 9.3 ppm	Azam et al. (2021)
Paddy	Rice polisher (JNMJ3, TZYQ)	–	**Polishing removes** Toxic elements: 44.4% As, 19.8% Cd, and 15.4% Cr Nutrients: 30.9% Cu, 31.5% Se, and 29.1% Zn	Yao et al. (2020)

(Continued)

TABLE 9.2 (*Continued*)
Importance of Polishing on Nutritional, Structural, Functional, and Safety Aspects of FRK

Rice Variety	Polishing Method/ Machine	Method of Fortification	Major Findings	References
Brown rice	LTJM160 rice milling machine **DOM** 1. 0% 2. 0%<DOM≤4.56% 3. 4.98%≤ DOM≤8.72% 4. (DOM≥9.13%).		**DOM≤6.56%** Whiteness, transmittance, and L*-value: Increased a* and b* values: Decreased to 180% and 58%, respectively Stage III: Better quality in appearance and morphological characteristics	Ren et al. (2021)
Brown rice and polished rice	Rice polisher	-	**Brown rice** Dietary fibre (3.08% – 3.68%), Oil absorption (116.0% – 139.0%) Emulsion stability (87.46% – 99.93%) Resistant starch content (6.80% – 9.00%) **Polished rice** Water absorption (102.0% – 122.0% Foaming capacity (8.00% – 13.63%), Apparent amylose (19.16% – 22.62%	Qadir and Wani (2023)
Broken rice	–	**Extrusion method (iron, folic acid, and vitamin B_{12})** Die head temperature (70°C–110°C), Extruder screw speed (40–80 rpm), Feeder screw speed (20–60 rpm) Feed moisture content (26%–34% w.b.)	Final moisture content: 10%–12% w.b. WAI: 3.05–4.34 Water solubility index: 3.34%–3.79%. Cooking time: 12.91–16.75 min, Cooking losses: 21.83%–58.51% Water absorption ratio: 0.38–2.64,	Dalbhagat and Mishra (2019)

(*Continued*)

TABLE 9.2 (*Continued*)
Importance of Polishing on Nutritional, Structural, Functional, and Safety Aspects of FRK

Rice Variety	Polishing Method/ Machine	Method of Fortification	Major Findings	References
Brown rice	Pressure-type rice polisher	–	Total starch content increased with DOP **Brown rice** Starch hydrolysis (65.68%) Lower glycaemic index (72.50) **Polished rice** Starch hydrolysis (89.74%–96.17%) Glycaemic index (84.95–92.90)	Tamura et al. (2022)
Raw and parboiled rice	–	**Extrusion method (iron, folic acid, and vitamin B_{12})** Die head temperature (70°C–10°C), Extruder screw speed (40–80 rpm), Feeder screw speed (20–60 rpm) Feed moisture content (26%–34% w.b.)	Polished rice final moisture: 11%–12% w.b. **Raw polished rice** Higher lightness: 77.31 (highest) Water absorption index: 2.05 Water solubility index: 2.43 Cooking time: 15.05 min **Parboiled polished rice** Water absorption index: 2.52 (highest) Water solubility index: 3.53 (highest) Cooking times: 17.90 min (highest) **Texture profile:** Cooked rice (raw): 32.49 N Cooked rice (parboiled): 54.09 N	Nithya et al. (2022)

NaFeEDTA, sodium iron ethylenediaminetetraacetate; DOM, degree of milling; DOP, degree of polished.

9.5.2 Polishing Methods

Traditionally, two types of polishing techniques are employed: abrasion-type and friction-type. The polishing process of FRK is shown in Figure 9.7. The choice of polishing method significantly influences the quality of rice. Understanding these methods and their impact is essential for optimizing rice processing to meet consumer preferences and industry standards.

Abrasive Polishing

Abrasion is a mechanical process characterized by repeated frictional stress between solid surfaces, resulting in material removal. In rice polishing, an abrasion polisher facilitates the displacement of the bran layer from the outer surface of rice kernels through continuous rubbing (Huang et al. 2020). During polishing, rice grains interact with steel rollers, leading to the gradual removal of bran layers from brown rice.

The morphology of brown rice plays a crucial role in determining the final characteristics of polished rice kernels. According to Rachmat et al. (2022), raw rice with a moisture content of 14% typically yields 18%–20% husk weight, 5%–15% bran, and 60%–65% polished rice. Adjustments to the normal force, grit number of the abrasive surface, and sliding distance in the polishing machine alter the abrasive characteristics of brown rice. Both normal load and grit number significantly influence the specific abrasion area and mass loss (Meng et al. 2024). Furthermore, the restoration of the nourishing embryo structure is more effective when rice kernels are polished using an abrasive surface. The application of moderate milling improves milling precision, minimizes fragmentation, and reduces nutritional loss, thereby enhancing the palatability of polished rice.

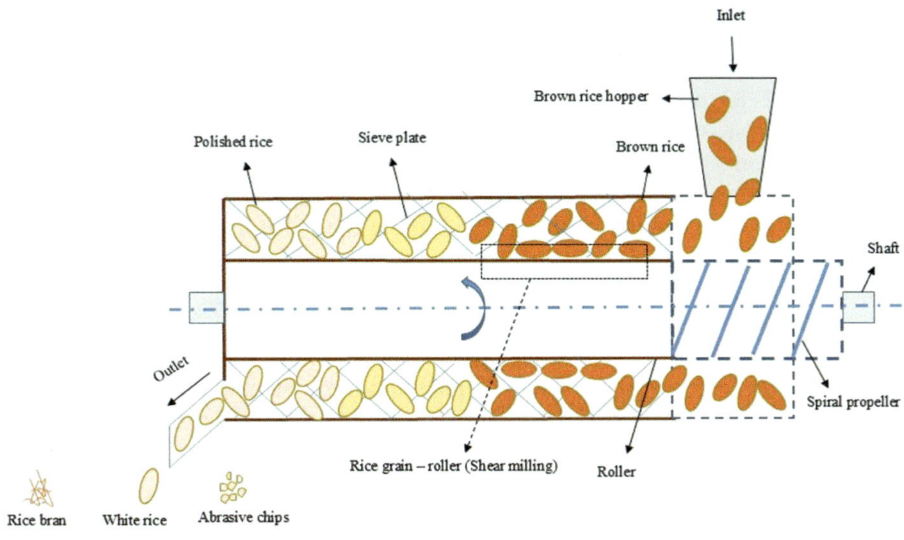

FIGURE 9.7 Polishing of FRK.

Friction Polishing

Friction is a fundamental force that opposes relative motion between two contacting surfaces. It is directly proportional to normal force and acts against movement (Schuderer et al. 2025). In the rice milling process, brown rice undergoes significant mechanical interactions, including pressure, shear forces, and friction during roller rotation, which increases the likelihood of rice breakage. The frictional milling process enhances the production of head rice with superior edible quality; however, it also increases the rate of broken rice due to the applied friction and pressure.

The frictional polishing mechanism can be categorized into four distinct stages: initial sliding, bran layer friction, transition friction, and core stable friction. These stages correspond to the structural composition of brown rice and influence the milling outcome (Yang et al. 2021). From an engineering perspective, the frictional polishing process affects rice porosity and its static and dynamic coefficients of friction but does not significantly alter its specific gravity. Consequently, frictional dynamics play a crucial role in determining the quality of polished rice. Polished rice typically undergoes substantial nutritional loss due to frictional processing, contributing to a growing preference for moderately processed rice. Brown rice with moderate processing retained its nutrients and had a better texture. The careful selection and implementation of these frictional milling methods contribute significantly to producing high-quality rice with improved texture, appearance, and nutritional retention (Yang et al. 2021).

9.6 IMPACT OF POLISHING ON QUALITY OF FRK AND MITIGATING STRATEGIES

9.6.1 Impact on Nutritional Quality of FRK

These nutritional components contribute to its value in food and feed applications. These nutrients are lost when the brown rice is subjected to a polishing process. Rice bran is a rich source of essential minerals, including calcium (198–397 mg/kg), iron (32–83 mg/kg), manganese (90–136 mg/kg), and zinc (43–73 mg/kg) (Wisetkomolmat et al. 2022). Polishing decreases micronutrient content, including thiamine, niacin, pyridoxine, and tocopherol, by around 75%–90%. The physical (size, shape), physicochemical (initial moisture content), and processing conditions (polishing time) of rice influenced the intensity of the polishing. The physical properties of rice grain minimize the loss of mineral content while polishing. Frictional polishing reduces 60%–80% of mineral elements including Mg, P, Fe, Mn, and K with the 8%–10% grain weight loss, and gentler polishing reduces minimal loss with 3%–5% (Hansen et al. 2012). Azam et al. (2021) examined iron and zinc fortification in polished rice, reporting significantly higher mineral content in fortified rice before cooking (97.72 ppm Zn; 106.31 ppm Fe) compared to unfortified rice (14.74 ppm Zn; 7.44 ppm Fe). Despite some nutrient loss during cooking, fortified rice retained more minerals post cooking (36.7 ppm Zn; 43.54 ppm Fe) than unfortified rice (9.3 ppm Zn; 4.24 ppm Fe). Essential micronutrients like Zn, Fe, folic acid, and vitamin B_{12} showed retention rates of 75%–100% after cooking (Kachhawa and Mogra 2023).

Fortification not only restores lost nutrients but also adds essential vitamins, including folic acid, cobalamin, and vitamin A. FRK closely resembles milled rice in size, shape, and colour while maintaining enhanced nutrient content even after soaking and boiling (FSSAI 2018).

9.6.2 Changes in Functional and Morphological Characteristics of FRK

Fortification enhances rice nutrition while ensuring stability during processing. During cooking, gelatinization modifies FRK structure, leading to starch granule rupture and protein dispersion. Extrusion further affects physicochemical properties, influencing water absorption, oil retention, emulsion stability, and pasting behaviour. A study by Qadir and Wani (2023) investigated that the polished rice exhibited a higher water absorption capacity (102%–122%) compared to brown rice (92%–109%). This difference is attributed to the elevated hydrophobic groups in brown rice, which contain higher oil content (12%–18.6%) and hinder starch hydration during cooking. Nithya et al. (2024) examined how varying amylose content (0.46%–23.89%) affects polished FRK functional properties. WAI increased with lower non-glutinous rice content but decreased with 100% glutinous rice. Higher amylopectin improved gelatinization and debranching. Extrusion temperature, feed moisture, and screw speed impacted WAI, while moisture acted as a hydrophilic plasticizer, enhancing gelatinization (Dalbhagat and Mishra 2019). WAI indicates starch transformation, while WSI measures starch disintegration during extrusion (Nithya et al. 2022).

9.7 TECHNOLOGICAL ADVANCES AND FUTURE DIRECTIONS

Advancements in food processing have significantly improved the drying and polishing of FRK, enhancing efficiency, quality, and nutrient retention while ensuring sustainability. Novel drying technologies optimize heat and mass transfer, overcoming challenges in conventional hot air drying like inefficiency and degradation. Methods such as superheated steam, vacuum, radio frequency, IR, fluidized bed, and microwave drying improve drying kinetics and energy use. Electromagnetic radiation-based drying ensures uniform thermal distribution with lower energy consumption. Hybrid RF-MW drying enhances efficiency while minimizing quality loss (Mahmood et al. 2022). Advanced polishing techniques refine FRK surface morphology, improving structural integrity and nutrient retention. AI-driven modelling and imaging further optimize processing for sustainability.

9.7.1 Research Gaps and Future Scope

Drying and polishing are essential post-harvest processes for fortified rice, influencing physicochemical integrity, nutrient bioavailability, and consumer acceptance. Optimizing drying methods is challenging due to limited studies on fortificant stability. In some cases, hot air, microwave-assisted, IR, and FBD had impact rice structure and nutrient retention differently. Improper drying leads to colour degradation and fissures, compromising quality (Dalbhagat and Mishra 2021a). Temperature, air velocity, and humidity variations affect oxidation and micronutrient loss. Future research should refine drying parameters, enhance nutrient retention, and integrate energy-efficient technologies. Standardized protocols will ensure consistent quality, supporting food security and micronutrient deficiency mitigation. Polishing enhances sensory attributes but may reduce nutrients. AI-driven monitoring improves precision, supporting sustainable rice fortification.

9.8 CONCLUSIONS

The drying and polishing of FRK are essential post-harvest operations that significantly impact their physicochemical stability, nutrient bioavailability, and consumer acceptance. Drying facilitates moisture removal while preserving nutritional retention, preventing microbial proliferation and enzymatic degradation, and maintaining kernel integrity. Advanced drying technologies such as vacuum drying, microwave-assisted drying, IR drying, and hybrid methods offer promising solutions for enhancing nutrient preservation and energy efficiency. Polishing, a critical finishing process, influences the texture, colour, and sensory attributes of fortified rice, contributing to its marketability. However, excessive polishing can result in considerable nutritional loss due to surface abrasion and thermal effects, necessitating alternative techniques that optimize aesthetic appeal and micronutrient retention. Standardizing drying and polishing protocols in industrial production is vital for product consistency and fortification efficacy. Real-time monitoring with AI and IoT sensors enhances precision control over moisture dynamics and nutrient stability, reducing variability and optimizing quality. Sustainability-driven approaches, including low-energy drying and polishing techniques, mitigate resource depletion. A multidisciplinary approach can advance fortified rice production for global food security.

REFERENCES

Abrahão, Fernanda Rezende, and Jefferson Luiz Gomes Corrêa. 2023. "Osmotic Dehydration: More Than Water Loss and Solid Gain." *Critical Reviews in Food Science and Nutrition* 63(17): 2970–2989.

Azam, M. Mohibbe, Sarla Padmavathi, R. Abdul Fiyaz, Amtul Waris, K. T. Ramya, and Chirravuri N. Neeraja. 2021. "Effect of Different Cooking Methods on Loss of Iron and Zinc Micronutrients in Fortified and Non-Fortified Rice." *Saudi Journal of Biological Sciences* 28(5): 2886–2894.

Chemat, Farid, and Muhammed Kamran Khan. 2011. "Applications of Ultrasound in Food Technology: Processing, Preservation and Extraction." *Ultrasonics Sonochemistry* 18(4): 813–835.

Chen, Xianghua, Fangman Chen, Shuo Sun, Yingwen Li, Yongxing Li, Hui Mo, Zhian Li, and Ping Zhuang. 2022. "Effect of Polishing on Lead and Cadmium Bioavailability in Rice and Its Health Implications." *Foods* 11(17): 2718.

Dalbhagat, Chandrakant Genu, A. Nithya, Shubham Mandliya, Siddharth Vishwakarma, and Hari Niwas Mishra. 2024. "Effect of Microwave Drying and Tempering on Color Attributes, Fissure Formation, and Cooking Characteristics of Fortified Rice Kernels." *Journal of Food Science and Technology* 61(4): 706–716.

Dalbhagat, Chandrakant Genu, and Hari Niwas Mishra. 2019. "Effects of Extrusion Process Conditions on System Parameters; Physicochemical Properties and Cooking Characteristics of Extruded Fortified Rice Kernels." *Journal of Cereal Science* 89: 102782.

Dalbhagat, Chandrakant Genu, and Hari Niwas Mishra. 2021a. "Drying Modeling, Cooking Characteristics, Pasting Properties, and Crystallinity of Fortified Rice Kernels." *Journal of Food Processing and Preservation* 45(6): e15579.

Dalbhagat, Chandrakant Genu, and Hari Niwas Mishra. 2021b. "Effect of the Drying Process on the Color Change, Fissure Development, and Morphology of Fortified Rice Kernels." *Journal of Food Process Engineering* 44(7): e13719.

Dalbhagat, Chandrakant Genu, Dipendra Kumar Mahato, and Hari Niwas Mishra. 2019. "Effect of Extrusion Processing on Physicochemical, Functional and Nutritional Characteristics of Rice and Rice-Based Products: A Review." *Trends in Food Science and Technology* 85: 226–240.

Duan, Xu, Min Zhang, Xinlin Li, and Arun S. Mujumdar. 2008. "Ultrasonically Enhanced Osmotic Pretreatment of Sea Cucumber Prior to Microwave Freeze Drying." *Drying Technology* 26(4): 420–426.

Eikevik, Trygve Magne, Odilio Alves-Filho, and Michael Bantle. 2012. "Microwave-Assisted Atmospheric Freeze Drying of Green Peas: A Case Study." *Drying Technology* 30(14): 1592–1599.

Food Safety and Standards Authority of India. 2018. *Food Safety and Standards (Packaging) Regulations, 2018*. Ministry of Health and Family Welfare, India. https://fssai.gov.in/upload/uploadfiles/files/Compendium_Packaging_01_02_2022.pdf.

Fu, Tiantian, Liya Niu, Leiyan Wu, and Jianhui Xiao. 2021. "The Improved Rehydration Property, Flavor Characteristics and Nutritional Quality of Freeze-Dried Instant Rice Supplemented with Tea Powder Products." *LWT* 141: 110932.

Hansen, Thomas H., Enzo Lombi, Melissa Fitzgerald, Kristian H. Laursen, Jens Frydenvang, Søren Husted, Chanthakhone Boualaphanh, et al. 2012. "Losses of Essential Mineral Nutrients by Polishing of Rice Differ Among Genotypes Due to Contrasting Grain Hardness and Mineral Distribution." *Journal of Cereal Science* 56(2): 307–315.

Hii, Chinglik, Sze Pheng Ong, Choonlai Chiang, and A. S. Menon. 2019. "A Review of Quality Characteristics of Solar Dried Food Crop Products." *IOP Conference Series: Earth and Environmental Science* 292(1): 012054. IOP Publishing.

Hotz, Christine, Khandaker A. Kabir, Sharifa S. Dipti, Joanne E. Arsenault, and Moniruzzaman Bipul. 2015. "Rice Fortification with Zinc During Parboiling May Improve the Adequacy of Zinc Intakes in Bangladesh." *Journal of the Science of Food and Agriculture* 95(2): 379–385.

Huang, Shiyu, Gunsriwiang Benchamas, and Gangliang Huang. 2020. "Whole Processing and Use of Rice Polishings." *Innovative Food Science and Emerging Technologies* 63: 102373.

Inyang, Uwem, Innoccent Oboh, and Benjamin Etuk. 2017. "Drying and the Different Techniques." *International Journal of Food Nutrition and Safety* 8(1): 45–72.

Kachhawa, Kavita, and Renu Mogra. 2023. "Large-Scale Fortification of Rice in India and Its Distribution in Public Distribution System." *Indian Journal of Public Administration* 69(4): 779–787.

Kamaleeswari, K., N. Karpoora Sundara Pandian, B. Murugan, P. Gokul, M. Pugazhmani, V. Nithyalakshmi, R. Arulmari, et al. 2025. "Impact of Osmotic Dehydration on the Physical, Mechanical, and Textural Properties of Paneer from Different Milk Sources." *Journal of Texture Studies* 56(2): e70013.

Kishore, S. Ganga, Madhuresh Dwivedi, Chandrakant Genu Dalbhagat, Niranjan Thota, Chingakham Ngotomba Singh, Saranya Saravanan, Giripogu Nikitha, N. Vaibhav Kumar, and V. Vignesh. 2024. "Development of Extruded Millet Analogue Rice: A Comparison of Rheological, Structural, Physiochemical, and Cooking Properties with Conventional Rice." *ACS Food Science and Technology* 4(10): 2504–2516.

Kumari, Ankanksha, Ujala Ahuja, Manish Kumar Pandey, and Anupam Roy. 2025. "Optimizing Whole Grain Rice Fortification Using Microwave-Assisted Screw Conveying Spraying and Drying Setup: Exploring Solution Absorption, Gelatinization, and Micronutrient Retention." *Food Chemistry* 468: 142455.

Kyritsi, Athanassia, Constantina Tzia, and Vaios T. Karathanos. 2011. "Vitamin Fortified Rice Grain Using Spraying and Soaking Methods." *LWT-Food Science and Technology* 44(1): 312–320.

Likitrattanaporn, Charun, and Athapol Noomhorm. 2011. "Effects of Simultaneous Parboiling and Drying by Infrared Radiation Heating on Parboiled Rice Quality." *Drying Technology* 29(9): 1066–1075.

Mahmood, Naveed, Yanhong Liu, Zeeshan Munir, Yue Zhang, and Badar Munir Khan Niazi. 2022. "Effects of Hot Air Assisted Radio Frequency Drying on Heating Uniformity, Drying Characteristics and Quality of Paddy." *LWT* 158: 113131.

Mandliya, Shubham, and Hari Niwas Mishra. 2021. "Effect of Different Drying Methods on Quality Attributes and Microstructure of Mycelium (Pleurotus Eryngii)." *Biology and Life Sciences Forum* 6(1): 10.

Mandliya, Shubham, Jayshree Majumdar, Sourav Misra, Monalisha Pattnaik, and Hari Niwas Mishra. 2023. "Evaluation of Dry Microwave and Hot Water Blanching on Physicochemical, Textural, Functional and Organoleptic Properties of Indian Gooseberry (Phyllanthus Emblica)." *Journal of Food Measurement and Characterization* 17(3): 2881–2891.

Mandliya, Shubham, Siddharth Vishwakarma, and Hari Niwas Mishra. 2022. "Modeling of Vacuum Drying of Pressed Mycelium (Pleurotus Eryngii) and Its Microstructure and Physicochemical Properties." *Journal of Food Process Engineering* 45(10): e14124.

Meng, Xiangyi, Yadong Zhu, Fuguo Jia, Guizhong Tian, Xiaoming Feng, Yongsheng Pei, and Zhongqiu Mu. 2024. "Test and Application of Abrasion Characteristics of Brown Rice." *Innovative Food Science and Emerging Technologies* 93: 103635.

Nanvakenari, Sara, Kamyar Movagharnejad, and Asefeh Latifi. 2022a. "Modelling and Experimental Analysis of Rice Drying in New Fluidized Bed Assisted Hybrid Infrared-Microwave Dryer." *Food Research International* 159: 111617.

Nanvakenari, Sara, Kamyar Movagharnejad, and Asefeh Latifi. 2022b. "Multi-Objective Optimization of Hybrid Microwave-Fluidized Bed Drying Conditions of Rice Using Response Surface Methodology." *Journal of Stored Products Research* 97: 101956.

Nithya, A., Chandrakant Genu Dalbhagat, and Hari Niwas Mishra. 2022. "A Comparative Study on the Physicochemical, Cooking and Textural Properties of Fortified Rice Kernels Prepared from Raw and Parboiled Rice." *International Journal of Food Science and Technology* 57(2): 1325–1332.

Nithya, A., Siddharth Vishwakarma, Chandrakant Genu Dalbhagat, and Hari Niwas Mishra. 2024. "Apparent Amylose Content Positively Influences the Quality of Extruded Fortified Rice Kernels." *Carbohydrate Polymers* 338: 122213.

Ojediran, John, Adejoke Adewumi, Samouel Olugbenga Oladele, Olumuyiwa Moses, Adeniyi Olayanju, and Churchill Okonkwo. 2021. "Evaluating the Drying Characteristics of Paddy Rice Using Superheated Steam Dryer." *IOP Conference Series: Materials Science and Engineering*, 1107(1): 012044. IOP Publishing.

Ojediran, John Olusegun, Clinton Emeka Okonkwo, Tajudeen Adeniyi Olayanju, Syed Zameer Hussain, Abiola Folakemi Olaniran, and Adejoke Deborah Adewumi. 2020. "Effect of Polishing Duration on Physical, Milling, Cooking, and Sensory Properties of a Novel Mix-Colored Nigerian Parboiled Rice." *Cereal Chemistry* 97(6): 1172–1182.

Palamanit, Arkom, Angelique Musengimana Sugira, Somchart Soponronnarit, Somkiat Prachayawarakorn, Patcharee Tungtrakul, Fatih Kalkan, and Vijaya Raghavan. 2020. "Study on Quality Attributes and Drying Kinetics of Instant Parboiled Rice Fortified with Turmeric Using Hot Air and Microwave-Assisted Hot Air Drying." *Drying Technology* 38(4): 420–433.

Palamanit, Arkom, Somchart Soponronnarit, Somkiat Prachayawarakorn, and Patcharee Tungtrakul. 2013. "Effects of Inlet Air Temperature and Spray Rate of Coating Solution on Quality Attributes of Turmeric Extract Coated Rice Using Top-Spray Fluidized Bed Coating Technique." *Journal of Food Engineering* 114(1): 132–138.

Pande, Ranjana, Srishty Tyagi, Siddharth Vishwakarma, and Shubham Mandliya. 2025. "Dry microwave and hot water blanching of underutilized star fruit (*Averrhoa carambola*): A comparative assessment of quality attributes and sensory evaluation using chemometric approach." *Food Physics*: 100056.

Prom-u-thai, Chanakan, Benjavan Rerkasem, Ismail Cakmak, and Longbin Huang. 2010. "Zinc Fortification of Whole Rice Grain Through Parboiling Process." *Food Chemistry* 120(3): 858–863.

Prom-u-thai, Chanakan, Longbin Huang, Ismail Cakmak, and Benjavan Rerkasem. 2011. "Simultaneous Fortification of Iron and Zinc in Parboiled Rice Kernel." *Food Chemistry* 126(3): 1119–1124.

Qadir, Nafiya, and Idrees Ahmed Wani. 2023. "Functional Properties, Antioxidant Activity and In-Vitro Digestibility Characteristics of Brown and Polished Rice Flours of Indian Temperate Region." *Grain and Oil Science and Technology* 6(1): 43–57.

Rachmat, Ridwan, Eka Rahayu, Sandi Darniadi, and Mulyana Hadipernata. 2022. "Enhancement of Quality and Milled Rice Yield Recovery Through the Improvement of Polishing Process and Configuration." *IOP Conference Series: Earth and Environmental Science*, 1024(1): 012008. IOP Publishing.

Ren, Haibin, Shengmin Qi, Lianhui Zhang, Lu Wang, Jin Huang, Haiqing Yang, Chengang Ren, and Wenchen Zhou. 2021. "Variations in the Appearance Quality of Brown Rice During the Four Stages of Milling." *Journal of Cereal Science* 102: 103344.

Saha, Sreyajit, and Anupam Roy. 2020. "Whole Grain Rice Fortification as a Solution to Micronutrient Deficiency: Technologies and Need for More Viable Alternatives." *Food Chemistry* 326: 127049.

Sakare, Priyanka, Niranjan Prasad, Nandkishore Thombare, Ranjit Singh, and Satish Chandra Sharma. 2020. "Infrared Drying of Food Materials: Recent Advances." *Food Engineering Reviews* 12(3): 381–398.

Saniso, Eleeyah, Thanit Swasdisewi, Somchart Soponronnarit, and Somkiat Prachayawarakorn. 2022. "Methods of Producing Parboiled Rice by Hot Air and Combined Microwave-Hot Air Fluidized Bed Drying." *Drying Technology* 40(10): 1980–1993.

Schuderer, Matthias, Georg Rill, Thomas Schaeffer, and Carsten Schulz. 2025. "Friction Modeling from a Practical Point of View." *Multibody System Dynamics* 63(1): 141–158.

Srisang, Naruebodee, Somkiat Prachayawarakorn, Somchart Soponronnarit, and Thatchapol Chungcharoen. 2021. "An Innovative Hybrid Drying Technique for Parboiled Rice Production Without Steaming: An Appraisement of the Drying Kinetics, Attributes, Energy Consumption, and Microstructure." *Food and Bioprocess Technology* 14: 2347–2364.

Srisang, Naruebodee, Warunee Varanyanond, Somchart Soponronnarit, and Somkiat Prachayawarakorn. 2011. "Effects of Heating Media and Operating Conditions on Drying Kinetics and Quality of Germinated Brown Rice." *Journal of Food Engineering* 107(3–4): 385–392.

Szadzińska, Justyna, Joanna Łechtańska, Stefan Jan Kowalski, and Marcin Stasiak. 2017. "The Effect of High Power Airborne Ultrasound and Microwaves on Convective Drying Effectiveness and Quality of Green Pepper." *Ultrasonics Sonochemistry* 34: 531–539.

Taechapairoj, Chaiyong, and Krittanai Kaewyot. 2020. "Effects of Hot-Air Fluidized-Bed Drying on Cooking Quality, Antioxidant Activity and Bioactive Compounds in Germinated Brown Rice." *Science, Engineering and Health Studies*: 62–72.

Tamura, Masatsugu, Chisato Kumagai, and Yukiharu Ogawa. 2022. "Influence of Structural Changes of Brown Rice by Precise Polishing on In Vitro Starch Digestibility of Cooked Rice Grain." *Food Hydrocolloids for Health* 2: 100077.

Thiruselvam, Nishaanthini, Siaw Wei Cheong, Jagan Mohan, Janet Paterson, and Jayashree Arcot. 2014. "Micronutrients Fortification of Rice by Parboiling: Lab Scale and Pilot Scale Studies." *Journal of Nutrition and Food Science* 4(281): 2.

Timm, Newiton da Silva, Gustavo H. Lang, Cristiano D. Ferreira, Ricardo S. Pohndorf, and Maurício de Oliveira. 2020. "Infrared Radiation Drying of Parboiled Rice: Influence of Temperature and Grain Bed Depth in Quality Aspects." *Journal of Food Process Engineering* 43(4): e13375.

Wisetkomolmat, Jiratchaya, Chaiwat Arjin, Apinya Satsook, Mintra Seel-Audom, Warintorn Ruksiriwanich, Chanakan Prom-u-thai, and Korawan Sringarm. 2022. "Comparative Analysis of Nutritional Components and Phytochemical Attributes of Selected Thai Rice Bran." *Frontiers in Nutrition* 9: 833730.

World Health Organization (WHO). 2025. *Food Fortification. Health Topics – Overview*. Dated: 30.04.2025. https://www.who.int/health-topics/food-fortification#tab=tab_1.

Xu, Zekun, Yijuan Xu, Xiaojing Chen, Lin Zhang, Haitao Li, Zhongquan Sui, and Harold Corke. 2021. "Polishing Conditions in Rice Milling Differentially Affect the Physicochemical Properties of Waxy, Low-and High-Amylose Rice Starch." *Journal of Cereal Science* 99: 103183.

Yang, Liu, Yuchao Fan, Shaoyun Song, Deguo Wang, Yonglin Zhang, and Meili Cao. 2021. "Frictional Behavior of Brown Rice Grain During Moderate Processing." *Tribology International* 162: 107123.

Yao, Bao-Min, Peng Chen, and Guo-Xin Sun. 2020. "Distribution of Elements and Their Correlation in Bran, Polished Rice, and Whole Grain." *Food Science and Nutrition* 8(2): 982–992.

Yılmaz Tuncel, Neşe, Havva Polat Kaya, Ali Emre Andaç, Yonca Karagül Yüceer, and Necati Barış Tuncel. 2025. "The Effect of Different Pre-Treatments on Sensory Characteristics of Unformulated Pulse-Based Beverages." *Journal of Food Science and Technology*: 1–13.

Yussof, Nur Yasmeen Syukrina Mohammed, Shyang Pei Hong, Eng-Tong Phuah, and Ummul Hasanah Hassan. 2025. "Effects of Enzymatic Treatments and Drying Methods on the Physical Properties and Rehydration Kinetics of Instant Rice." *IOP Conference Series: Earth and Environmental Science* 1470(1): 012003.

Zielinska, Magdalena., Piotr Zapotoczny, Odilio Alves-Filho, T. M. Eikevik, and W. Blaszczak. 2013. "Microwave Vacuum–Assisted Drying of Green Peas Using Heat Pump and Fluidized Bed: A Comparative Study Between Atmospheric Freeze Drying and Hot Air Convective Drying." *Drying Technology* 31(6): 633–642.

10 Fortified Rice
Blending of Fortified Rice Kernels in Rice

Chandrakant Genu Dalbhagat, Shubham Mandliya, Siddharth Vishwakarma, and Hari Niwas Mishra

10.1 INTRODUCTION

Micronutrient deficiencies are a major health problem worldwide, affecting nearly 2 billion population. Iron, vitamin A, iodine, and zinc are key elements. Low intake of micronutrients aggravated the problem and caused serious issues for the vulnerable population. Micronutrient deficiency is also termed a 'hidden hunger' as it mainly affects the cognitive development, immunity, and overall productivity of the individual. Iron-deficiency anaemia (IDA), a problem due to a deficiency of iron, also affects a significant portion of the population, approximately 30% while India is facing an alarming situation due to the prevalence of anaemia among children (67.1%) and women (57.0%) (International Institute for Population Sciences (IIPS) 2021). In both global and Indian contexts, the primary causes of IDA include inadequate dietary intake of iron, cultural practices, and underlying health conditions. In India, socioeconomic factors, including poverty and inadequate sanitation, further exacerbate the problem, particularly among women and children. This indicates a pressing need for targeted interventions to address the nutritional deficiencies that contribute to these vulnerable groups. The problem of micronutrient deficiency is tackled using different approaches such as dietary diversification, supplementation, food fortification, and biofortification. Food fortification is a cost-effective and sustainable technique that offers no change in dietary habits and provides a unique opportunity to deliver essential micronutrients to the affected population.

Food fortification refers to the addition or enrichment of a particular food (e.g., grain flour, milk, oil, rice, and salt) with essential minerals and vitamins, irrespective of their presence in food, to achieve the recommended level declared by the competent authority or government body/organisation. It does not or minimally alter the product characteristics like taste, flavour, texture, etc. Rice fortification is one of the most adopted techniques for supplying the desired micronutrients to the affected population. This strategy does not require changing the dietary patterns and culinary practices (Forsman et al. 2014). Studies are available in the literature showing that rice fortification has improved the anaemic status of school-going children and women (Ara et al. 2019; Angeles-Agdeppa et al. 2008).

DOI: 10.1201/9781003661245-10

Rice fortification involves the preparation of fortified rice (FR), which is a blended product of fortified rice kernels (FRK) and normal rice (raw or parboiled). FRK is an extruded product containing micronutrients such as iron, folic acid, and vitamin B_{12} (Nithya, Dalbhagat, and Mishra 2022; Dalbhagat and Mishra 2019). FRK is designed to resemble normal rice (NR) in size (length, width, and thickness), shape, and density, ensuring that the FR will be indistinguishable from the unfortified normal rice in texture, appearance, and taste. This advantage allows the acceptability of FR among consumers without changing their dietary habits and maximises the impact on public health.

The blending operation is inevitable as the FRK contains very high vitamin and mineral content, which enhances the micronutrient status in the unfortified normal rice. FRK and NR are blended in specified ratios such as 1:100 or 1:200. The resulting blend is generally called FR. The motive for this blending is to reduce the production cost and handling of the FRK and FR (Dalbhagat and Mishra 2021). Although the blending operation cuts the price of the FR, it also adds certain challenges to the manufacturing process. This chapter discusses the principle of blending, different methods, and homogeneity evaluation in FR manufacturing.

10.2 PRINCIPLE OF BLENDING

Blending, also known as mixing, is a fundamental process involving mixing two or more ingredients or materials to form a homogeneous mass, mixture, or a single component. It typically involves dry and wet ingredients. Mixing can be categorised as solid–solid, solid–liquid, liquid–liquid, or solid–liquid–gas mixing (Chou et al. 2017).

10.2.1 Purpose

FR is a product of solid–solid mixing in which FRK and NR are the base ingredients. The primary purpose of the blending process is to distribute the FRK in the FR homogeneously to ensure the delivery of vital micronutrients to consumers in a uniform, safe, and acceptable manner. It aims to mix the FRK smoothly with NR to create a uniform product FR. This approach not only preserves the sensory properties of FR but also makes it convenient to scale up FR production for public health nutrition interventions through schemes like the Public Distribution System (PDS), Poshan Abhiyan, etc. It also facilitates the industrial-level mass production of FR, providing operational ease and better-quality products, particularly when automated equipment is used.

10.2.2 Blending Ratio and Homogeneity

The homogeneity of the FRK in the FR to achieve the uniform distribution of micronutrients is quite challenging because of the quality of the FRK. It is recommended that, while FRK is being manufactured, the physical properties of the FRK must be as close as possible to the NR in which it is intended to be mixed. Therefore, the blending depends on the physical size and shape, density, moisture content, surface and flow characteristics of FRK and NR, and the type of mixer.

Blending ratio refers to the proportion of FRK mixed in NR to ensure the delivery of recommended doses of the micronutrients to the affected population. The typical blending ratio ranges from 1:50 to 1:200, depending on the concentration of vitamins and minerals in the FRK and guidelines of the recommended doses of micronutrients. According to BIS (2021a), the recommended blending ratio in India is 1:100 or 1:200. The ratio 1:100 indicates that 1 part of FRK and 100 parts of FR are involved. The 100 parts of FR means 1 part of FRK plus 99 parts of NR. So, in a simple way, it can be said that 1 part of FRK is mixed with 99 parts of NR. The blending ratio is essential to calculate the batches or flow rates of the feeder while manufacturing the FR. For example, if the FR manufacturer has 44 kg of FRK batch, the FR can be made of 4,400 kg, which includes 44 kg of FRK (1 part) and 4,356 kg of NR (99 parts). NR quantity is calculated by multiplying the FRK quantity by 99.

Homogeneity is the uniform distribution of FRK in the FR batch. It ensures that in each batch of FR, the quantity of micronutrients is within the recommended range and consistent if the different samples are drawn for quality analysis. The blending ratio and homogeneity are interdependent and largely influenced by the properties of FRK and NR, the blending technique, and the types of blenders.

10.2.3 Factors Affecting the Homogeneity

Several factors influence the blending ratio and homogeneity, which are discussed below.

Physical Dimensions and Weight

The physical dimensions (length, width, and thickness) and weight of FRK should closely resemble those of NR. The deviation in the dimensions is allowed up to ±10% (BIS 2021a). The mismatch in dimension and weight causes uneven distribution and leads to segregation. A consistent size of FRK and its matching with NR reduces the segregation and enhances the homogeneity of FRK during the handling and storage.

Density

Matching the bulk and true densities of FRK and NR ensures the uniform distribution of FRK in FR. The density matching depends upon the physical dimensions and weight of individual kernels. The difference in densities causes the settling of the denser FRK at the bottom, leading to the uneven distribution of micronutrients. The physical dimensions, weight, and density of FRK can be controlled by optimising the extrusion process parameters and formulation.

Flowability

Good flow properties are essential for uniform mixing. Flowability is influenced by moisture, surface smoothness, particle size, and shape. Ensuring matching dimensions, surface smoothness, and uniform moisture in FRK improves its homogeneity in FR.

Handling and Transportation

Post-blending packaging, handling, and transportation may also disturb the homogeneity of the FRK in FR. Vibrations during transportation may settle or separate the FRK if the densities of FRK and NR are significantly different. Proper packaging, cushioning, and minimal disturbances are necessary to maintain a proper blending ratio.

10.3 BLENDING METHODS

The blending of FRK in FR can be done in two ways: batch and continuous (dynamic) blending. The selection of the blending process depends on processing scale, equipment availability, precision, and operational flexibility. As the target of blending is the uniform distribution of FRK in the FR, its physical properties must be matched with those of NR. The critical physical properties include size (physical dimensions) and shape, individual weight of kernels, density, colour, etc.

10.3.1 Batch Blending

Batch blending (Figure 10.1) is a controlled process in which a fixed, known amount of FRK and NR is mixed in a specific blender to create a homogeneous mixture of the desired blending ratio. It is widely used in small- to medium-scale blending units where the production of FR is done in smaller quantities. The pre-weighed quantities of FRK and NR are taken in small batches to provide high accuracy and traceability.

Batch blending involves weighing and loading FRK and NR in the blender and mixing them at predetermined operating parameters. The weighing and loading process can be manual and automatic.

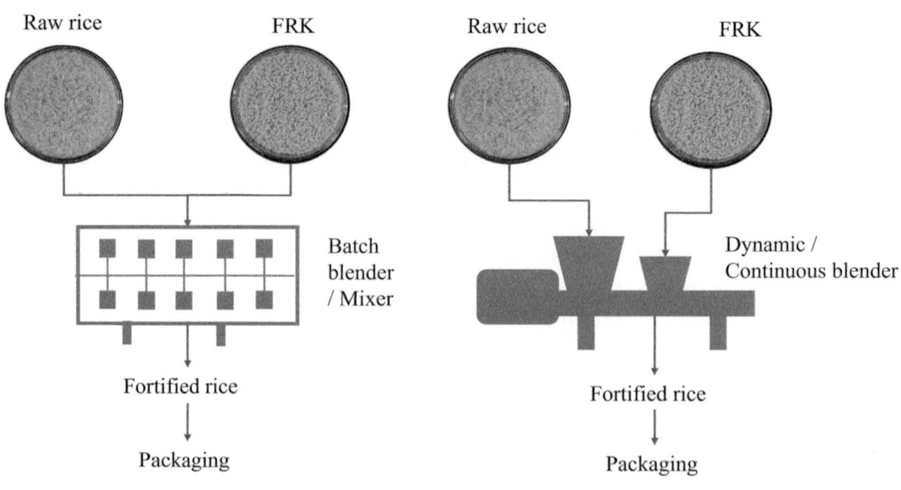

FIGURE 10.1 Types of blending: (a) batch blending and (b) continuous blending.

Weighing

It is the first step in batch blending, wherein the precise weight of the FRK and NR must be taken carefully to achieve the desired blending ratio. The distribution of FRK in the NR also depends on the accuracy of the weight taken while starting the batch blending process. The weighing balance/machine must be calibrated to avoid mistakes in weights. Automatic load cells can also be used to weigh the FRK and NR and store the data for further processing. For example, to produce FR of 100 kg, 1 kg of FRK and 99 kg of NR are required to achieve the blending ratio 1:100.

Loading

The pre-weighed FRK and NR are loaded into the batch blender to produce a calculated quantity of FR. The loading can be done manually or automatically using the different feeding systems. The loading of FRK and NR should be carried out using the optimised and validated blending protocol. The FRK and NR in the batch blender can be loaded using different methods, such as the top-down, bottom-up, and middle approaches. The top-bottom approach means the FRK is loaded at the top as the upper layer, while the NR is at the bottom as the bottom layer.

Mixing

The batch blenders are generally paddle or ribbon blenders in which the central shaft rotates in the mixing chamber. The shaft carries the mixing element, such as paddles or ribbons. When the shaft rotates, the mixing element rotates to lift, tumble, and agitate the material (FRK and NR). The optimum shaft speed and mixing time should be calculated to achieve the desired blending ratio with minimum relative standard deviation (RSD).

Discharge

The mixing in the blender should be finished within the stipulated mixing time, which is optimised for a specific blender, batch size, type of NR, and FRK. Once the FR batch is produced, it can be removed by tilting the mixing chamber on one side or opening the discharge gate at the bottom using pneumatic valves.

10.3.2 Continuous/Dynamic Blending

For large-scale commercial production, the limitation of batch blending is overcome by continuous blending (Figure 10.1). It is used to produce FR continuously to meet the demand of the distributor/consumer/market. These systems can handle the large batches of FRK and NR and consistently produce FR unless interruptions occur. Therefore, it is also called as dynamic blending.

Loading

It involves separate feeding units for the FRK and NR, which continuously feed them to the mixing unit. The feeding system may be equipped with vibratory or screw feeders. The calibration is performed by measuring the discharge rate (weight of FRK/NR per unit time) for a specific vibrational/rotational speed. However, in some advanced blenders, the load cell is also available. The feed rate is controlled using the control panel or programmable logic controller (PLC) to maintain the proper blending ratio.

Mixing and Discharge Mechanism

The final mixing is done in the mixing chamber, which may have a screw conveyor or rotating cylinder that gently conveys or tumbles the FRK and NR. The pre-set operating parameters, such as mixing speed and time, distribute the FRK in NR to produce FR. The FR is discharged and packaged in suitable bags. Dynamic blending provides an advantage of speed, automation, and consistency in FR, but requires high capital investment and technical support. It is ideal for the large-scale production of FR required for government or commercial rice fortification programmes.

10.4 EQUIPMENT

As discussed in the previous section, there are two types of blending processes: batch and dynamic. Like blending methods, the equipment is categorised as batch and continuous blenders.

10.4.1 Batch Blenders

Batch blenders are used to handle the smaller quantities of the FR batches. It is suitable for small- to medium-scale rice mills. The most commonly used batch blenders include the paddle and ribbon blenders. In these blenders, the feeding system allows the precise metering of FRK and NR in the blending chamber. Feed rates of the feeding system are maintained well to obtain the desired blending ratio.

Ribbon Blender

A ribbon blender is a U-shaped mixing chamber equipped with a rotating shaft. On the rotating shaft, the helical ribbons are installed. The ribbons are an inner helical ribbon and an outer ribbon agitator. The mixing of FRK and NR depends on the movement of these ribbons. The outer ribbon moves the material inwards towards the centre of the blending chamber, while the inner ribbon pushes the material outwards towards the wall. The continuous movement of material mixes the FRK in NR uniformly, ensuring homogeneous mixing. The ribbon's design is such that the formation of a dead zone is avoided. To prevent the breakage of FRK and NR, the speed of the shaft must be controlled and optimised. The mixing time, batch size, loading approach, and shaft speed must be optimised to produce FR for commercial purposes.

Paddle Blender

Like a ribbon blender, a paddle blender (Figure 10.2) also consists of a U-shaped blending chamber with a central rotating shaft on which the paddles are mounted. The paddles are designed in various shapes, such as rectangular, triangular, or square, with short cylindrical rods. The cylindrical rod is welded or fitted with a central shaft using a nut, bolt, and coupling. These paddles are oriented at a specific angle to the central shaft to promote the desired flow of FRK and NR during the mixing process.

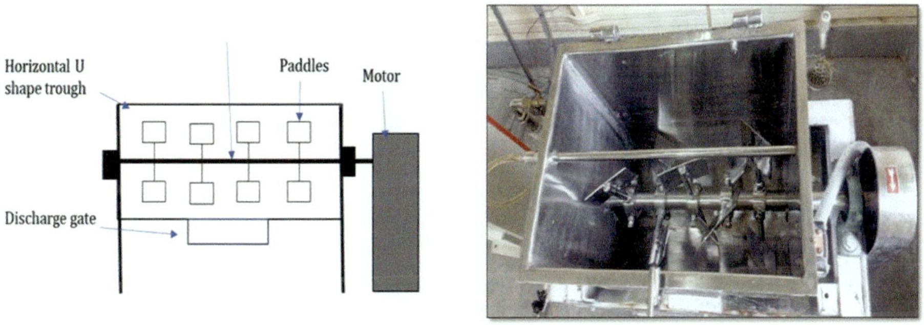

FIGURE 10.2 Paddle blender. (Courtesy: Iron Fortified Rice Manufacturing Unit, Agricultural and Food Engineering Department, IIT Kharagpur.)

The paddle angle may be fixed in welded paddles or changed using nut-bolt tightening. The paddles lift and fluidise the material in radial and axial directions to achieve the desired blending ratio. At a proper central shaft speed, the grains rise and become suspended before settling due to gravity, creating a tumbling effect resulting in homogeneous mixing of FRK in NR. The paddle blender may have twin rotating shafts in a double U- or W-shaped chamber, which is also called a fluidised bed mixer. In this mixer, the material moves upwards towards the centre and remains in a suspended/fluidised state above the central shaft, improving the blending efficiency.

Operation and Maintenance

Batch blenders require a calibrated feeding rate, i.e., vibration per unit time for the vibratory feeder, as well as a standardised mixing speed and time with a proper discharge mechanism for uniform mixing of FRK in FR. Proper standard operating procedures must be followed to prevent segregation and ensure consistent production of FR. Routine maintenance must be done to check wear and tear in screws or blades, housings, seals, bearings, gaskets, and lubrication in the moving parts. The cleaning and sanitation procedure should be employed to avoid cross-contamination and ensure consistent production of FR.

10.4.2 Dynamic Blenders

Operation

Dynamic blenders (Figure 10.3) continuously produce a large quantity of FR by feeding FRK and NR at a predefined feed rate. The blending ratio is adjusted by controlling the feed rates of FR and NR to the main mixing chamber. Feed rates are calculated using the rotational or vibrational speed of the feeders. In most dynamic blenders, screw feeders are used, whose feed rate is standardised using the rotational speed and delivery of material. The feeding capacity of both NR and FRK feeders decides the production capacity of the dynamic blender. Screw feeders operate on the principle of volumetric feeding of material. Therefore, feed rates have to be calibrated for different rice varieties and FRK, due to the working mechanism of the screw. The main mixing chamber consists of a central shaft having paddles or a

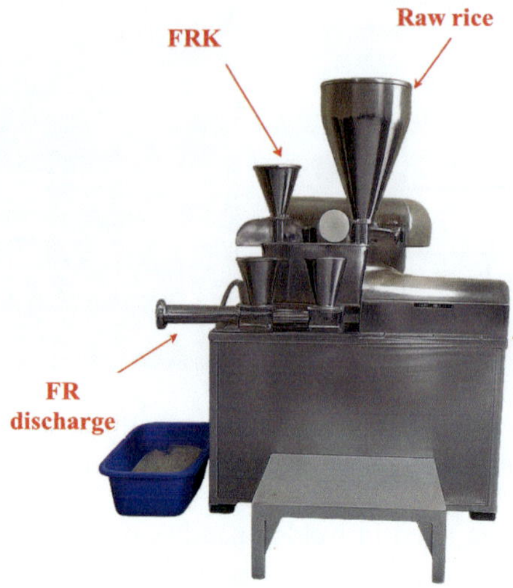

FIGURE 10.3 Continuous FRK and normal rice blending unit. (Courtesy: Iron Fortified Rice Manufacturing Unit, Agricultural and Food Engineering Department, IIT Kharagpur).

rotating screw that mixes and conveys the blend to the discharge end. The blending mechanism ensures uniform distribution of FRK in NR to produce FR.

Feeders

Feeders play a vital role in batch and dynamic blending systems as they decide the blending ratio in the FR. The precise and consistent delivery of FRK and FR in the mixing chamber mainly depends on the feeding mechanism. Feeders may be classified as gravimetric or volumetric feeders. Gravimetric feeders measure the actual weight of FRK and NR using an online load cell and deliver the material accurately, while volumetric feeders dispense the material based on volume. The density and volume of material are used to calculate the feed rate in volumetric feeders. The commonly used feeders in the blending system may be screw and vibratory feeders.

Screw feeders consist of a screw housed in a pipe and an electric motor equipped with a reduced gearbox. The screw rotates in the pipe with a predefined rotational speed, conveying the material in the forward direction. It delivers the material volumetrically, and the capacity is dependent on screw design, rotational speed, density, and volume of material. It is a precise delivery method once properly standardised. Consistent and controlled screw speed discharges material continuously at a set feed rate, preventing overfeeding or underfeeding. Vibratory feeders work on the principle of vibration, in which the material experiences horizontal and vertical acceleration, causing its tossing and sliding action. As a gentle slope is given to the vibratory feeder, the material moves forward due to vibrations, and gravity decides the direction of flow.

Maintenance

Dynamic blenders must be monitored for motor alignment, screw wear, residue buildup, and feeder calibration. Sensors and PLC units should be regularly checked and recalibrated. Maintenance schedules should be automated for preventive alerts and downtime planning.

10.4.3 Control System

Control over operational parameters such as feeding rate, mixing speed, and time is essential as they decide the blending ratio and homogeneity of FRK in FR. Modern blending equipment integrates sensors such as load cells, encoders or proximity switches, and variable frequency drives (VFD) with PLC and Human-Machine Interfaces (HMI) to control the blending efficiency and accuracy. The control panels monitor feed rates, blending time, and FRK distribution in FR. Alarms, feedback loops, and automatic corrections may also enhance the homogeneity of FRK in FR and reduce human error.

10.5 BLENDING TIME AND HOMOGENEITY

10.5.1 Blending Time

Blending time is a critical parameter for mixing FRK in NR. Under- or overmixing may cause the segregation or uneven distribution of FRK in the FR. Moreover, excessive mixing may lead to the breakage of FRK and NR. Therefore, the blending time should be standardised for uniform mixing, depending on the blender type, batch size, and physical properties of the FRK and NR. The blending time may be defined as the time required to distribute the FRK in NR such that whenever the sample is drawn from the FR batch for analysing the blending ratio, it will be 1:100 with a permissible deviation of ±10%. The blending time can be calculated using a graphical method, i.e., a plot of RSD and time (Figure 10.4). The mixing time varies according to the blender design, characteristics of FRK and NR, and operational parameters.

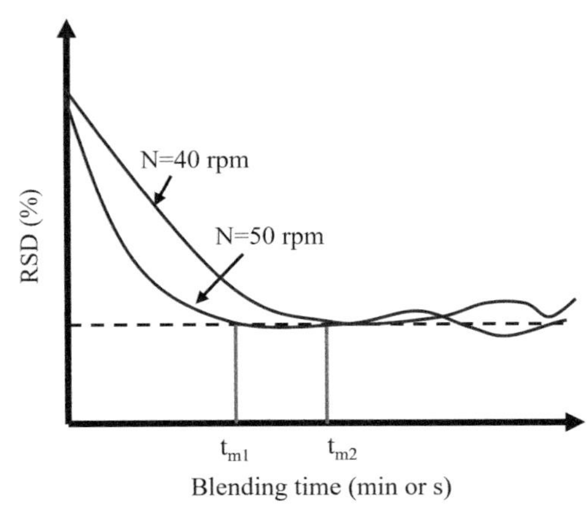

FIGURE 10.4 Blending time estimation using RSD (%) vs blending time.

10.5.2 Homogeneity of FRK in FR

During blending, the FRK is mixed in the NR in a specific ratio, such as 1:100. The blending ratio is predefined to achieve the desired micronutrient level in the FR. Therefore, the homogeneous distribution of FRK in the NR is always desired during the blending. In the homogeneity calculation, the targeted concentration of FRK in the FR is always the blending ratio. For example, a blending ratio 1:100 indicates 1% concentration of FRK in 100 kg FR. Depending on regulatory compliance or process requirements, the blending ratio may vary from 1:50 to 1:200. Consider a batch of FR is being produced with any specified blending ratio. Assume x is the overall targeted concentration of FRK in the FR, meaning any random sample drawn from the FR batch would theoretically give FRK concentration x%. However, practically, there may be a deviation from the targeted concentration of FRK. This deviation is calculated using a percentage RSD. Suppose a batch of FR is being prepared in a batch/dynamic blender for a particular shaft speed and mixing time. At the end of the mixing process, samples are collected from different positions within the blend, and the concentration of FRK (x_i) is measured. Simultaneously, the average $\bar{x}$ concentration of FRK is also recorded. The minimum difference between x_i and $\bar{x}$ indicates more uniform blending ($|x_i - \bar{x}| \to 0$). The variance of the FRK concentration (σ^2) in the FR can be calculated using the Eq. 10.1.

$$\sigma^2 = \frac{1}{N-1} \sum_{i=1}^{N} (x_i - \bar{x})^2 \tag{10.1}$$

Where N is the number of random samples collected to calculate FRK concentration. $\bar{x}$ and x_i are the FRK's average concentration and concentration in the ith random sample, respectively.

The percentage RSD (Eq. 10.2) indicates variability in the concentration of FRK relative to the overall targeted concentration of FRK (x). It is calculated using the values of σ^2 and x (Yaraghi et al. 2018).

$$\text{RSD}\ (\%) = \frac{\sqrt{\sigma^2}}{x} \times 100 \tag{10.2}$$

A perfect mixing occurs when the concentration of FRK in a randomly drawn sample at any point in the blender gives the desired concentration of FRK. The variance in the perfectly blended FR is supposed to reach zero, but it is practically impossible (Weinekötter 2016). Therefore, as much as a low percentage RSD is expected during the blending. According to the Indian Standard for FR, the permitted variation in FRK concentration is allowed up to ± 10% (BIS 2021b). The RSD values can be used to calculate the blending time. The graph of % RSD for the corresponding blending speed can be plotted against blending time (Figure 10.4). The minimum cutoff of percentage RSD can be set to determine the optimum blending time. In the blending process, overmixing can lead to segregation of FRK, causing changes or an increase in percentage RSD.

The concentration of FRK is calculated using the weight of FRK and the sample of FR drawn. In this method, FRK are separated manually from the drawn sample and the weight of FRK is recorded. Consider 'w' as the total weight of the FR sample (g) drawn from which 'm' quantity of FRK (g) is obtained, and then the concentration of FRK can be calculated using the Eq. 10.3.

$$\text{FRK concentration } (\%) = \frac{m}{w} \times 100 \tag{10.3}$$

The iron spot and iodine-povidone test can also be used to qualitatively identify FR, i.e., the presence of FRK in FR.

10.6 CHALLENGES AND FUTURE SCOPE

Despite the advancement in blenders, several challenges persist in achieving the uniform distribution of FRK in FR. The challenges and future scopes are listed below.

10.6.1 Challenges

- Physical mismatch of the FRK and NR, i.e. size, shape, and density, leads to segregation and uneven distribution of FRK in the FR.
- Limitations in equipment designs and operational flexibility cause problems in precise blending.
- Lack of robust control panels and a quality monitoring system for real-time evaluation of homogeneity and its validation.
- FRK and NR breakage during mechanical blending affect the product quality.
- Post-blending segregation during handling, transportation, and storage reduces the consistency and quality of FR.
- Limited adoption of automation and artificial intelligence and machine learning (AI-ML)-based control systems in batch or dynamic blenders.

10.6.2 Future Scope

- Improve FRK quality using advanced extrusion systems that mimic the physical properties of NR to improve compatibility. AI-ML-based estimation of extrusion proves that parameters may also improve the quality of FRK, ultimately improving blending efficiency.
- Development of AI-ML and sensor-based technologies for real-time monitoring of the homogeneity of FRK and validation.
- Design and development of portable, low-cost batch or dynamic blenders for rural and small-scale operations.
- Implement standardised training/seminar/workshop programmes for equipment/machinery operators and quality assurance personnel.
- Promotion of community awareness and consumer acceptance of FR through education campaigns.

10.7 CONCLUSIONS

The blending operation plays a key role in the rice fortification process, which ensures the success of this programme in combating IDA or micronutrient malnutrition. The blending process is simple and easy, but requires proper attention to properly mix FRK in the NR. Effective blending ensures the distribution of micronutrients such as iron, folic acid, and vitamin B_{12} across each produced batch of FR as per the recommended regulatory guidelines. Therefore, the blending process is crucial for attaining the desired blending ratio. Both batch and dynamic blending methods have advantages and disadvantages. Therefore, the types of blenders (paddle/ribbon/dynamic), capacity, feeding system, and control mechanism must be considered while selecting a blender for commercial use. The blending efficiency also depends on the quality of FRK, which has to match the NR in which FRK is intended to be mixed. The key parameters include the physical dimensions, individual weight of kernel, bulk, true densities, and flowability, which need to be controlled during FRK production. Moreover, the homogeneity of FRK in FR is affected by the operating parameters of blenders such as shaft speed, mixing time, feed rates, and paddle angles. These operational parameters are critical in distributing the FRK throughout the NR during the blending process. Modern blending units include PLC, HMI, and sensor-driven technologies, enhancing efficiency and precision. They can gather data and store it for further analysis. Therefore, with the growing demand for FR, research and development should be expanded to different rice varieties to develop FRK and innovate blending technology to improve efficiency and accuracy in FR quality.

REFERENCES

Angeles-Agdeppa, Capanzana, Barba, Florentino, and Takanashi. 2008. "Efficacy of Iron-Fortified Rice in Reducing Anemia Among Schoolchildren in the Philippines." *International Journal for Vitamin and Nutrition Research* 78 (2): 74–86. https://doi.org/10.1024/0300-9831.78.2.74.

Ara, Gulshan, Mansura Khanam, Ahmed Shafiqur Rahman, Zhahirul Islam, Shahriar Farhad, Kazi Istiaque Sanin, Sihan Sadat Khan, Mohammad Mahbobor Rahman, Herma Majoor, and Tahmeed Ahmed. 2019. "Effectiveness of Micronutrient-Fortified Rice Consumption on Anaemia and Zinc Status among Vulnerable Women in Bangladesh." *PLoS ONE* 14 (1): 1–16. https://doi.org/10.1371/journal.pone.0210501.

Bureau of Indian Standards (BIS). 2021a. "Fortified Rice-Specification (IS 17780: 2021)." https://standardsbis.bsbedge.com/BIS_SearchStandard.aspx?Standard_Number=17780&id=0

Bureau of Indian Standards (BIS). 2021b. "IS 17782 : 2021: Fortified Rice Kernels Specifications." https://standardsbis.bsbedge.com/BIS_searchstandard.aspx?Standard_Number=IS+17782&id=34819

Chou, Shih-Hao, Yue-Lou Song, and Shu-San Hsiau. 2017. "A Study of the Mixing Index in Solid Particles." *KONA Powder and Particle Journal* 34 (34): 275–81. https://doi.org/10.14356/kona.2017018.

Dalbhagat, Chandrakant Genu, and Hari Niwas Mishra. 2019. "Effects of Extrusion Process Conditions on System Parameters; Physicochemical Properties and Cooking Characteristics of Extruded Fortified Rice Kernels." *Journal of Cereal Science* 89 (October 2018): 102782. https://doi.org/10.1016/j.jcs.2019.05.016.

Dalbhagat, Chandrakant Genu, and Hari Niwas Mishra. 2021. "Effect of the Drying Process on the Color Change, Fissure Development, and Morphology of Fortified Rice Kernels." *Journal of Food Process Engineering* 44 (7). https://doi.org/10.1111/jfpe.13719.

Forsman, Carmen, Peiman Milani, Jill A. Schondebare, Dipika Matthias, and Christophe Guyondet. 2014. "Rice Fortification: A Comparative Analysis in Mandated Settings." *Annals of the New York Academy of Sciences* 1324 (1): 67–81. https://doi.org/10.1111/nyas.12453.

International Institute for Population Sciences (IIPS). 2021. "National Family Health Survey (NFHS-5) India 2019–21." *Mumbai: International Institute for Population Sciences* I:1–714. https://www.nfhsiips.in/nfhsuser/publication.php

Nithya, A., Chandrakant Genu Dalbhagat, and Hari Niwas Mishra. 2022. "A Comparative Study on the Physicochemical, Cooking and Textural Properties of Fortified Rice Kernels Prepared from Raw and Parboiled Rice." *International Journal of Food Science and Technology* 57 (2): 1325–32. https://doi.org/10.1111/ijfs.15529.

Weinekötter, Ralf. 2016. "Mixing of Solid Materials." In *Production, Handling and Characterization of Particulate Materials*, edited by Henk G. Merkus, and Gabriel M.H. Meesters, 291–326. https://doi.org/10.1007/978-3-319-20949-4_9.

Yaraghi, Amirsalar, Mohammadreza Ebrahimi, Farhad Ein-Mozaffari, and Ali Lohi. 2018. "Mixing Assessment of Non-Cohesive Particles in a Paddle Mixer through Experiments and Discrete Element Method (DEM)." *Advanced Powder Technology* 29 (11): 2693–2706. https://doi.org/10.1016/j.apt.2018.07.019.

11 Packaging, Storage and Shelf-Life Evaluation of Fortified Rice Kernels and Fortified Rice

Sujosh Nandi, Indrajit Dalai, Saurabh Kumar, and Hari Niwas Mishra

11.1 INTRODUCTION

Rice is a staple food for nearly half of the human population (~45%), supplying 20% and 80% of daily dietary energy requirements for the human population worldwide and Asia, respectively (Razzaq et al. 2020). India is one of the largest cultivating countries in the Southeast Asian region, covering around 44 million ha, which is nearly 14% of the total geographical area of the country and around one-fourth of the cultivable area. Rice feeds more than 60% of the Indian population and is mainly consumed by people in the eastern and southern regions (A Status Note on Rice in India 2015). Moreover, rice and rice-based products are the primary dishes for groups of people who come under medium to lower and lower levels of income. However, with an increase in income, they shift their diet more towards high-value foods such as protein-rich or enriched with vitamins and minerals and shift away from rice.

The rice grain constitutes 12% water, 75%–80% starch, 7% protein and minerals like calcium (Ca), magnesium (Mg) and phosphorus (P) along with some traces of iron (Fe), copper (Cu), zinc (Zn) and manganese (Mn). Additionally, rice grain comprises Vitamin B_1 (~0.18 mg), Vitamin B_2 (~0.05 mg), Vitamin B_3 (~5.09 mg), Vitamin B_5 (~1.49 mg), Vitamin B_6 (0.51 mg) and Vitamin B_9 (~20 µg) per 100 g (Verma and Srivastav 2017). However, during milling and polishing to produce white milled rice, a substantial amount of nutrients is washed away, amounting to 67% of vitamin B_3, 80% of vitamin B_1, 90% of vitamin B_6, half of manganese, half of phosphorus, 60% of iron and all of the dietary fibre and essential fatty acids.

Micronutrient malnutrition (MNM) is referred to as a deficiency of vitamins and minerals in the daily diet. MNM is a silent epidemic spreading worldwide, especially in developing countries like India, affecting their economic growth. In India, the prevalence of anemia is at an alarming stage. According to the National Family Health Survey 2015–2016, 58.6% of children (6–59 months), 50.4% of pregnant women (15–49 years) and 53.1% of all women are anaemic (Dalbhagat and Mishra 2019). Therefore, fortifying staple foods such as rice with essential micronutrients can address this issue. Moreover, when enriched with essential vitamins and

DOI: 10.1201/9781003661245-11

minerals, rice can offer a sustainable and cost-effective solution to improve public health and well-being. Furthermore, appropriate measures have been implemented to deliver fortified rice to the vulnerable population by creating a robust supply chain management (Public Distribution System). However, post-manufacturing operations, including storage and packaging, have been neglected over the years, and thus, very limited research is conducted on the shelf life of fortified rice.

The shelf life of fortified rice is related to its nutritional stability, colour, microbiological safety and sensory properties, and thus, creating a robust system is a practical necessity in the successful implementation of rice fortification programmes in public distribution systems, school feeding programmes and emergency food assistance programmes in developing countries like India. Micronutrients like vitamins and minerals are sensitive to processing, packaging and environmental conditions, resulting in quick degradation of micronutrients during storage and the supply chain (Saffarionpour and Diosady 2021). Environmental factors, such as light, oxygen, humidity and temperature, affect the stability of micronutrients in fortified rice. Similarly, the storage stability of fortified rice kernels (FRK) manufactured through the extrusion process can be compromised if extrusion processing parameters are improperly maintained. Moreover, qualities of the package, such as water vapour transmission rate, oxygen transmission rate and opacity, play crucial roles in the shelf life of fortified rice and fortified rice kernels. For example, ferric pyrophosphate can readily oxidize to ferrous iron due to exposure to oxygen and light during storage, which consequently lowers its bioavailability. Similarly, rice fortified with Vitamin A is highly susceptible to light and oxygen, and thus, continuous exposure can diminish its availability while being stored (Hemery et al. 2018). Moisture content is another essential factor that affects the shelf life of FRK and fortified rice (FR) during storage and transportation. Maintaining adequate moisture content of FRK (<12% w.b.) and FR (<15.5% w.b.) throughout the storage periods can restrict oxidative reactions as well as the growth of toxin-generating microbes, particularly yeasts and moulds. Therefore, it is necessary to consider an appropriate packaging system to protect FRK and FR from detrimental effects when exposed to and used in an inadequate external environment (Groh et al. 2019). It is worth mentioning here that in addition to protective purposes, packages can be used as a medium of communication and convenience (Schaefer and Cheung 2018). Packages may come in different sizes and shapes as per the requirements, and they also act as an interface that is convenient and easy to use for users. The labelling on packages about product nutrient quality, shelf life and manufacturing details makes consumers aware of the product's details and traits. Therefore, in this chapter, an overview of the packaging and storage requirements for FRK and FR, along with various factors influencing the shelf life of FRK and FR, has been discussed.

11.2 FACTORS AFFECTING THE QUALITY OF STORED FR AND FRK

The quality of FR and FRK during storage is a cumulative effect of processing, handling, storage and management of the kernels. During storage, kernels are generally exposed to various abiotic and biotic factors such as moisture content, temperature and relative humidity of the stored environment, initial conditions of the grain and insects and moulds, which influence the alteration of physicochemical and physiological properties. Further, the adaptation of improper manufacturing techniques of FRK can also lead to deterioration of quality of FRK during storage. Hence, comprehensive understanding of these factors is imperative to diminish the deteriorative effects on FR and FRK during storage and maintain desirable quality.

11.2.1 Abiotic Factors

Moisture Content

The stability and safety of FRK and FR can be significantly affected by moisture content (MC) because it can promote undesirable chemical reactions such as oxidation, which can change the physicochemical characteristics and promote degradation of starch during storage of FRK and FR (Romruen et al. 2024). Further, high moisture creates favourable conditions for growth of mould, insects and mites, resulting in decreased quality and safety and a shorter shelf life of FR and FRK. Optimal MC for storage of FR is below 15.5% (BIS 2021a), while for FRK moisture content should be below 12% (BIS 2021b).

Temperature

Besides moisture content, controlling the temperature of stored FR and FRK is important to safeguard their quality attributes. Temperature during storage can be affected by air temperature and heat produced due to respiration of grain, insects and moulds. High temperature can cause detrimental effects on nutritional value of FR and FRK by reducing micronutrient content. In our experimental study, a significant reduction in folic acid and vitamin B_{12} was noticed when storage temperature increased by 20°C (Figure 11.1). Further maintaining low temperature (<17°C) of the stored grain is effective in diminishing insect infestation by saw-toothed grain beetle, grain weevil and other granivorous insects.

Grain Metabolism

Grain metabolism is another important factor that can affect the quality of fortified rice during storage. Grain metabolism is closely related to the respiration of rice grain and its accompanying microorganisms. During the respiration process,

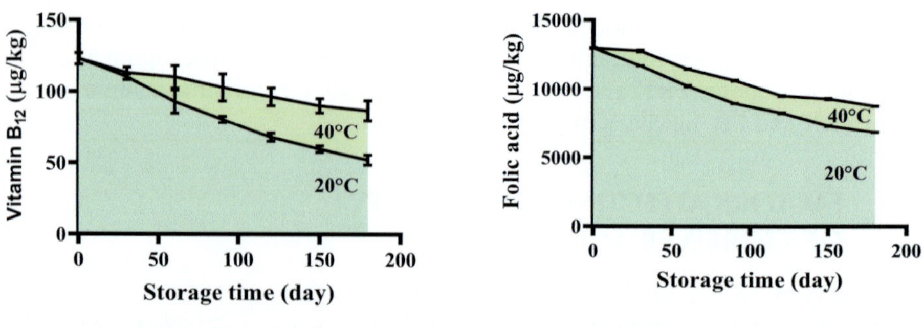

FIGURE 11.1 The effect of temperature on vitamin B_{12} and folic acid content of FRK.

rice grain consumes energy by breaking down starch, protein and lipid through the oxidation process, causing losses of dry matter and production of carbon dioxide, heat and water. Therefore, the respiration process is positively correlated with storage temperature and relative humidity, influencing the cooking properties, colour and digestibility of FR. Controlling storage temperature by installing an appropriate aeration system in the storage premises is essential to reduce the effect of heat by maintaining adequate temperature. The recommended storage temperature for FR and FRK is 25°C or below.

11.2.2 Biotic Factor

Biotic factor, particularly insect infestation, is not a major problem for storage of rice grain for less than 10 months if stored under appropriate conditions (Befikadu 2014). However, in long-term storage, insect activity can cause severe qualitative and quantitative losses in FR, which may lead to 10% of storage loss. Two categories of insects are primarily involved in damaging FR, namely primary pests and secondary pest (Table 11.1). Primary pest can infest undamaged and sound FR, while secondary pests can attack damaged and broken grains. Therefore, mitigating this problem, it is necessary to store FR below MC of 15% and storage temperature of 25°C with

TABLE 11.1
Possible List of Primary and Secondary Pests to Attack FR During Storage

S. No	Insects Common Name	Insects Scientific Name	Insects Image
		Primary Insects	
1.	Rice weevil	*Sitophilus oryzae*	
2.	Lesser grain borer	*Rhyzopertha dominica*	
3.	Red flour beetle	*Tribolium castaneum*	

(*Continued*)

TABLE 11.1 (*Continued*)
Possible List of Primary and Secondary Pests to Attack FR During Storage

S. No	Insects Common Name	Insects Scientific Name	Insects Image
4.	Rice moth	*Corcyra cephalonica*	
5.	Khapra beetle	*Trogoderma granarium*	
		Secondary Insects	
6.	Saw-toothed grain beetle	*Oryzaephilus surinamensis*	
7.	Long-headed flour beetle	*Latheticus oryzae*	
8.	Flat grain beetle	*Cryptolestes minutus*	
9.	Grain mite	*Acarus siro*	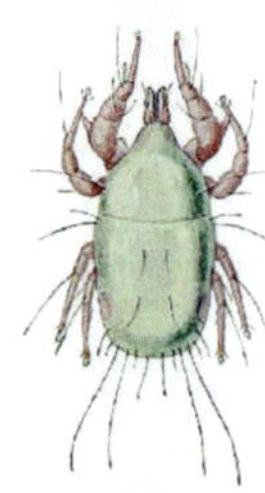
10.	Grain lice	*Liposcelis divinitorius*	

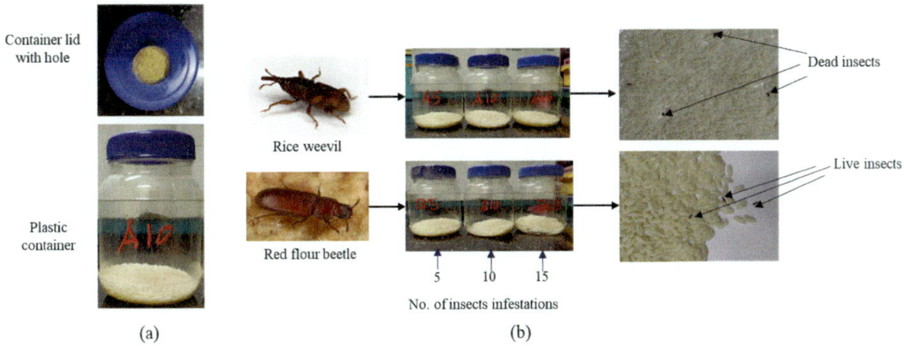

FIGURE 11.2 Insect infestation in FRK: (a) plastic container with lid and (b) experimental details of insect infestation in FRK.

proper aeration system. Further, pre-storage fumigation is recommended to prevent early infestation in FR during storage. Additionally, curative treatment using aluminium phosphide tablet (3 g per metric tonne for an exposure period of 7 days) and prophylactic treatment using deltamethrin and malathion spray (50 EC 10 mL/L with @ 3 L of spray fluid/100 m^2 over the bags) should be carried out at regular intervals to prevent insect infestation in FR during storage. On the other hand, FRK is less susceptible to insect attack owing to their compact and dense microstructure, which an insect fails to penetrate. In an experimental study conducted to understand the growth of insects in FRK, two types of rice insects, viz., rice weevil (*Sitophilus oryzae*) and red flour beetle (*Tribolium castaneum*), were kept in a ventilated plastic container with FRK. An interesting observation was noted with dead rice weevils and red flour beetles after 30 and 50 days of storage. On the contrary, when FR was infested with same insects, eggs and pupae developed in the rice, while for up to 45 days, insects grew (Figure 11.2). It is worth mentioning that insect infestation in FR can be restricted by vacuum and CO_2 packaging with high gas barrier polymeric film, but it would be costly.

11.2.3 Exposure to Light

The stability of micronutrients present in FR and FRK can be greatly affected by visible light and infrared radiation. A continuous exposure to visible light and UV radiation degrades folic acid and cyanocobalamin present in FRK and FR, resulting in a sharp decrease in their nutritional value and quality (Kwon et al. 2018). Furthermore, it has been reported that just 5% of the exposure to the Sun's UV light can break down several chemical bonds, including C—C, C—O and C—H, causing significant damage to the physicochemical properties of rice grains (Nikvarz, Khayati, and Sharafi 2020). In addition, exposure to UV radiation damages packaging materials (plastic), decreasing their capacity to protect and possibly exposing contents to increased environmental harm. UV exposure may also lead to negative changes in FRK and FR, such as a decrease in weight and molecular structure, as well as a loss of glossiness and softness. Photo-oxidation, a process where light energy produces reactive oxygen species (ROS) that harm vitamins and minerals, can be triggered by light, especially UV light (Ashooriyan et al. 2023). Therefore, adequate UV and visible light protection is necessary to prevent qualitative losses of FRK and FR during storage.

11.2.4 Processing Factors

In addition to adopting appropriate storage conditions, processing conditions have an impact on the shelf life of FRK and FR. For instance, the particle size of rice flour and the uniform mixing of rice flour and vitamin-mineral premix ensure the quality of FRK after the extrusion process. Bigger particle size (>250 μm) and poor mixing index cause uneven distribution of the micronutrients in the FRK. Bigger particle size, on the other hand, may end up producing poor quality FRK, particularly distorted extrudate kernels, causing breakage of FRK during transportation, storage and handling.

Along with mixing, conditioning of the blend of rice flour and vitamin-mineral premix affects the extrusion process, which consequently influences micronutrient stability during storage. A consistent and equal moisture distribution throughout the mixture ensures uninterrupted and steady material flow through the extruder and adequate gelatinization of the rice dough, producing superior quality FRK (Nithya et al. 2024). On the contrary, poor conditioning of the mixture can produce FRK with poor structural integrity, which consequently promotes the quick degradation of micronutrients during storage.

In the coating method, binding agents are used to apply micronutrients in a thin coating on the surface of rice kernels. This method provides little defence against environmental influences, increasing the susceptibility of nutrients to loss during storage, cooking and washing (Espinoza-Espinoza et al. 2024). Cold Extrusion involves combining micronutrients and rice flour at room temperature to produce kernels without using much heat. Despite minimizing nutrient loss during processing, the lack of heat results in decreased shelf life and structural stability (Bilgi Boyaci et al. 2012). On the other hand, in hot extrusion, dense, compact FRK are often produced by extruding the rice flour and nutrition mixture at a high temperature and pressure. Higher nutrient incorporation within the matrix is made possible by the thermal and mechanical treatments, which also provide superior protection from light, oxygen and moisture (Boonyasirikool and Charunuch 2000).

Drying is a crucial step, as it has a major impact on the micronutrients' stability and structural integrity of FRK. Mechanical breaking of the kernels, micronutrient' degradation and microbial contamination might result from either too much or too little drying. For instance, high drying temperatures can create fissures in FRK, enhancing the probability of breakage during storage and transportation. Moreover, high drying temperature and longer drying time can degrade heat-sensitive vitamins, producing poor quality FRK, which causes reduction in shelf life. Inadequate drying, on the other hand, keeps high moisture, making it susceptible to mould growth and hydrolytic breakdown. Therefore, an optimum drying temperature of 40°C is preferable to produce high-quality FRK and enhance its shelf life (Dalbhagat and Mishra 2021).

Adopting an adequate blending technique for the production of FR by mixing FRK with normal rice at a ratio of 1:100 or 2:100 has a significant effect on micronutrient stability and efficacy during storage. This will provide accurate micronutrient delivery per serving. Due to machine constraints, human mistakes or variations in kernel size, shape or density, improper blending may result in an uneven distribution, which could cause micronutrient concentrations to be either too high or too low. Moreover, if blending takes place in a humid environment, FRK may absorb moisture, increasing the possibility of microbial development and nutrient deterioration (WFP 2018).

11.3 EFFECTS OF STORAGE ON PROPERTIES OF FR AND FRK

During storage of FR and FRK, several quality attributes, including physicochemical, cooking, nutritional and textural properties, change due to modifications in structural orientation and interaction among starch, protein, lipids and incorporated micronutrients. Although, the mechanism behind these modifications has not been explored yet, it is ascertained that these changes are related to storage time, temperature and relative humidity. Further, it is assumed that the effect on properties of FR during storage is similar to normal rice as 99% of kernels in FR are normal rice kernels.

11.3.1 Effect of Storage on Cooking

Cooking properties of rice are related to amylose content and its molecular structure and interactions of non-starchy components (protein and lipids) with starchy components. It is believed that storage does not moderate the content of these three macromolecules in the rice, but changes their molecular distribution and structure (Figure 11.3), resulting in changes in cooking properties.

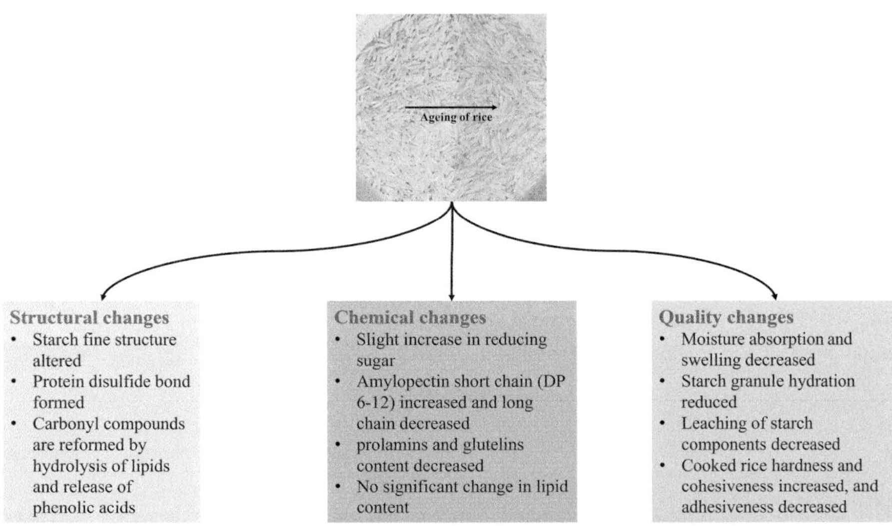

FIGURE 11.3 Structural, chemical and quality changes in rice during storage.

Cooking Time and Water Absorption Ratio

Cooking time of rice is associated with solubility of starch and protein. Thus, changes in solubility of these macronutrients during storage can affect the cooking time (Zhou et al. 2002). During storage, water solubility of starch decreases due to alteration in intra- and inter-molecular interactions of amylose–amylopectin and starch–protein (Ziegler et al. 2018), and these interactions were also evident in FRK and FR during storage, causing an elevated cooking time. Certainly, an increasing trend was generally observed for water absorption ratio (WAR) of FR during increasing storage duration. However, interestingly, no changes for WAR were noticed for FRK during 6 months of storage (Table 11.2). The mechanism behind decreasing WAR in FRK could be the cleavage of starch molecules and denaturation of protein molecules during extrusion process and subsequent enzymatic activity during storage (Ziegler et al. 2018). Moreover, the changes were significant when FR and FRK were at low and high temperatures (20°C and 40°C), which may be attributed to the reduction of leaching of amylose with changes in temperature and time (Park et al. 2012).

Cooking Loss

The estimation of cooking loss during storage of FR and FRK is crucial to maintain nutritional quality over storage time. With increased storage time of FR, cooking loss diminished, which may be attributed to increase in insolubility of protein and starch (Table 11.2). While FRK followed a divergent trend with significant increase in cooking loss due to hydrolysis of amylopectin and formation of weak lipid–amylopectin complex during storage (Mohamed 2021).

Textural Properties

Texture of cooked FR and FRK is a major attribute for consumer acceptability. Among various textural qualities, hardness, cohesiveness and chewiness are important parameters to be evaluated. Researchers reported that hardness, cohesiveness and chewiness of freshly harvested rice are lower compared to stored rice. Similar results were obtained in our experiment when FR was stored for 6 months at different

TABLE 11.2
Changes in Cooking Properties of FR and FRK During 6 Months of Storage at 20°C and 40°C

Storage Temperature (°C)	Cooking Time	Cooking Loss (%)	WAR	Textural Properties		
				Hardness	Cohesiveness	Chewiness
Fortified Rice						
40	Increased	Decreased	Increased	Increased	No changes	Increased
20	No changes	Decreased	Increased	Increased	No changes	Decreased
Fortified Rice Kernel						
40	Increased	Increased	Decreased	Decreased	No changes	Decreased
20	No changes	Increased	Decreased	Decreased	No changes	Decreased

temperatures, except for cohesiveness, where no significant variation was observed. However, FRK followed an opposite trend with decreasing hardness and chewiness during storage. Further, the effects of storage temperature were noticed when the FR and FRK were stored at both 40°C and 20°C. FR and FRK stored at high temperatures showed maximum variations in the properties compared to lower temperatures.

11.3.2 Effect of Storage on Appearance of FR and FRK

Colour changes during storage are predominant qualitative changes of rice. Researcher reported that the lightness (L^*) value of rice decreased and the reddish (a^*) and yellowish (b^*) values increased with storage time due to the formation of disulfide bonds between protein and starchy components of rice (Ziegler et al. 2017). Sirisoontaralak and Noomhorm (2007) described that yellowness of rice was enhanced during storage due to breakdown of glycosidic and peptide bonds, which consequently form carbonyl and amino compounds and undergo Maillard reactions. Similar phenomenon can be expected to occur in FR during storage (Figure 11.4) as the L^* value decreased and the a^* and b^* values increased during 6 months of storage. For FRK also, colour changes were prominent and became reddish with the storage time. Interestingly, changes in colour values of FRK are affected by the permeability

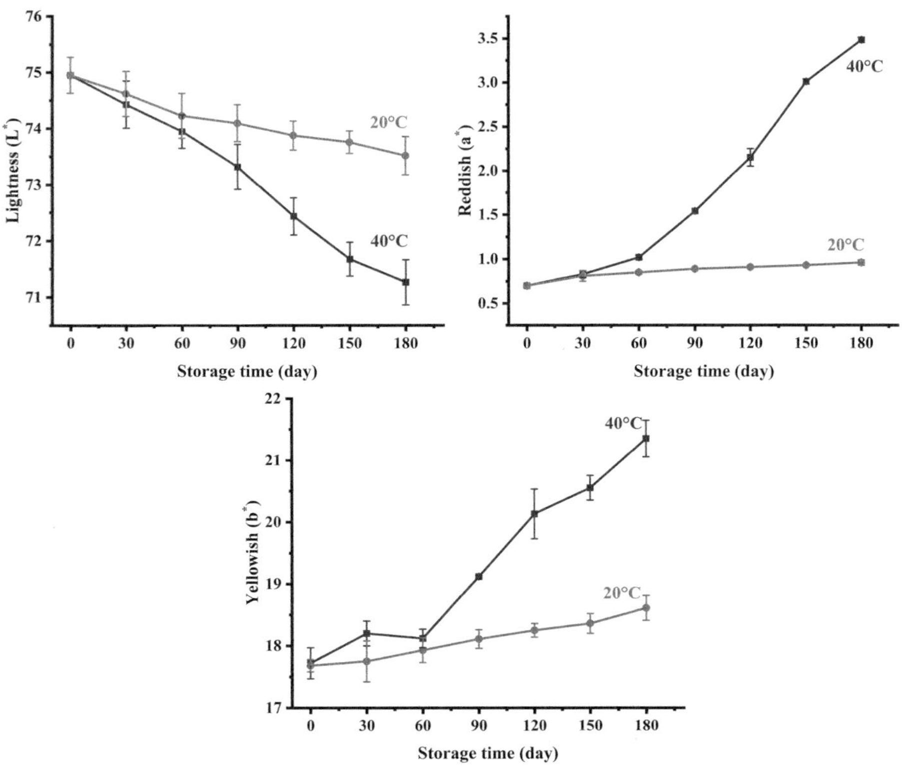

FIGURE 11.4 Changes in colour properties of FRK during storage at 40°C and 20°C.

of oxygen and moisture through packaging materials, as observed in our study. This pattern suggests that micronutrients present in FRK underwent oxidative degradation during storage and eventually made darker colour FRK. Further, changes in colour in the kernels can happen due to storage temperature, moisture content and insect infestation (for FR only) (Haydon and Siebenmorgen 2017; Ahmad et al. 2017).

11.3.3 Effect of Storage on Micronutrient Contents in FR and FRK

Estimating degradation of micronutrients is essential for evaluating shelf life of FR and FRK. It is well known that micronutrients, particularly vitamins, degrade during processing and storage. Therefore, adding vitamins in excess of the recommended level can ensure the required level of vitamins when reaching the consumers. However, this added extra cost to product, and sometimes, excessive consumption of certain vitamins is not healthy. Therefore, understanding the cause of degradation of added micronutrients in FRK and FR and its degradation patterns during storage is important for FRK manufacturers, blenders, policymakers and consumers.

FRK are generally incorporated with iron, folic acid and vitamin B_{12}. Iron acts as a pre-oxidant in fortified cereal foods and thus largely affects sensory attributes in long storage time (Akhtar, Anjum, and Anjum 2011). Iron shows better stability in various storage conditions compared to the vitamins, but losses occur due to exposure to heat, light and air (Sanni 2012). Folic acid is heat sensitive, and its degradation escalates in the presence of oxygen and light. However, it is found stable in dry products in airtight containers (Burch 2011). On the other hand, vitamin B_{12} is susceptible to oxidative and reducing agents and light. Interestingly, its stability is affected by sugars, iron and vitamin C (Yamada et al. 2008).

In our experiment on stability of iron, folic acid and vitamin B_{12} in FRK and FR at different temperature and packaging conditions, it was observed that retention of the micronutrients in laminates of PET/LDPE was significantly higher than woven polypropylene bags, owing to the low oxygen and water vapour permeability of the laminates. Moreover, it is worth mentioning that temperature and relative humidity played a significant role in degrading micronutrient contents. An increase in storage temperature and relative humidity accelerated the degradation of vitamins and iron by 60% and 5%, respectively (Figure 11.5). Previously, Hemery et al. (2018) reported similar degradation patterns of vitamin A in fortified wheat flour. They observed 45%–85% reduction in vitamin A content with an increase in storage temperature from 25°C to 40°C and relative humidity from 65% to 85%. Further, the effect of light on the degradation behaviour of micronutrients was observed by illuminating the storage space artificially and results showed a sharp decrease in folic acid and vitamin B_{12} content from around 7%–56% at 20°C storage temperature. In the case of fortified rice stored in jute bag, iron degradation rate was noticed comparatively higher than the FRK due to higher oxygen transmission rate of the jute bag compared to woven polypropylene bag and PET/LDPE laminate. On the other hand, degradation of folic acid and vitamin B_{12} was found to be significantly lower for FR compared to FRK.

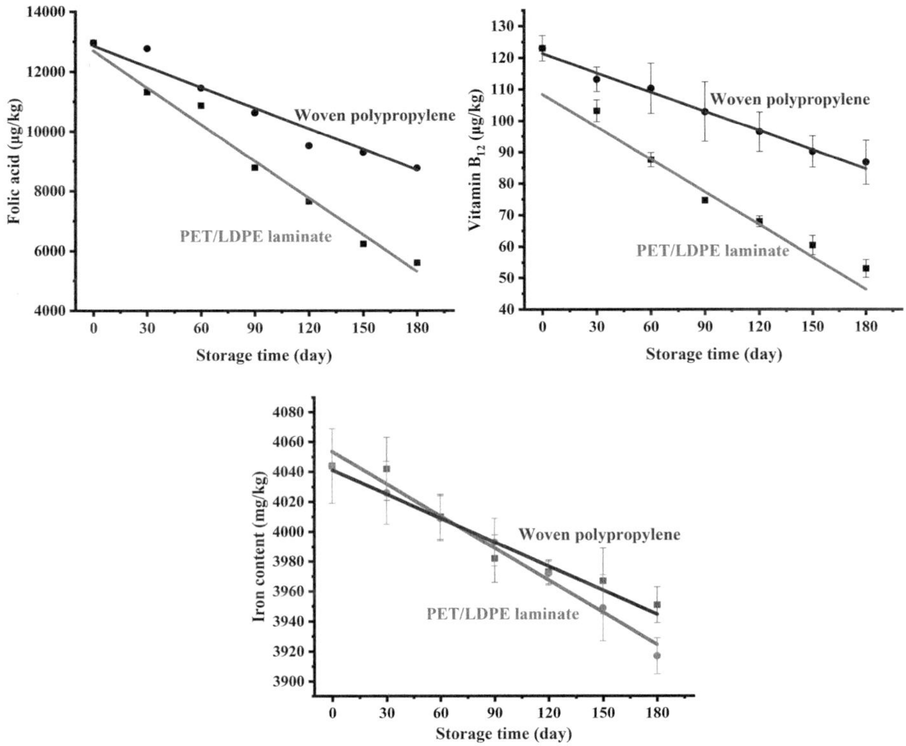

FIGURE 11.5 Degradation of micronutrient in FRK during storage at different packaging materials.

11.4 PACKAGING REQUIREMENTS FOR STORAGE OF FRK AND FR

Selection of appropriate packaging for storage of FRK and FR is crucial to provide protection against adverse storage conditions, insect infestation and physical damage during handling. Further, packaging should act as a medium of communication to the consumers about brand identity, nutritional facts, cooking instructions and expiry date. Therefore, before choosing packaging materials for FR and FRK, the following points should be considered:

- Packaging material for FR and FRK must be food grade and non-reactive as per FSSAI rules and regulations.
- Packaging materials for FRK must have very low water vapour permeability, preferably below 2 g/m^2/day, and be impermeable to oxygen and carbon dioxide.
- Package materials for FR and FRK must have complete resistance to light transfer.
- If jute bag is used for storage of FR, the bag should have heat-cut mouth with a single-fold bottom and closed by double stitching with suitable thread.

- If high-density polyethylene (HDPE) or polypropylene (PP) is used for FR storage, it has to be ensured that bags are flexible and puncture-proof. Additionally, water vapour transmission rate (WVTR) and oxygen transmission rate (OTR) should be 20–50 g/m^2/day and 8–1.9 cc/m^2/day, respectively.
- Packaging materials must have the capability to prevent insect infestation.
- Packaging materials must have enough strength to withstand external forces during handling and transportation.
- Packaging materials for both FRK and FR must be easily available and economically viable.

11.5 PACKAGING MATERIALS FOR STORAGE OF FRK AND FR

In order to preserve physicochemical and nutritional quality, recommended packaging material for FRK is coextruded laminates, while for FR, jute bags are preferred. In addition to these, the below-mentioned packaging materials with appropriate water vapour, oxygen gas and light barrier properties can be considered for storage of FRK and FR. Figure 11.6 illustrates current trends in packaging system for FRK and FR.

11.5.1 Polyethylene (PE)

HDPE and low-density polyethene (LDPE) are the most commonly used food packaging materials due to their affordability, mouldability and lightweight (Lim 2023). Both LDPE and HDPE possess good water and gas barrier properties, with HDPE having a slightly higher value. In addition, HDPE has higher light barrier properties compared to LDPE, making HDPE more suitable for use as primary packaging materials for FRK. However, LDPE can be considered as lining materials or secondary packaging materials for the storage of FRK where primary packaging materials possess inadequate water and gas barrier properties. Furthermore, good mechanical

FIGURE 11.6 Current trends in packaging system of (a) FRK and (b) FR.

properties of both the polyethene can provide substantial protection against physical damage during transportation and handling.

11.5.2 Polypropylene (PP)

Polypropylene (PP) is widely used for the packaging of rice owing to its strength, durability and tear resistance properties (da Silva et al. 2019). Two types of PP bags, namely woven and non-woven, can be utilized for the packaging of FRK and FR. Biaxially oriented polypropylene (BOPP) is another form of PP, which is currently preferred by the rice millers to pack FRK due to its superior protection against external environments (light, water and gases) and good printability.

11.5.3 Paper Bags with Polyethene Lining

Paper bags are made of interwoven materials of cellulosic fibres derived from wood or other cellulosic materials by using sulphite or sulphate. It has good mechanical properties, printability and biodegradability. However, poor water and oxygen barrier properties and porous internal structure are the major drawbacks to use as sole packaging materials for FRK. Hence, paper bag with LDPE lining can be considered for short-term storage of FRK.

11.5.4 Multilayered Film Packaging

Multilayered film packaging is an emerging film-forming technique that creates packaging film by integrating superior properties of various polymers. In this technique, polymers are selected and conjugated in such a manner that they can manufacture a film with enhanced barrier against water vapour, gases and light. Although synthetic polymers have been used to develop multilayered packaging films due to low cost and good structural characteristics, recently, biodegradable polymers are replacing those synthetic polymers to minimize environmental pollution. For the packaging of FRK and FR, aluminium/PET/LDPE or EVOH laminates and PP with metalized polyester can be preferred to ensure the safety of kernels from the external environment during storage and transportation.

11.6 SHELF-LIFE EVALUATION OF FR AND FRK

The shelf life of FR and FRK is an intriguing concept because it merges scientific and technological issues with economic, regulatory and consumer behaviour. In general, the shelf life can be denoted as a finite time during which a product maintains its certain quality parameters at a required or defined level under adequate storage conditions after manufacturing and packaging. Therefore, it is crucial to select those quality indicators (I) and monitor their losses with storage time (t) for FR and FRK. The mathematical expression showcases the changes of quality indicators with time, which can be expressed as $t = f(I, \lambda)$, where λ is the kinetic parameter. When FR and FRK are stored at different environmental conditions (temperature and relative humidity) and packaging materials, the rate of decay of each quality indicators will follow a different reaction kinetics. Therefore, it is crucial to assess that reaction kinetics along with the kinetic parameters for evaluation of shelf life of FRK. The step-by-step experimental process for estimating shelf life is described in Figure 11.7, and important quality indicators with packaging and environmental factors responsible for quality deterioration are given in Table 11.3.

From this table, it is evident that the shelf life of FR and FRK is an interactive effect of Q_i, P_i and E_i, and thus, an integrating approach needs to be designed and adopted to estimate the shelf life. However, the estimating the shelf life of FR and FRK considering the interactive effect of all the variables is not an easy task. Mostly, it is observed that researchers are calculating shelf life, particularly for accelerated

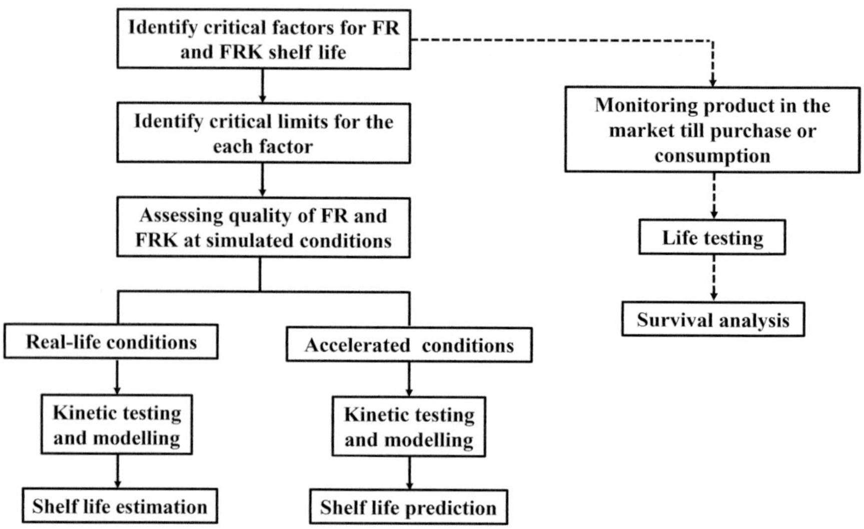

FIGURE 11.7 Flowchart of experimental process for shelf-life estimation of FR and FRK.

TABLE 11.3
Specific Quality Indicator, Properties of Packaging Materials and Environmental Conditions Expected to Affect the Shelf Life of FR and FRK

Quality Indicator (Q_i)	Packaging Material Properties (P_i)	Environmental Factors (E_i)
Iron content	Gas barrier property	Temperature
Folic acid content	Water vapour barrier property	Relative humidity
Vitamin B_{12} content	Light barrier property	Light
Water activity	Tensile strength	
Total plate count	Elongation at break	
Yeast and mould count		
Aflatoxin (for FR only)		

shelf-life testing, considering a few variables or ignoring the interactive effect of a variable. For instance, oxygen permeability of a package depends on temperature, and thus, eliminating this factor in shelf-life testing may end up predicting erroneous shelf life. This may have happened for FR and FRK as adequate barrier properties of packaging materials are important for stability of vitamins, as mentioned in Section 11.3.3. However, for simplification of shelf-life calculation procedure, one should remember the definition of shelf life, mentioning that products should be packed in predefined packaging materials to estimate their shelf life. Hence, the factor P_i can be considered constant for FR and FRK by adopting BIS-recommended packaging materials. For further simplification of the process, physical characteristics and microbiological characteristics (slower growth rate compared to degradation rate of vitamins) of FR and FRK should be considered constant. Therefore, the stability of micronutrients in FR and FRK becomes solely dependent on environmental factors like temperature and relative humidity.

This section of this chapter is curated to provide an overview of modelling the degradation reactions of micronutrients in FR and FRK at accelerated conditions. The term "accelerated shelf-life testing" is denoted here as the accelerated predictive stability (APS) study of FR and FRK. In the APS, degradation process of the micronutrients present in FR and FRK is dependent on temperature and relative humidity. In this context, a modified form of the Arrhenius equation was proposed by Genton and Kesselring (1977) and Waterman and Adami (2005) by incorporating relative humidity (RH) into the temperature dependence equation 11.1.

$$k = A \times \exp\left(-\frac{E_a}{RT}\right) \times \exp(B \times RH) \tag{11.1}$$

$$\ln k = \ln A + C \times \left(\frac{1}{T}\right) + B \times RH \tag{11.2}$$

Where k is the reaction rate, A is the pre-exponential factor related to temperature, E_a is the activation energy of the reaction (typically measured in kJ/mol or kcal/mol), R is the universal gas constant, 8.314 J/K mol, $C = -E_a/R$ and B is the humidity-sensitivity parameter.

To predict the shelf life of FR and FRK based on iron, folic acid and vitamin B_{12} contents, linear regression analysis of Eq. (11.2) is performed and parameters like A, B and E_a are calculated. Equation 11.3 is used to calculate the shelf life of FR and FRK:

$$\text{Shelf life } (t) = \frac{Q_{i0} - Q_{it}}{\exp\left(C \times \left(\frac{1}{T}\right) + B \times RH + \ln A\right)} \tag{11.3}$$

Using the aforementioned approach, the following model equations were calculated for FR and FRK stored in jute bag and LDPE/PET laminates (Table 11.4).

TABLE 11.4
Modified Arrhenius Equations Calculated for FR and FRK

	FR*	FRK*
Iron	$\ln(k) = -1860\beta_1 + 1.439\beta_2 - 7.545$	$\ln(k) = -3180\beta_1 + 5.767\beta_2 - 24.31$
Folic acid	$\ln(k) = -1204\beta_1 + 0.3752\beta_2 - 4.488$	$\ln(k) = -662.1\beta_1 + 0.3887\beta_2 - 4.857$
Vit B_{12}	$\ln(k) = -1932\beta_1 + 0.9016\beta_2 - 3.628$	$\ln(k) = -710.8 + 0.3909\beta_2 - 4.718$

* $\beta_1 = \frac{1}{T}$ and $\beta_2 = RH$

However, it should be noted that parameters *A*, *B* and *C* may be varied based on FRK manufacturing conditions, quality of raw materials and packaging conditions.

11.7 CONCLUSIONS

Proper storage conditions for FRK and FR are essential in preserving the quality, safety and nutritional value of the product. Proper storage prevents nutrient degradation, lipid oxidation and moisture-related issues that cause detrimental effects on the sensory attributes of the kernels. By controlling temperature, avoiding contaminants and adhering to set procedures, the shelf life of FRK and FR can be optimized, contributing to cost savings, waste reduction and regulatory compliance. These guidelines are not only economically important but also consumer-centric, ensuring that rice maintains high-quality standards and consumer trust. Additionally, optimal storage practices promote resource efficiency, aligning with sustainability goals in the food production chain.

FRK and FR should be stored in ventilated warehouses with controlled temperatures of 25°C or below and relative humidity of less than 60%. These environmental conditions are important in avoiding the nutrients deterioration and moisture accumulation, which also lowers the growth of insects and mould. Airtight packaging should be used to prevent oxidation and damage. In addition, the use of pest control measures, including frequent fumigation and the placement of rodent traps and pest-o-flash devices, should be implemented to prevent contamination risks from pests, insects and rodents. Also, FRK and FR should be protected from direct sunlight as light can break down the quality of nutrients and negatively affect the sensory attributes of rice. Jute bag and laminates of LDPE/PET can be considered packaging of FR and FRK, respectively, with adequate properties.

REFERENCES

A Status Note on Rice in India, 1, 2015. https://www.nfsm.gov.in/StatusPaper/Rice2016.pdf. Retrieved on 18/06/2025.

Ahmad, Usman, Luis Alfaro, Millicent Yeboah-Awudzi, Emmanuel Kyereh, Bennett Dzandu, Franklin Bonilla, Alexander Chouljenko, and Subramaniam Sathivel. 2017. "Influence of Milling Intensity and Storage Temperature on the Quality of Catahoula Rice (*Oryza Sativa L.*)." *LWT* 75: 386–92. https://doi.org/10.1016/j.lwt.2016.09.014.

Akhtar, Saeed, Faqir M. Anjum, and Muhammad Akbar Anjum. 2011. "Micronutrient Fortification of Wheat Flour: Recent Development and Strategies." *Food Research International* 44 (3): 652–59. https://doi.org/10.1016/j.foodres.2010.12.033.

Ashooriyan, Payam, Maedeh Mohammadi, Ghasem Najafpour Darzi, and Maryam Nikzad. 2023. "Development of Plantago Ovata Seed Mucilage and Xanthan Gum-Based Edible Coating with Prominent Optical and Barrier Properties." *International Journal of Biological Macromolecules* 248: 125938. https://doi.org/10.1016/j.ijbiomac.2023.125938.

Befikadu, Dubale. 2014. "Factors Affecting Quality of Grain Stored in Ethiopian Traditional Storage Structures and Opportunities for Improvement." *International Journal of Sciences: Basic and Applied Research* 18 (1): 235–57. https://gssrr.org/index.php?journal=JournalOfBasicAndApplied.

Bilgi Boyaci, Berna, Jae Yoon Han, Mustafa Tugrul Masatcioglu, Erkan Yalcin, Sueda Celik, Gi Hyung Ryu, and Hamit Koksel. 2012. "Effects of Cold Extrusion Process on Thiamine and Riboflavin Contents of Fortified Corn Extrudates." *Food Chemistry* 132 (4): 2165–70. https://doi.org/10.1016/j.foodchem.2011.12.013.

Bureau of Indian Standards (BIS). 2021a. "Fortified Rice-Specification (IS 17780: 2021)." https://archive.org/details/gov.in.is.17780.2021.

Bureau of Indian Standards (BIS). 2021b. "IS 17782: 2021: Fortified Rice Kernels Specifications." https://archive.org/details/gov.in.is.17782.2021.

Boonyasirikool, Pracha, and Chulaluk Charunuch. 2000. "Development of Nutritious Soy Fortified Snack by Extrusion Cooking." *Agriculture and Natural Resources* 34 (3): 355–65.

Burch, Rachel. 2011. "The Stability and Shelf Life of Vitamin-Fortified Foods." In *Food and Beverage Stability and Shelf Life*, edited by David Kilcast and Persis Subramaniam, 743–54. Woodhead Publishing Limited. https://doi.org/10.1533/9780857092540.3.743.

Dalbhagat, Chandrakant Genu, and Hari Niwas Mishra. 2019. "Effects of Extrusion Process Conditions on System Parameters; Physicochemical Properties and Cooking Characteristics of Extruded Fortified Rice Kernels." *Journal of Cereal Science* 89: 102782. https://doi.org/10.1016/j.jcs.2019.05.016.

Dalbhagat, Chandrakant Genu, and Hari Niwas Mishra. 2021. "Drying Modeling, Cooking Characteristics, Pasting Properties, and Crystallinity of Fortified Rice Kernels." *Journal of Food Processing and Preservation* 45 (6): 1–9. https://doi.org/10.1111/jfpp.15579.

Espinoza-Espinoza, Luis Alfredo, Henry Daniel Muñoz-More, Juliana Maricielo Nole-Jaramillo, Luis Alberto Ruiz-Flores, Nancy Maribel Arana-Torres, Luz Arelis Moreno-Quispe, and Jaime Valdiviezo-Marcelo. 2024. "Microencapsulation of Vitamins: A Review and Meta-Analysis of Coating Materials, Release and Food Fortification." *Food Research International* 187. https://doi.org/10.1016/j.foodres.2024.114420.

Genton, D., and U. W. Kesselring. 1977. "Effect of Temperature and Relative Humidity on Nitrazepam Stability in Solid State". *Journal of Pharmaceutical Sciences* 66 (5): 676–80. https://doi.org/10.1002/jps.2600660517.

Groh, Ksenia J., Thomas Backhaus, Bethanie Carney-Almroth, Birgit Geueke, Pedro A. Inostroza, Anna Lennquist, Heather A. Leslie, Maricel Maffini, Daniel Slunge, Leonardo Trasande, A. Michael Warhurst, and Jane Muncke. 2019. "Overview of Known Plastic Packaging-Associated Chemicals and Their Hazards." *Science of the Total Environment* 651: 3253–68. https://doi.org/10.1016/j.scitotenv.2018.10.015.

Haydon, Kathryn N., and Terry J. Siebenmorgen. 2017. "Impacts of Delayed Drying on Discoloration and Functionality of Rice." *Cereal Chemistry* 94 (4): 683–92. https://doi.org/10.1094/CCHEM-10-16-0257-R.

Hemery, Youna M., Arnaud Laillou, Laura Fontan, Vincent Jallier, Regina Moench-Pfanner, Jacques Berger, and Sylvie Avallone. 2018. "Storage Conditions and Packaging Greatly Affects the Stability of Fortified Wheat Flour: Influence on Vitamin A, Iron, Zinc, and Oxidation." *Food Chemistry* 240:43–50. https://doi.org/10.1016/j.foodchem.2017.07.084.

Kwon, Seongyoung, Aungkana Orsuwan, Nattinee Bumbudsanpharoke, ChanSuk Yoon, Jungwook Choi, and Seonghyuk Ko. 2018. "A Short Review of Light Barrier Materials for Food and Beverage Packaging." *Korean Journal of Packaging Science and Technology* 24 (3): 141–48. https://doi.org/10.20909/kopast.2018.24.3.141.

Lim, Matthew. 2023. "Comparison of Plastics Bags and It's Most Common Alternatives." *The National High School Journal of Science,* 1–9.

Mohamed, Ibrahim O. 2021. "Effects of Processing and Additives on Starch Physicochemical and Digestibility Properties." *Carbohydrate Polymer Technologies and Applications* 2. https://doi.org/10.1016/j.carpta.2021.100039.

Nikvarz, Navid, Gholam Reza Khayati, and Shahriar Sharafi. 2020. "Preparation of UV Absorbent Films Using Polylactic Acid and Grape Syrup for Food Packaging Application." *Materials Letters* 276: 128187. https://doi.org/10.1016/j.matlet.2020.128187.

Nithya, A., Siddharth Vishwakarma, Chandrakant Genu Dalbhagat, and Hari Niwas Mishra. 2024. "Apparent Amylose Content Positively Influences the Quality of Extruded Fortified Rice Kernels." *Carbohydrate Polymers* 338: 122213. https://doi.org/10.1016/j.carbpol.2024.122213.

Park, Chan Eun, Yun Sook Kim, Kee Jai Park, and Bum Keun Kim. 2012. "Changes in Physicochemical Characteristics of Rice during Storage at Different Temperatures." *Journal of Stored Products Research* 48: 25–29. https://doi.org/10.1016/j.jspr.2011.08.005.

Razzaq, Abdul, Arfan Ali, Luqman Bin Safdar, Muhammad Mubashar Zafar, Yang Rui, Amir Shakeel, Abbad Shaukat, Muhammad Ashraf, Wankui Gong, and Youlu Yuan. 2020. "Salt Stress Induces Physiochemical Alterations in Rice Grain Composition and Quality." *Journal of Food Science* 85 (1): 14–20. https://doi.org/10.1111/1750-3841.14983.

Romruen, Orapan, Pimonpan Kaewprachu, Samart Sai-Ut, Passakorn Kingwascharapong, Thomas Karbowiak, Wanli Zhang, and Saroat Rawdkuen. 2024. "Impact of Environmental Storage Conditions on Properties and Stability of a Smart Bilayer Film." *Scientific Reports* 14 (1): 23038. https://doi.org/10.1038/s41598-024-74004-4.

Saffarionpour, Shima, and Levente L. Diosady. 2021. "Multiple Emulsions for Enhanced Delivery of Vitamins and Iron Micronutrients and Their Application for Food Fortification." *Food and Bioprocess Technology* 14 (4): 587–625. https://doi.org/10.1007/s11947-021-02586-2.

Sanni, Silifat. 2012. "Stability of Iron-Fortified Gari during Storage in Different Packaging Materials." *Journal of Food Processing and Preservation* 36 (3): 207–13. https://doi.org/10.1111/j.1745-4549.2011.00579.x.

Schaefer, Dirk, and Wai M. Cheung. 2018. "Smart Packaging: Opportunities and Challenges." *Procedia CIRP* 72: 1022–27. https://doi.org/10.1016/j.procir.2018.03.240.

Silva Barbosa Ferreira, Eduardo da, Caio Henrique do Ó Pereira, Edcleide Maria Araújo, Elieber Barros Bezerra, Danilo Diniz Siqueira, and Renate Maria Ramos Wellen. 2019. "Properties and Morphology of Polypropylene/Big Bags Compounds." *Materials Research* 22: 1–8. https://doi.org/10.1590/1980-5373-MR-2018-0850.

Sirisoontaralak, P., and Athapol Noomhorm. 2007. "Changes in Physicochemical and Sensory-Properties of Irradiated Rice during Storage." *Journal of Stored Products Research* 43 (3): 282–89. https://doi.org/10.1016/j.jspr.2006.06.008.

Verma, Deepak Kumar, and Prem Prakash Srivastav. 2017. "Proximate Composition, Mineral Content and Fatty Acids Analyses of Aromatic and Non-Aromatic Indian Rice." *Rice Science* 24 (1): 21–31. https://doi.org/10.1016/j.rsci.2016.05.005.

Waterman, Kenneth C., and Roger C. Adami. 2005. "Accelerated Aging: Prediction of Chemical Stability of Pharmaceuticals." *International Journal of Pharmaceutics* 293 (1–2): 101–125.

WFP. 2018. "Rice Fortification in Senegal. Landscape Analysis." Edited by WFP Senegal.

Yamada, Keiko, Michiko Shimodaira, Seiko Chida, Noriko Yamada, Norio Matsushima, Morimichi Fukuda, and Shoji Yamada. 2008. "Degradation of Vitamin B12 in Dietary Supplements." *International Journal for Vitamin and Nutrition Research* 78 (4–5): 195–203. https://doi.org/10.1024/0300-9831.78.45.195.

Zhou, Zhongkai, Kevin Robards, Stuart Helliwell, and Christopher Blanchard. 2002. "Ageing of Stored Rice: Changes in Chemical and Physical Attributes." *Journal of Cereal Science* 35 (1): 65–78. https://doi.org/10.1006/jcrs.2001.0418.

Ziegler, Valmor, Cristiano Dietrich Ferreira, Jorge Tiago Schwanz Goebel, Shanise Lisie Mello El Halal, Gabriela Soster Santetti, Luiz Carlos Gutkoski, Elessandra da Rosa Zavareze, and Moacir Cardoso Elias. 2017. "Changes in Properties of Starch Isolated from Whole Rice Grains with Brown, Black, and Red Pericarp after Storage at Different Temperatures." *Food Chemistry* 216: 194–200. https://doi.org/10.1016/j.foodchem.2016.08.045.

Ziegler, Valmor, Cristiano Dietrich Ferreira, Jessica Fernanda Hoffmann, Fábio Clasen Chaves, Nathan Levien Vanier, Maurício de Oliveira, and Moacir Cardoso Elias. 2018. "Cooking Quality Properties and Free and Bound Phenolics Content of Brown, Black, and Red Rice Grains Stored at Different Temperatures for Six Months." *Food Chemistry* 242: 427–34. https://doi.org/10.1016/j.foodchem.2017.09.077.

12 Plant and Machinery for Fortified Rice Kernels Manufacturing and Blending

Siddharth Vishwakarma, Yuvraj Bhosale, Rakesh Kumar Raigar, Shubham Mandliya, Chandrakant Genu Dalbhagat, and Hari Niwas Mishra

12.1 INTRODUCTION

Fortified rice kernels (FRK) are rice-shaped kernels enriched with iron, folic acid, and vitamin B_{12}, designed to combat micronutrient deficiencies. The fortified rice (FR) is the blend of FRK and NR in the specified ratio (Dalbhagat and Mishra 2019). In India, rice fortification has become a national priority to address anemia and malnutrition in vulnerable population. The fulfilment of this priority is backed by strong policy support: The Food Safety and Standards Authority of India (FSSAI) has established fortification standards and mandated the use of a +F logo on fortified foods. Likewise, the Bureau of Indian Standards (BIS) introduced specifications for FRK quality and for the machinery used in FRK production (BIS 2021a). Mechanization plays a pivotal role in the success of rice fortification. Producing FRK at the scale required for national programs, thousands of tons distributed to millions of households would be impractical without advanced machinery and automation. Modern FRK production is carried out using extrusion technology: rice flour is mixed with a vitamin-mineral premix (VMP) and water, and the dough is extruded through a die to form rice-shaped kernels, which are then dried and cooled. These fortified kernels are blended with milled rice at typically 1:100 ratio to produce FR for consumption (BIS 2021b). Key components of an FRK plant include mixers for flour and micronutrients, feeders and dosing equipment, a twin-screw extruder (for hot extrusion of kernels), dryers and coolers, automatic blending machines for mixing FRK with normal rice, and packaging systems. Each of these components must be precisely controlled to ensure that the final product meets nutrient specifications and quality standards. Mechanization and automation help maintain this precision: for example, extruders allow exact regulation of temperature and screw speed for gelatinization of starch, and automatic blending machines ensure a uniform 1:100 distribution of FRK in rice with minimal breakage. In short, advanced machinery underpins the

 DOI: 10.1201/9781003661245-12

efficiency, scalability, and quality control of FRK manufacturing, which in turn enables India's fortification program to reach its public health goals. In this chapter, the plant and machinery used for the rice fortification is discussed.

12.2 PLANT LAYOUT

Efficient plant layout is fundamental for the hygienic, economical, and scalable production of FRK. It ensures logical sequencing of operations, minimizes material handling, reduces cross-contamination risk, facilitates equipment maintenance, and complies with regulatory norms set by FSSAI and BIS.

12.2.1 Process Flowchart

The standard process flow for FRK production via extrusion technology is shown in Figure 12.1. It includes manufacturing and blending, employing extrusion technology and handling systems. Firstly, the broken rice is fed into a micropulverizer and milled into fine flour. This flour is then directed to a cyclone separator, which removes dust and fine particulates. The de-dusted flour enters a mixer-cum-conditioner, where Reverse Osmosis (RO) water and a pre-calibrated micronutrient premix (typically containing iron, folic acid, and vitamin B_{12}) are added. The resulting conditioned dough is transferred to a twin-screw extruder, where it undergoes gelatinization and shaping under controlled temperature and shear to form FRK. These extruded FRK are then dried in a multilayer dryer to reduce moisture to safe storage levels (~12%). The dried FRK are conveyed to an automatic blending machine, where they are mixed at a precise 1:100 ratio with milled raw rice. The resulting FR is finally packed using an automated packaging system that complies with FSSAI guidelines with "+F" logo. This streamlined and hygienic process ensures consistent nutrient

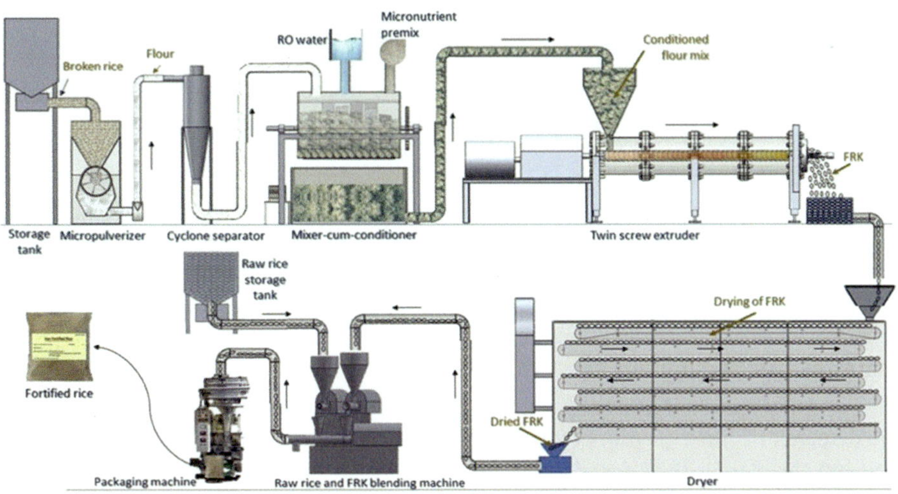

FIGURE 12.1 Design of continuous FRK plant. (Designed by Food Chemistry and Technology Laboratory, IIT Kharagpur.)

delivery, operational efficiency, and regulatory -compliance across FRK production facilities (BIS 2021a; BIS 2022a).

Each unit operation is connected through food-grade conveying systems (e.g., screw conveyors, bucket elevators, and pneumatic lines), ensuring hygienic and efficient material transfer.

12.2.2 Building Layout

The building layout for an FRK production unit should accommodate all process equipment, storage areas, utilities, and worker facilities, while maintaining unidirectional flow of materials and personnel to prevent contamination. The layout is divided into various sections:

Raw Material Section

The raw material section mostly includes the raw rice flour storage silos or sacks, micronutrient premix storage, and material inspection area.

Processing Section

The processing section contains grinding, mixing, and extrusion. The grinding zone includes micropulverizer and cyclone with bag filter. The mixing area contains ribbon or paddle mixer. The extrusion chamber contains twin-screw extruder placed on vibration-dampening platform, with an adjacent control panel.

Post-Processing Section

The post-processing section mainly contains the drying section and the quality control lab. The drying unit room is enclosed and equipped with dryer along with the exhaust system. The quality control lab is for batch-wise nutrient assay and moisture testing.

Packaging and Blending Area

The FRK and Normal rice blending can be done in the packaging area. Some FRK can go directly for the packaging and some can go for the blending section.

Utilities and Support

It includes the compressor and blower house for pneumatic conveying systems and an electrical control room with PLC/SCADA panels. Regular equipment maintenance is facilitated by a maintenance workshop and tool room. A staff hygiene area with handwashing and gowning facilities also provides sanitary operations.

Material and Personnel Flow

The layout of the plant must allow linear material flow from raw material receiving to the dispatch of finished products with minimal cross-contamination. Segregation of raw and clean areas is required for hygiene. Individuals must pass through designated hygiene stations (shoe wash, gowns), with fire exits and ventilation designed as per BIS and FSSAI regulations (BIS 2021a).

12.3 FRK MANUFACTURING EQUIPMENT

12.3.1 Cleaners and Graders

Cleaners and graders are used at the start of the FRK manufacturing line to prepare the raw material (typically broken rice grains) by removing impurities and ensuring a uniform input size. Cleaning equipment often includes vibrating sieve separators, de-stoners, and magnetic separators. These remove dust, chaff, stones, metals, and other foreign matter from the broken rice before further processing (Dalbhagat and Mishra 2020). For example, BIS guidelines specify that broken rice should be initially cleaned for metals, stone, plastic or paper, etc., (BIS 2021a).

After basic cleaning, grading may be performed to ensure consistent particle size of the rice input. In practice, since the broken rice is milled into flour, size grading of intact kernels is less critical than in whole grain milling; however, removing extremely large or undersized fragments can improve grinding efficiency. A classifier or scalping screen may be used to separate out any oversize particles/materials or fine sand/silts that passed through the initial sieve. Ensuring uniform raw material size and purity helps to produce a consistent rice flour and, subsequently, uniform fortified kernels (BIS 2021a; BIS 2021b).

12.3.2 Grinders

After cleaning, the broken rice is milled using industrial mills capable of producing a very fine and consistent particle size. The BIS standards for FRK specify that the rice flour should have an average particle size of less than 250 μm, with a recommendation of achieving even finer size (e.g., "preferably < 150 microns") for best results (BIS 2021a; BIS 2021b). Fine particle size ensures a homogeneous mixture with micronutrient premix and aids in the extrusion process. In practice, many FRK manufacturers grind to 100–120 mesh, roughly 125 μm, which is within the recommended range. Commonly, a hammer mill or pin mill is used in FRK plants. Pin mills can achieve very fine and uniform particle size, while hammer mills can handle larger throughput and slight variations in feed moisture.

Following rice flour milling, the particulate-laden airstream typically exits the grinding chamber at high velocity. The airflow conveys both the wanted fine rice flour and suspended ultrafine dust particles that have been generated by impact and attrition (Dalbhagat and Mishra 2019). FRK factories generally employ a two-stage air-solid separation system composed of a cyclone separator and a baghouse (fabric) filter

Cyclone separators use centrifugal force to separate particles. Dust-laden air enters tangentially, creating a spiral flow that pushes heavier rice flour particles (<250 μm) to the walls, where they settle and exit through the bottom. Cleaner air rises and exits through the central vortex. In FRK production, this step recovers over 90%–95% of flour before secondary filtration. This enhances product recovery and also reduces the dust load on the subsequent bag filter, extending its service life.

The cyclone exhaust still contains fine particulate matter, notably <10 μm particles. To eliminate them, a high-efficiency filtration barrier is used, which comprises a series of cylindrical filter bags made of woven or non-woven fabric (e.g., polyester and PTFE-coated felt). As air passes through the material, dust is deposited on the outer surface of the bags. Clean air flows out through the inner core of the bags. The trapped dust can be periodically cleaned by mechanical shaking (Dhankhar 2014).

12.3.3 Mixer

The mixer blends rice flour with the VMP, additives, and water to produce a conditioned dough for extrusion. Achieving a homogeneous blend of micronutrients in the rice flour is critical, as it ensures that each kernel will have the intended nutrient content. In FRK production, horizontal batch mixers such as ribbon blenders or paddle mixers are commonly used (BIS 2021a). These mixers consist of a trough with internal elements (ribbon spirals or paddles) that rotate to tumble and fold the ingredients. According to the BIS guidelines, the mixing process is two-stage: first, a dry mix of rice flour and premix (and any additives like binders) is done, followed by wetting/conditioning (BIS 2021a). In FRK production, water is sprayed into the mixer to raise blend moisture to ~25%–30% for extrusion. Mixers use spray nozzles and flow meters, and the dough rests for 5–10 min to equilibrate moisture. A well-designed mixer prevents micronutrient segregation and "hot spots." Ribbon blenders are efficient, achieving uniform mixing in 3–5 min, with slow to moderate agitation followed by water spraying and continued mixing. The target is a cohesive, slightly moistened meal that will feed consistently into the extruder. A good practice is to ensure that all mixer parts are tight and there are no worn pieces; mixers often have magnetic traps at discharge points as an extra safety step.

12.3.4 Extruder

After conditioning, the fortified dough is cooked and shaped into rice-shaped kernels using extruder. In the extruder, conditioned dough is propelled by screws under controlled heat and pressure through a die to form rice-shaped kernels that are cut to a specified size. The process also partially gelatinizes starch, ensuring proper hardness and cooking properties (Dalbhagat and Mishra 2019).

For FRK, twin-screw extruders (TSE) are generally that employ two intermeshing screws providing better mixing and more uniform cooking of the dough. TSE deliver intensive mixing, which ensures that micronutrients are evenly distributed in every kernel and can handle higher mineral/vitamin loads without issues like segregation or clogging. They also allow more precise control of temperature and shearing. Typical configurations used in India include co-rotating TSE with multiple heating zones and a die head before cut.

A typical industrial FRK extruder in India could have a throughput of 200–500 kg/h of kernels, consuming around 50–100 kW of power for the main drive. TSE have the advantage of consistent, high-quality output, the kernels are very uniform in shape and density, and nutrient distribution is assured (Dalbhagat and Mishra 2021b). The disadvantages are primarily higher cost and complexity.

The single-screw extruders (SSE) can also be used for FRK, and some smaller manufacturers or pilot plants use them due to lower cost and easier maintenance. SSE can produce kernels, but achieving uniform mixing of the micronutrients is more challenging and the throughput is generally lower for the same barrel size. To address nutrient uniformity, producers may recycle extrudate or use a high-shear screw profile, but overall, the trend is toward TSE for any large-scale fortification program. There are also cold extrusion methods (sometimes called a pelletizer approach) where the dough is pressed through a die at a lower temperature (~60°C) to minimize heat exposure of vitamin (Ganachari et al. 2022). Cold extrusion requires binders and careful drying, producing uncooked kernels with altered cooking properties. In contrast, hot extrusion, which partially cooks starch, is preferred commercially for yielding kernels with cooking time and texture similar to milled rice.

12.3.5 Dryer

The freshly extruded FRK are moist and hot and thus need to be dried to a stable moisture content (~12% moisture) to ensure shelf stability and proper texture. Drying is a delicate step: it must remove moisture uniformly without causing cracks or deforming the kernels, and without degrading the added vitamins (Dalbhagat and Mishra 2020). This typically achieved using lower drying air temperatures of 40°C–60°C in many cases, as higher temperatures can damage the product.

Dryers used in FRK manufacturing can be batch or continuous. In smaller setups, a tray dryer (hot air oven) is used: wet kernels are spread in thin layers on trays, and hot air (at e.g. 40°C–60°C) is circulated for several hours. Larger commercial facilities use continuous belt dryers or multilayer dryers, where the kernels are fed onto a moving belt (or a series of belts in tiers). For example, an industrial belt dryer might have 6–7 layers of mesh belt, with air blowers circulating hot air across the kernels. In practice, the kernels are usually dried in a single pass to reach the target moisture (e.g., from ~30% down to 11%–12% moisture per hour). To ensure even drying, the BIS code suggests spreading the FRK in a thin layer (preferably a single kernel thick), which prevents the kernels at the bottom of a pile from staying wet while the top ones over-dry.

A critical quality issue during drying is the formation of fissures or cracks in the kernels if drying is too rapid or uneven. For instance, Dalbhagat and Mishra (2021a) found that FRK dried at 40°C remained intact, while drying at ≥50°C caused fissures and disintegration on cooking. Gentle drying preserves structure but is slow, so a two-stage process, mild initial drying followed by tempering is used to balance quality and efficiency (Dalbhagat et al. 2024). Another strategy is fluidized bed drying in which the warm air quickly reduces kernel moisture from ~30% to 18%–20%, followed by slow drying to 12% in a tray or belt dryer. This two-stage method, common in parboiled rice processing, helps to prevent stress cracks.

In terms of design parameters, drying air temperature, airflow velocity, and humidity are monitored. A gentle airflow (e.g. 0.5–1 m/s) is used. If the dryer is multi-zone, the first zone might be set around 60°C and later zones at 50°C. The relative humidity control drying can be useful to check "case hardening," and some advanced dryers recirculate a portion of air to maintain humidity (BIS 2021a; Dalbhagat and Mishra 2019).

12.3.6 Polisher

Once the fortified kernels are dried and cooled, these are passed through a polishing machine. The goal of polishing is to improve the appearance and surface finish of the kernels so that they closely resemble the sheen of milled rice.

In FRK manufacturing, the polisher used is often a modified rice polisher or whitener. Since FRK has no bran, the polisher here serves to *buff* the kernel. It usually consists of a steel polishing chamber with an internal rotating roll and an outer screen. Kernels enter, are rubbed against each other and the roll, and a very fine layer of surface material is scrubbed off as powder. The BIS manufacturing protocol describes this step: the dry FRK (moisture $\leq$12%) is fed slowly into the polisher to maintain a uniform feed rate, and the machine is run for only about 15–20 s for smoothening the surface of the FRK.

Key equipment parameters include the type of polishing roll, the mesh size of the screen (to allow fine powder out but not let kernels escape), and the pressure or feed rate which controls. Some polishing machines might allow a small amount of water mist or oil to be added to enhance gloss; however, with FRK, this is usually avoided and dry polishing should be employed. Polishers risk contamination from emery dust and loose metal parts; BIS guidelines mandate checks and metal detection. Over-polishing can also damage kernels by excessive grinding or breakage.

12.3.7 Packaging Machine

Packaging, the final stage, involves accurately weighing and packing FRK to protect against moisture, light, pests, and damage (Mishra et al. 2024). Modern FRK plants use automatic Form-Fill-Seal (FFS) machines that form, fill, and seal pouches, typically with electronic weighers or cup feeders ensuring precise fills (e.g., 1 kg). Recommended packaging is co-extruded laminate—polyethylene inside with polyester or metallized film outside—providing food-grade, non-reactive, and moisture- and gas-barrier protection to preserve vitamins. Packs are hermetically heat-sealed.

For bulk packs (25–50 kg), semi-automatic fillers with weighing and stitching/sealing units are used—typically filling woven polypropylene or BOPP-laminated bags, stitched securely (preferably double). Accurate weighing is essential to maintain fortification ratios, so fillers require regular calibration.

12.3.8 Handling Equipment

Throughout the FRK production process, material handling equipment is employed to convey raw materials, intermediates, and finished kernels between the various machines. Efficient handling systems reduce manual labor, increase throughput, and protect the product from spillage or contamination (Wimberly 1983). The main types of handling equipment used are pneumatic conveyors, belt conveyors, bucket elevators, screw conveyors, and vibrating conveyors (or feeders). The selection of conveyor type is based on factors like capacity required, distance and height to cover, need for gentleness (to avoid kernel breakage), and layout constraints. Below, each type of handling equipment is discussed in the context of an FRK manufacturing setup, highlighting its function, advantages, and any limitations.

Pneumatic Conveyors

Pneumatic conveyors use air flow, via pressure or vacuum, to move powders and grains through enclosed pipes, ensuring hygiene and dust control. They can transport materials vertically and over long distances, with some systems combining suction and pressure for multi-stage transfer. For FRK, pneumatic conveyors offer the advantage of being gentle if operated in "dilute phase" at moderate speeds; kernels are suspended in air, so there's minimal mechanical stress. The systems are generally enclosed, preventing any dust of vitamin premix or rice flour from escaping into the work environment. On the downside, pneumatic conveyors tend to consume significant energy and they can be noisy (Wimberly 1983).

Belt Conveyors

Belt conveyors are one of the most common and straightforward methods of moving materials horizontally (or at a gentle incline). They consist of a continuous looped belt (made of rubber, fabric, or modular plastic) that runs over pulleys, driven by a motor, carrying material on its surface. In FRK manufacturing, belt conveyors can be used to carry dry kernels from one unit to another gently. For example, after drying and cooling, a belt conveyor might transport the finished kernels to the packaging hopper (Wimberly 1983).

The main advantage of belt conveyors is their gentle handling of products. The kernels sit stationary on the moving belt and experience very little agitation. Belt conveyors are engineered to minimize impact damage to fragile materials. This is crucial for FRK because the kernels can develop cracks, and a belt conveyor imparts almost no breakage. Belts can also handle high capacities (tonnes per hour) and can be made in various widths to accommodate the flow. The speed of the belt is adjustable; slower speeds can be used if needed to be extra gentle or to allow manual inspection (Mohidem et al. 2022). Belt conveyors also offer high capacity, adjustable speeds, food-grade materials, and easy cleaning, with design options like side rails, trough shapes, and limited inclines (10°–20°) to reduce spillage.

Bucket Elevators

A bucket elevator uses a series of buckets on a belt or chain running in vertical loops to lift material from a lower point to a higher discharge. In FRK plants, they save floor space while moving broken rice to grinders or dried kernels from dryers to screening or packaging. Their main advantage is high-capacity vertical transport in a compact footprint, capable of lifting materials several meters. Common in rice mills, they feature closely spaced buckets, a loading leg (boot) for feeding, and a discharge head where kernels are released by gravity or centrifugal force.

Screw Conveyors

Screw conveyors (also known as augers) consist of a helical screw blade (flight) rotating within a trough or tube, which drags material along. They are widely used in many industries for moving powders or granular products horizontally or at a small incline. In an FRK plant, screw conveyors can be used for short-distance transfers or for feeding equipment. For example, a screw feeder is typically used to feed the rice flour blend from a hopper into the extruder steadily. Screw conveyors might also transport premix or flour from one mixer to another or unload finished kernels from a bin to the packaging machine (Chongchitpaisan and Sudsawat 2022).

The advantage of screws is their simplicity and controlled flow. They can serve not only to convey but also to meter the material by adjusting screw speed, one can control the flow rate quite precisely.

Vibrating Conveyors

Vibrating conveyors use rapid oscillations to move material in small jumps along a trough. In FRK production, they ensure uniform feeding (e.g., flour to the extruder) and gentle handling (e.g., cooling and spreading kernels). With adjustable frequency and amplitude, they provide precise flow control, making them ideal for dosing from hoppers (Wimberly 1983).

12.4 FRK AND NR BLENDING

Blending or mixing two components, such as FRK and normal rice, is a fundamental process that requires careful consideration of various factors. Precision and control are essential to achieve a homogeneous mixture. The blending needs uniform FRK distribution throughout the fortified rice, and its physical properties must be matched with the regular rice. The important physical properties include size and shape, weight of kernels, density, and color. The commercial blending of FRK is done in batch and continuous mode, depending upon the equipment availability.

12.4.1 Batch Blenders

Batch mixing of solids refers to combining dry or solid materials in a controlled manner to achieve a homogeneous mixture. This is a standard process in various industries such as pharmaceuticals, food and beverage, and chemicals, where the precise blending of solid components is crucial to produce consistent and high-quality products (Basinskas and Sakai 2016). Batch blenders typically consist of two hoppers, one for FRK and other for normal rice connected with dosing systems. The dosing systems play a pivotal role in ensuring the accuracy of FRK and normal rice proportions, a critical factor in achieving the desired blending ratio. The blender consists of a "U"-shaped trough with a central shaft and ribbons/paddles mounted on it. The motion of paddles/ribbon mix the FRK homogeneously in normal rice. A twin shaft fluidized paddle mixer can also be employed for mixing, which creates less shear than other blender.

12.4.2 Continuous Blending

To meet demand, capacity can be increased by scaling batch blending or shifting to continuous blending at the source. In continuous systems, FRK and rice are fed at controlled rates into ribbon or paddle mixers, with automated dosing ensuring the correct ratio. The blended output is discharged continuously for further processing or packaging.

12.5 AUTOMATION AND CONTROL SYSTEMS

Modern FRK manufacturing facilities in India increasingly employ automation and control systems to ensure consistent product quality, traceability, and efficiency. Key automation technologies include programmable logic controllers (PLCs) for machine control, supervisory control, and data acquisition (SCADA) systems for monitoring and oversight, and emerging Industrial Internet of Things (IoT) devices for real-time data collection. These technologies collectively enable precise control of the extrusion and blending processes, while also ensuring processing quality.

12.5.1 PLC and SCADA

PLCs are important in modern FRK production lines, executing real-time control of equipment such as mixers, extruders, dryers, and blending machines. PLCs are rugged industrial computers programmed to maintain process parameters within specified ranges and to sequence the operations. For example, a PLC on an extruder will regulate barrel zone temperatures, screw speed, and feeder rate to shape the FRK correctly. Many Indian-made FRK blending machines come equipped with PLCs and human-machine interface (HMI) touchscreens, making them user-friendly to operate and calibrate even by novice operators. PLC automation reduces human error and variability in the production process, which is critical for meeting fortification standards batch after batch (Kondakci and Zhou 2017).

SCADA provides a supervisory layer over PLCs, enabling remote monitoring of process parameters (e.g., extrusion temperatures, moisture, and motor loads) via graphical interfaces. It logs data for traceability, triggers alarms or shutdowns on deviations, and supports root-cause analysis and audits. In India, SCADA records also help demonstrate compliance with fortification standards (Režek Jambrak et al. 2021).

12.5.2 IoT and Decision Support Systems (DSS)

Internet of Things (IoT) integration in FRK plants enables real-time monitoring and remote access to process data. Smart sensors track parameters like extruder temperature, torque, die pressure, dryer humidity, and product moisture, transmitting data to cloud dashboards. Alerts are triggered near critical limits, allowing timely intervention (e.g., detecting rising extrusion pressure before a blockage) (Dadhaneeya, Nema, and Arora 2023).

Decision Support Systems (DSS) use IoT and SCADA data to guide optimal operations in FRK plants. They can recommend drying conditions, optimize blending, or adjust FRK ratios when nutrient levels vary. By analyzing real-time and historical data, DSS enables fine-tuning that is difficult to calculate manually, ensuring consistent fortification quality (Gupta et al. 2018). For example, if lab tests show that the vitamin content in a batch of FRK is at the lower end of the acceptable range, the DSS might advise blending a bit more than 1% FRK to ensure that the final FR still meets nutrient requirements.

12.6 REGULATIONS AND STANDARDS

The Indian Standard IS 17853 (BIS 2022a) provides the requirements of the equipment used in the production of FRK. The material requirement specifications for key components of FRK manufacturing equipment are shown in Table 12.1. The standard prescribes the detailed specifications for micropulverizer, mixing unit, extruder, and dryer, which are the main components of equipment used in manufacturing of FRK.

12.6.1 Requirements for Micropulverizer

The constructional requirements of the micropulverizer ensure the proper design and functionality of the rice grinding equipment. Features like the hopper, grinding chamber, cyclone, control panel, blower, and motor specifications are meticulously designed to facilitate efficient and safe operation. The hopper regulates feed intake and incorporates a magnet for metal impurity removal, crucial for product quality and machinery integrity. The grinding chamber should have arrangements for installation of wear-resistant screen and hammers with proper clearance. Dust filters prevent cyclone emissions, and performance standards ensure final product quality. The average particle size of rice flour must be $< 250\,\mu m$, preferably $< 150\,\mu m$, ensuring fine texture and consistency. Additionally, the rice flour must be free from any metal or foreign impurities post pulverization, guaranteeing food safety and quality (BIS 2021a; BIS 2022a).

TABLE 12.1
Material Specifications for Key Components of FRK Manufacturing Equipment

Equipment	Component/Part	Material Specification
Micropulverizer	Hopper, grinding chamber, screen, hammer, cyclone, Pipes and fittings, supporting frame	Stainless steel grade X04Cr19Ni9
Mixer	a. Mixing chamber, cover/lid, Shaft, paddle, helical ribbon, supporting frame and structure	Stainless steel grade X04Cr19Ni9
	b. Gasket	Food-grade rubber
Extruder	a. Barrel, extruder screw, and die-head	Food-grade nitriding steel of grade EN41B or 38CrMoAl or any similar material resistant to wear, abrasion, and corrosion
	b. Die cutter, body cover, supporting frame and structure	Stainless steel grade X04Cr19Ni9
Dryer	Belt of dryer	Made of food-grade material

12.6.2 Constructional Requirements for Mixing Unit (Powder-Mixer)

The mixer must be designed for efficient, standardized, and hygienic blending in fortified food production. Key requirements include a suitable capacity and type, a well-designed mixing chamber, water spray system, and lump-breaking mechanism to ensure uniform mixing without clumping. Paddles and lids with food-grade gaskets prevent spillage and contamination. A water dosing pump with automatic flour control should provide uniform moisture distribution, supported by integrated dosing, feeding, and control systems for precision and efficiency. Compliance with motor, material, and metal impurity standards ensures safety, while the mixer must uniformly disperse nutrients such as iron, folic acid, and vitamin B_{12} (BIS 2021a; BIS 2022a).

12.6.3 Constructional Requirements for Extruder

The extruder must be designed for efficient and precise FRK production. The barrel, square or circular with heating elements, should maintain optimal extrusion temperature, while a wear-resistant screw and precise feeder control ensure smooth conveyance and uniform output. A properly designed hopper must deliver wet material to the screws. Die head sensors for temperature and pressure safeguard product consistency, while specific die and cutter designs adapted to rice varieties prevent sticking and allow adjustable cutting speeds. A control panel with SCADA enables real-time monitoring of torque, die pressure, and barrel/die temperatures, improving efficiency. Compliance with material and motor standards ensures safety and reliability. Performance requirements emphasize uniform kernel size and weight to meet quality standards (BIS 2021a; BIS 2022a).

12.6.4 Constructional Requirements for Dryer

The constructional requirements for the dryer are pivotal for maintaining the quality of FRK. The standard stipulates the use of continuous electrical type dryer with multilayer vibratory arrangement to ensure efficient drying, meeting agreed capacity demands. Temperature and humidity control through provision of suitable heating and cooling arrangement, along with precise airflow management through installation of blower of suitable size, prevent undesirable outcomes like color alteration and fissure formation, preserving product integrity. The integration of sensors for temperature and relative humidity (RH), an air flowmeter, and a control panel allows for meticulous monitoring and adjustment of crucial parameters, enhancing operational precision. FRK obtained after drying should not have more than 12% moisture, ensuring proper preservation. Additionally, the absence of fissures and color change in FRK ensures its visual and structural quality (BIS 2021a; BIS 2022a).

12.6.5 Requirements of Equipment for Manufacture of Fortified Rice

The Indian Standard for specifications of equipment for the manufacture of Fortified Rice, IS 17854 (BIS 2022b) permits both blending unit batch or continuous type depending upon the capacity and requirement. Blender may be of the following types offering flexibility in blending methods: (a) continuous screw/ribbon blender; (b) drum blender; (c) continuous drum blender; (d) paddle blender; and (e) octagonal blender. Each type is designed to handle specific blending tasks efficiently. The hopper, lid screw/ribbon, paddle, containers, and agitator of blender shall be made of stainless steel of grade $X_{04}Cr_{19}Ni_9$. The standard also stipulates that the agitator should be capable of uniform mixing and not damage either FRK or raw rice. The blender is required to be provided with a suitable magnet to trap the metal impurities. The blending unit should have a control system that is capable of feeding the FRK and rice at a preadjusted rate with a display to control process. The blended output should meet the standards set by IS 17780, ensuring compliance with regulatory requirements and maintaining product integrity (BIS 2021b).

12.7 CHALLENGES IN FRK MANUFACTURING AND BLENDING

Despite significant advancements in technology and supportive policies, manufacturers of FRK face numerous challenges in practice. These challenges can be broadly categorized into technical issues inherent in the fortification process, regulatory compliance demands, and gaps in quality control on the ground. This section discusses these hurdles in detail:

12.7.1 Technical Challenges

Fortification of rice via extrusion and blending is technically demanding. Maintaining consistent product quality and nutrient levels requires careful control, yet in real-world operations, variability and practical issues often arise. Some of the major technical challenges include process variability and consistency, nutrient loss during extrusion, micronutrient premix integration and segregation, blending uniformity and equipment calibration and equipment reliability, and scale-up issues (Table 12.2).

FRK quality depends on flour properties, extrusion, and drying, which affect size, density, and structure. Consistency is essential for blending and nutrient retention, with BIS standards requiring kernels to match local rice. This demands precise extruder control and measures against brittleness.

Hot extrusion embeds micronutrients, but degrades heat-sensitive vitamins like folic acid and vitamin B_{12} through heat and shear. Overages and process optimization help meet targets, yet losses may occur during storage or cooking. Poor encapsulation also increases leaching, highlighting the need for improved extrusion and encapsulation methods.

Uniform micronutrient distribution requires thorough VMP–flour blending, aided by mixers, binders, and strict process control. FRK must visually match rice and blend at 1%, but large-scale uniformity is challenging. Automatic blending machines with metered feeders support continuous dosing but need frequent calibration

TABLE 12.2
Common Technical Challenges in FRK Manufacturing and Potential Mitigations

Technical Challenge	Potential Mitigation Strategies
Process variability in extrusion	Strict process controls (PLC/SCADA monitoring), standardized operating procedures; use of preconditioners to stabilize input; frequent sampling of kernel quality
Vitamin degradation at high temperature	Optimize extrusion parameters (lower temp, shorter time); use vitamin overages; incorporate heat-stable vitamin forms or protective coatings; switch to warm extrusion if feasible
Premix segregation in mixing	Use high-efficiency mixers; add binders or moisture to improve adhesion; ensure premix particle size compatibility with rice flour; minimize transfer steps to avoid settling
Blending non-uniformity (1:100 ratio)	Employ BIS-certified automatic blenders with load cell feedback; perform hourly blending efficiency tests; calibrate feeders regularly with test runs; use kernels that mimic rice size/density to prevent sorting
Equipment calibration and wear	Schedule routine calibration of weighing systems; implement preventive maintenance (e.g., periodic replacement of extruder screws and cutter blades); train operators for the early detection of mechanical issues (noise, heat, etc.)
Small-scale production constraints	Explore smaller, cost-effective extruders for cluster-level production; share equipment among miller cooperatives; seek government subsidies or financing for machinery upgrades

due to drift from humidity. SOPs in India mandate hourly blending checks, while BIS-standardized blenders (IS 17854:2022) enhance accuracy.

Operational issues include extruder wear under high torque, dryers struggling in humid climates, and mechanical stress during large-scale mixing. High-precision equipment costs also remain a major barrier for small millers.

12.7.2 Regulatory Compliance

FR production in India must comply with FSSAI and BIS regulations, covering nutrient levels, premix quality, packaging, labeling, traceability, and audits. FSSAI's Fortification of Foods Regulations specify nutrient ranges, for example, 28–42 mg iron, 75–150 μg folic acid, and 0.75–1.5 μg vitamin B_{12} per kg rice. Since FRK is usually blended at 1:100, producers must formulate kernels so the final rice meets these standards, which are verified by lab testing and adjusted with overages for stability. In 2023, the FSSAI added stricter premix standards to ensure consistent quality.

BIS has issued standards for both FRK and equipment: IS 17782:2021 (FRK quality), IS 17780:2021 (fortified rice), IS 17853:2022 (extruders), and IS 17854:2022 (blending machines). The Food Corporation of India (FCI) has mandated that the FRK producers using BIS-marked equipment can supply FR and FRK to government schemes like PDS/ICDS. A 2022 directive required all producers to transition to certified machinery, with compliance ensured through licensing, inspections, and conformity tests.

12.8 FUTURE SCOPE

As India expands rice fortification to combat micronutrient deficiencies, future FRK production will move toward smarter, decentralized, and nutrient-responsive technologies. Building on current extrusion systems, the next generation will integrate AI and IoT for closed-loop control, using real-time sensor data to adjust screw speed, barrel temperature, and moisture feed. These predictive systems will minimize manual error, optimize vitamin retention, and ensure consistent quality across batches.

Alongside automation of processes, digital traceability using blockchain technology holds vast potential. Every batch of FRK and FR can be assigned a tamper-proof digital identity with production, blending, testing, and distribution data throughout the value chain. With cloud-powered SCADA, along with mobile dashboards, not just this would ensure regulatory transparency, but enable even consumers to verify nutrition quality by scanning a retail or PDS pack's QR code. Blockchain integration with biosensors and decision-support systems can enable real-time nutrient profiling in the near future when a handheld portable device can deliver instant data on iron and vitamins, enabling quality assurance right up to remote locations without lab infrastructure. Putting intelligence into machines, making supply chains traceable, and injecting precision into nutrition delivery not only can allow India to strengthen its public health reports, but also can make it a global leader in fortification technology.

12.9 CONCLUSIONS

FRK production is a key technology intervention for controlling micronutrient deficiency, i.e., iron-deficiency anemia and folic acid and vitamin B_{12} deficiency in India. Rice fortification, through policy directives, FSSAI and BIS norms, and pan-India roll-out through government programs, has emerged as a viable, large-scale, and sustainable public health intervention. The key to its success lies in utilizing specialist, standardized, and automated plant machineries. These machines ensure uniform delivery of nutrients, food safety processing requirements, and large-scale production at a national level. FRK production in the future would be guided by artificial intelligence trends, smart diagnostics, real-time nutrient analysis, and blockchain traceability. All these would aid in ensuring consistency in quality, allowing operational autonomy, and improving access to fortification, particularly in rural and decentralized settings. A food processing technology and digital technology integration would enable India to lead the world in fortified food production. Ultimately, robust plant design, new technology, and data-driven process control are a solid basis for a successful FR program, so each grain that makes it to market can contribute to a healthier, more nourished populace.

REFERENCES

Basinskas, G., and M. Sakai. 2016. "Numerical Study of the Mixing Efficiency of a Batch Mixer Using the Discrete Element Method." *Powder Technology* 301: 815–829.

Bureau of Indian Standards (BIS). 2021a. "IS 17782: 2021: Fortified Rice Kernels Specifications." https://standardsbis.bsbedge.com/BIS_searchstandard.aspx?Standard_Number=IS+17782&id=34819

Bureau of Indian Standards (BIS). 2021b. "Fortified Rice-Specification (IS 17780: 2021)." https://standardsbis.bsbedge.com/BIS_SearchStandard.aspx?Standard_Number=17780&id=0

Bureau of Indian Standards (BIS). 2022a. "Equipment for Manufacture of Fortified Rice Kernel — Specification (IS 17853 : 2022)." https://standardsbis.bsbedge.com/BIS_searchstandard.aspx?Standard_Number=IS+17853&id=35289

Bureau of Indian Standards (BIS). 2022b. "Equipment for Manufacture of Fortified Rice — Specification (IS 17854: 2022)." https://standardsbis.bsbedge.com/BIS_searchstandard.aspx?Standard_Number=IS+17854&id=35288

Chongchitpaisan, P., and S. Sudsawat. 2022. "A Review on Screw Conveyors for Bulk Materials in Various Applications." *Ladkrabang Engineering Journal* 39 (2): 1–12.

Dadhaneeya, H., P. K. Nema, and V. K. Arora. 2023. "Internet of Things in Food Processing and Its Potential in Industry 4.0 Era: A Review." *Trends in Food Science and Technology* 139: 104109.

Dalbhagat, C. G., and H. N. Mishra. 2019. "Effects of Extrusion Process Conditions on System Parameters; Physicochemical Properties and Cooking Characteristics of Extruded Fortified Rice Kernels." *Journal of Cereal Science* 89: 102782.

Dalbhagat, C. G., and H. N. Mishra. 2020. *Studies on Manufacture of Micronutrient Fortified Rice Kernels Using Extrusion Technology*. Indian Institute of Technology Kharagpur.

Dalbhagat, C. G., and H. N. Mishra. 2021a. "Drying Modeling, Cooking Characteristics, Pasting Properties, and Crystallinity of Fortified Rice Kernels." *Journal of Food Processing and Preservation* 00: e15579.

Dalbhagat, C. G., and H. N. Mishra. 2021b. "Effect of the Drying Process on the Color Change, Fissure Development, and Morphology of Fortified Rice Kernels." *Journal of Food Process Engineering*: e13719.

Dalbhagat, C. G., A. Nithya, S. Mandliya, S. Vishwakarma, and H. N. Mishra. 2024. "Effect of Microwave Drying and Tempering on Color Attributes, Fissure Formation, and Cooking Characteristics of Fortified Rice Kernels." *Journal of Food Science and Technology* 61 (4): 706–716.

Dhankhar, P. 2014. "Rice Milling." *IOSR Journal of Engineering* 4 (5): 34–42.

Ganachari, A., U. Nidoni, S. Hiregoudar, K. T. Ramappa, N. Naik, S. Vanishree, and P. F. Mathad. 2022. "Development of Rice Analogues Fortified with Iron, Folic Acid and Vitamin A." *Journal of Food Science and Technology* 59 (9): 3474–3481.

Gupta, A., A. Deokar, L. Iyer, R. Sharda, and D. Schrader. 2018. "Big Data and Analytics for Societal Impact: Recent Research and Trends." *Information Systems Frontiers* 20 (2): 185–194.

Kondakci, T., and W. Zhou. 2017. "Recent Applications of Advanced Control Techniques in Food Industry." *Food and Bioprocess Technology* 10 (3): 522–542.

Mohidem, N. A., N. Hashim, R. Shamsudin, and H. Che Man. 2022. "Rice for Food Security: Revisiting Its Production, Diversity, Rice Milling Process and Nutrient Content." *Agriculture* 12 (6): 741.

Režek Jambrak, A., M. Nutrizio, I. Djekić, S. Pleslić, and F. Chemat. 2021. "Internet of Nonthermal Food Processing Technologies (IoNTP): Food Industry 4.0 and Sustainability." *Applied Sciences* 11 (2): 686.

Wimberly, J. E. 1983. *Technical Handbook for the Paddy Rice Postharvest Industry in Developing Countries*. Los Baños, Laguna,: International Rice Research Institute.

13 Supply Chain Management of Fortified Rice Kernels and Fortified Rice

Sachin Kumar, Bindu Bala, and Sujosh Nandi

13.1 INTRODUCTION

Milling practices that are conventional remove essential micronutrients, which make fortification mandatory for iron, zinc, and vitamins. This is important for the people in Asia who are dependent on rice as their primary source of nourishment (Nguyen et al. 2024). To reduce "hidden starvation" (Nguyen et al. 2024; Nandeep et al. 2024), rice should be transformed with some kind of micronutrients such as iron, folic acid, and vitamins B_1, B_2, B_6, and B_{12}. Both fortified rice (FR) and fortified rice kernels (FRK) are effective solutions for a society where micronutrient deficiency is a major concern. The two main components of the production of FR are the rice itself and a premix of micronutrients that are added to it. To enhance the nutritional value of FR, premix is critical, but a major concern is the bioavailability of micronutrients during processing, storage, and distribution. For the FR schemes to be successfully adopted in developing countries like India, many stakeholders including government departments, private players, owners of rice mills, and logistics service providers need to collaborate. The major roles of government are policy enforcement, regulatory supervision, and welfare scheme-based distribution. Other than this, FCI has the responsibility of supplying along monitoring food safety measures uninterruptedly (Nandeep et al. 2024). It is also the private sector that provides food fortification enhancement with the help of novel techniques, building capacity, and developing biofortification for iron enrichment via refined breeding methods (Singh et al. 2023). According to Nandeep et al. (2024), logistics providers should ensure that the supply chain is integral and intact by transporting FR to the users in an efficient way as much as possible. Thus, communication between these partners is very important in resolving challenges such as the allocation of resources and regulatory compliance.

Though promising, the FR value chain is saddled with major challenges like traceability problems, inefficient management of waste, and logistics issues. Fragmented data structures obstruct the assurance of transparency other than with immediate suppliers, and jurisdictional variations create regulatory difficulties, requiring harmonized standards (Reynolds 2024). The expenditure on implementation and scalability issues also discourages the usage of traceability-improving technologies like

DOI: 10.1201/9781003661245-13

blockchain (Reynolds 2024). Secondly, the reduction of waste measures is dependent on enhancing coordination in the supply chains of retail, where a good flow of information can diminish food loss (Lemos and Castro 2021). To remove these kinds of challenges, the contemporary fortification supply chain has adopted digital transformation through artificial intelligence (AI), blockchain, and the Internet of Things (IoT). Through predictive analytics and live data processing, AI increases operational effectiveness, which contributes to a 20% improvement in supply chain visibility and a 10% increase in the satisfaction level of customers (Nweje and Taiwo 2025). Blockchain increases the traceability of the supply chain with immutable records that minimize inefficiencies and help mitigate fraud risks, as exemplified by Walmart's blockchain pilot on vendor management and risk minimization (Kumar et al. 2024). According to Faiyaz et al (2024), real-time monitoring and data-informed decision-making, which enhance responsiveness to market changes, can be achieved by combining blockchain with Industry 4.0 technologies like IoT and cloud computing. Research and adjustment are needed for scalability, but regulatory limits and technology adoption resistance remain challenges (Quttainah et al. 2024).

Therefore, the effectiveness of FR programs depends on a well-organized, transparent, and resilient supply chain. However, the integration of FRK and FR into public distribution systems has been done, and challenges in areas of regulatory compliance, traceability, and waste management remain to affect overall efficiency. With advanced technologies and improvement in stakeholder coordination, supply chain visibility and operation efficiency can be greatly improved. Closing these gaps with innovation, policy action, and partnership will be essential for enhancing food security and the long-term sustainability of FR fortification programs. This includes overcoming difficulties such as how to ration resources and abide by rules and policies.

This chapter illustrates the entire overview of processes, stakeholder engagement, issues, and technological innovations in the supply chain of FRK and FR, as well as projects the key steps to be eligible for future supply chain improvement.

13.2 RAW MATERIAL PROCUREMENT AND LOGISTICS

In both the FRK and FR supply chains, buying raw materials is a crucial asset. It directly influences how efficient, reliable, and successful the business is. Companies secure a steady supply of rice and essential micronutrient premixes when they plan their purchase, which helps them avoid delays and disruptions. Good planning also helps to cut costs and improve overall performance. For example, at PT XYZ, smart procurement strategies led to a Perfect Order Fulfillment rate of 88.36% (Sinaga et al. 2023).

As per Dong (2022), for reducing risks like supply shortages, price changes, and delivery delays, good procurement also plays a big role. Management of these risks is critical, especially in industries where running out of raw materials can entirely break the production flow. The House of Risk method is a convenient tool that aids in recognizing and dealing with procurement-related risks effectively (Maulana and Arvitrida 2022). In addition, real-time tracking systems are helpful in improving the coordination among different stages of supply chain, which, therefore, makes it easier to identify problems early and act accordingly. However, the food processing sector encounters various challenges, including complex contracts, poor infrastructure, and tough regulations that make procurement harder (Dua and Sahu 2024). Therefore, it is necessary to tackle these issues with robust planning and supply chain strategies to develop a more reliable procurement process.

13.2.1 Supplier Selection and Quality Control

Maintaining rules and regulations associated with food safety is essential in supply chain of rice because about 15% of the rice produced worldwide is affected by mycotoxins such as aflatoxins and fumonisins (Chandravarnan et al. 2024). Therefore, robust safety measures before and after harvesting of rice are needed. In this context, selection of supplier becomes major challenge for government and rice millers in India. Researchers have proposed various statistical tools like FUZZY MCDM, AHP, and TOPSIS to select supplier after assessing the quality of supplied rice, reliability, and supplier's previous background of supplying rice (Magableh 2024; Prasetya and Donoriyanto 2024). Another technique, namely stable isotope ratio analysis, can also be used to evaluate types of pesticides and their quantity (Chi et al. 2024). In the case of FRK and FR, regular audits of FRK manufacturing facilities and testing of quality of FRK and FR can be useful to identify purity and stability of ingredients. Therefore, careful implementation of these techniques can be helpful in the selection of right supplier (Mejia et al. 2019; O'Dwyer and Vegiraju 2020) with better quality FRK and FR.

13.2.2 Procurement Challenges and Risk Management

There are a lot of procurement challenges, such as contract complexity, supply chain inadequacy, and lack of infrastructure, that are continuously disrupting the supply chain in the food processing sector (Dua and Sahu 2024), and the FR sector is not an exception. The major problem in this sector is insufficient and costly raw materials for production of vitamin-mineral premix and FRK, which consequently halt the production and supply of FR to the end consumer. Therefore, the implementation of layered risk management strategies from the very early stages of FR supply chain is necessary (Maulana and Arvitrida 2022) to distribute high-quality FR among vulnerable population. Further, the implementation of new statistical tools namely blockchain, can make the buying process more transparent. On the contrary, automated inventory systems can be utilized more effectively to keep raw materials stocked constantly (Reynolds 2024). With these steps in place, the FR industry can reduce weak points in the supply chain and keep production running smoothly.

13.2.3 Logistics and Supply Chain Optimization for Raw Materials Procurement

For transporting raw materials and FR to production services and distribution points, well-organized logistics and transportation systems are critical. Predictive modeling methodologies, such as Long Short-Term Memory (LSTM) neural networks, enhance the effectiveness of transportation through real-time route adaptation according to demand variance and environmental conditions (Wang and Li 2024). Even delivery performance can be improved by balancing travel time, cost, and congestion and by finally eliminating inefficiencies (Papară and Schirliu 2024) for which a multi-objective optimization model can be used. Practices like Total Quality Management (TQM) even minimize delays by tackling fleet maintenance and transportation bottlenecks, enhancing responsiveness in the supply chain. Notwithstanding these developments, challenges such as high costs of implementation, technology adaptation, and variability in the market remain (Anaba et al. 2025). Therefore, to enhance the long-term efficacy, sustained investment in automation, predictive analytics, and intelligent supply chain solutions is required.

13.2.4 Manufacturing and Processing Integration

For the nutritional quality and economic viability in the production of FR, a perfect integration of manufacturing and processing depends on technological innovation, resource utilization, and supplier partnerships. Extrusion, parboiling, and impregnation are some modern processing techniques that enhance the storage and handling of rice while also retaining its micronutrients. Integrated Resource-Efficient (IRE) rice mill established complexes to utilize the by-products of rice. It separates raw material (broken rice) and generates energy (Lim et al. 2013).

These complexes reduce waste, maximize water and energy use, and ultimately minimize environmental effects. They also enhance the sustainability and cost-effectiveness of FR production. Effective employment depends on infrastructure investment and industrial uptake. For the development of products, quality checks, and process optimization, supplier participation is required. There are limitations in the socioeconomic sector such as limited accessibility to advanced technologies and gaps in capacity building. These limitations halt integrated collaboration between suppliers, millers, and processing units. These issues can be overcome by targeted programs initiatives, better regulatory systems, and stronger ties with industry, which are important for successfully expanding rice fortification production.

13.2.5 Coordination between FRK Manufacturing and Rice Blending Units

The coordination of FRK production and rice blending unit is important to provide a uniform, high-quality FR product. The guarantees have been produced that nutrients are evenly spread in the final product based on the recommended mixing ratio of 1:100. The automated blending equipment promises accurate mixing during large-scale manufacture and enhances the distribution and consistency of nutrients. Segregation of micronutrients is avoided in effective blending methods so that the nutritional values of FR are preserved during cooking and storage.

13.3 STORAGE AND INVENTORY MANAGEMENT

13.3.1 Handling and Warehousing of FRK and FR

For maintaining quality handling of FRK, FR plays a critical role in maintaining quality. It guarantees nutritional stability and minimizes post-harvest losses. Appropriate inventory administration is required to guarantee the availability of products and the reduction of loss from spoilage, pests, and climatic factors. Moreover, automated rice discharge systems increase working efficiency effectively through decreased manual labor and efficient inventory flow (Holah 2023; Müller et al. 2022). Supply and demand are optimized by an efficient inventory management system that optimizes supply and demand while keeping costs low.

Economic Order Quantity (EOQ) is commonly applied to establish maximum order quantities to minimize excess stock and storage expenses (Tatuh et al. 2015). Alternatively, real-time monitoring systems allow for the monitoring of stock levels, avoidance of waste, and time-efficient replenishment (Mustafa et al. 2024).

13.3.2 Automation in Storage and Inventory Management

Inventory management transformation has been done with automation, which boosts efficiency, minimizes loss, and boosts compliance with protocols. For example, robotic arms and conveyor belts of Automated Storage and Retrieval Systems (AS/RS) in stores fetch rice effectively, which aids in minimizing physical work and human error (Müller et al. 2022). IoT-based Warehouse Monitoring systems can be executed in godowns for the monitoring of temperature, humidity, and inventory levels to enhance transparency and predictive maintenance (Zaidi et al. 2024). Sharma et al. (2023) showed that the blockchain for inventory traceability can guarantee tamper-proof data records and fulfill food safety regulations. Nonetheless, there are impediments to complete automation such as high investment costs, cybersecurity threats, and training needs for workers. With the growing attention of global supply chains toward intelligent storage solutions, overcoming these issues is critical to achieving mass adoption of automation in FR and FRK supply chains.

13.3.3 Inventory Control Strategies

The supply of FR is ensured by optimal inventory management. It also reduces the chances of surplus inventory, saves costs, and enhances business proficiency. However, monitoring stock ages manually is time-consuming and laborious. For this problem, computerized inventory systems can be used to enhance the accuracy and effectiveness of inventory management plans by implementing different state-of-the-art technologies like Just-In-Time (JIT), EOQ, First-In-First-Out (FIFO), batch tracking, and real-time inventory monitoring (Roosevelt et al. 2024). Furthermore, AI improvement, predictive analytics, and digital monitoring systems improve inventory processes, which are made smooth with seamless demand-supply matching. According to Malakooti (2013), a 30% reduction in wastage was done by FIFO while keeping products fresh. Tracking and traceability of batch and products are equally effective methods for maintaining efficient inventory control by documenting production and distribution data for every batch. This method can be utilized for FR and FRK supply chains because it was found effective for product recalls and quality control in food processing (Mittal and Shah 2016).

13.3.4 Real-Time Inventory Monitoring

Visibility, over-stock balance control, movements, and replenishment cycles can be traced by real-time inventory monitoring (Mittal and Shah 2016). IoT sensors and cloud computing can be used to monitor inventory conditions in real time and enhance the productivity of warehouses (Roosevelt et al. 2024). On the other hand, predictive analytics technique along with AI can maximize demand forecasting and warehouse space utilization (Granillo-Macías 2020). Therefore, incorporation of these techniques in real-time inventory management of FR and FRK can be helpful to improve the overall supply chain system. However, a continuous increase in dependency on digital inventory solutions poses cybersecurity threats, and thus, developing a strong data protection system is required to avert breaches (Roosevelt et al. 2024).

13.4 DISTRIBUTION AND TRANSPORTATION CHALLENGES

13.4.1 Ensuring Integrity of Fortified Rice during Transit

The transportation of FRK and FR is challenging considering their high chances of breakage and contamination. Therefore, innovative packaging and handling methods need to be implemented to reduce losses and maintain product integrity. One of such key methods used includes 360-degree revolving support mechanisms alongside fixing disks, which provide support to various delicate food products in place during transportation irrespective of the alignment of their packaging. This system also minimizes internal collision, breakages, and contamination risk within the containers. Moreover, the application of controlled elasticity cushioning materials can absorb shock and excess vibrations to avoid cracks and deformations. Apart from the physical protection, appropriate environmental conditions (temperature and relative humidity) must be maintained during the supply chain of FRK and FR. For instance, temperature- and light-sensitive FR needs to be transported in temperature-controlled containers to restrict degradation of nutrients (Maleš 2020). For protecting rice and keeping nutrients intact from extreme temperature fluctuations, advanced packaging technologies are being used, which prevent environmental and mechanical damage (Singh and Singh 2018). It is very important to use a good support system to minimize bulk packaging losses. During movements, damage happening to packages can be reduced by using soft padding and strong straps. Monitoring of products by tracking systems can fix transportation problems in real time (Singh and Singh 2018), but these methods need to be cost-effective.

13.4.2 Cold Chain Requirements and Bulk Logistics

The stability of FR throughout bulk transportation and long-term storage required stable storage conditions. It has been observed in various studies that the shelf life and micronutrient viability greatly increased under controlled environmental conditions. Therefore, the controlled temperature played an important role in minimizing biochemical deterioration in FRK and FR.

A recent FRK and FR storage study conducted at IIT Kharagpur highlighted that the temperature above 25°C speed up the micronutrient losses; therefore, the desired temperature for storage of FRK and FR should be below 25°C. Another study showed that 40°C±5°C temperatures lead to the extensive deterioration of vitamin A, while at the moderate temperature of 25°C±5°C, essential micronutrients such as iron and zinc remain intact more effectively (Kuong et al. 2016). Another factor such as humidity should also be taken into consideration. The recommended relative humidity for storage space is between 60% and 70% since values above 75% lead to the growth of mold and nutrient degradation (Kuong et al. 2016). Aeration systems and dehumidifiers keep the storage area at an accurate temperature and humidity (Müller et al. 2022). These techniques help to prevent spoilage, but nutrient breakdowns still happen over time. Therefore, monitoring at regular intervals of time and smart control are required to keep rice safe and maintain its nutrients intact.

13.5 REGULATORY COMPLIANCE IN SUPPLY CHAIN OPERATIONS

13.5.1 Alignment with Food Security and Fortification Policies

In supply chain operations, regulatory compliance is crucial for the maintenance of consumer safety, product efficacy, and alignment with national food security and fortification policies. A supply chain that is well-regulated helps in strengthening stakeholder trust and food production sustainability. Safe and healthy food is ensured by rules and standards that identify the quality levels (Shi 2024). Food Safety and Standards Authority of India (FSSAI) in India oversees fortified policies and checks whether FR meets the required nutrition and safety limits. This quality check is done by the Food and Drug Administration (FDA) in the United States. At the worldwide level, the World Health Organization (WHO) and the Food and Agriculture Organization (FAO) set global standards and rules for countries. These organizations deliver guidance through systems like the Codex Alimentarius, which acts like a global book of food rules (Cormick et al. 2020).

Apart from regulatory compliance in the quality and safety of FR, regulatory compliance on adaptation in digital traceability is another serious factor. Hazard Analysis and Critical Control Points (HACCP) and Good Manufacturing Practices (GMP) are used by businesses to comply with global food safety standards (Kumar et al. 2024).

Furthermore, in the case of contamination, traceability strengthens recall mechanisms. Businesses can rapidly recognize affected batches, reducing health risks and financial losses (Wu and Zou 2023). Case studies from OECD countries done by Charlebois et al. in (2024) demonstrate how active traceability systems increase recall management and reduce risks related to food safety.

However, in developing countries where the infrastructure is limited and implementation costs are high, these traceability technologies face challenges (Arunfred and Bini Marin 2023). Financial assistance in digital infrastructure and effective policies is required to expand access to traceability systems, refining global food safety and post-harvest management.

13.6 PUBLIC DISTRIBUTION SYSTEM (PDS) AND PRIVATE MARKET LINKAGES

13.6.1 Supply Chain Integration with Government Food Programs

Among vulnerable groups, the Indian government implements welfare programs such as the PDS, mid-day meals, and maternal nutrition programs to distribute FR to combat micronutrient deficiency. But there are certain issues such as ineffective supply chains, corruption, and poor-quality food grains, which raised doubts about its proficiency. Poor grain quality, irregular food supply, and illegal sales cause dissatisfaction among beneficiaries and weaken the system's integrity (Pawade et al. 2023). Multiple intermediaries in the supply chain create complexity in the monitoring processes and increase the risk of fraud (Pawade et al. 2023). Therefore, effective solutions like incorporating private market linkages are effective to mitigate these inefficiencies. Involvement of the private players can help in increasing service delivery, supply chain efficiency, and minimizing operational wastages (Prakash 2018). Blockchain technology can be used to increase transparency and accountability in the supply chain of FRK and FR. It also helps in minimizing wastage and better auditing mechanisms (Valarmathi 2024; Pawade et al. 2023). The COVID-19 pandemic facilitated the use of digital tools in the PDS and simplifying the system and reducing wastage (Kumar 2021). Along with transparency, the improvement in distribution efficiency is also important. Researchers indicated K-Means clustering can be adopted for route planning, which consequently minimizes the delivery cost and increases the service efficiency (Wang 2025). Although innovative technologies play their role very well, problems like limited infrastructure and their expansion must be taken into consideration to improve the PDS supply chain.

13.6.2 Role of Private Players in Distribution

The involvement of the private sector in the distribution of FR increases distribution efficiency by enhancing its availability through organizational buying, commercial outlets, and digital platforms. This participation is important for the regions where government efforts alone are insufficient to meet demand. Thus, government can establish sustainable systems for the production and delivery of FR along with private players (Forsman et al. 2014). Previously, Indian private sector-driven value chains have demonstrated the capability to supply fortified food, but their success largely depended on strong institutional support to ensure distribution to low-income households (Parasar and Bhavani 2018). Therefore, a strong collaboration between private industry and public institutions is required with immediate effect to make a big success of the food fortification initiatives. Programs like Technoserve's SAPFF in Africa showcase how capacity-building initiatives and technical assistance can empower the industry to meet national fortification standards effectively (Durotoye et al., 2022). Moreover, private investment in rice fortification across Southeast Asian countries has brought in capital, advanced technology, and managerial expertise, which further improves supply chain efficiency and reduces post-harvest losses (Shepherd et al. 2011).

Besides, private companies can play a crucial role in distributing FR among malnourished child, pregnant women, and other micronutrient deficient people. In this context, a field experiment was conducted in Indonesia, which showed that the participation of large number of private industries could reduce markups and improve product quality. In addition, competition among private distributors further reduces costs and improves service quality (Banerjee et al. 2016). However, private sector-controlled distribution requires proactive government oversight and regulation to ensure that nutritionally vulnerable populations are getting adequate quantity of FR. Therefore, it is essential to maintain a balance between commercial objectives and public health priorities to maintain quality and assure that the benefits of fortification are distributed across all segments of the population (Parasar and Bhavani 2018).

13.7 CONCLUSIONS

The management of supply chains for FRK and FR's is vital for reducing micronutrient deficiencies, strengthening food security, and maintaining product quality throughout the entire process from procurement to final delivery. Adequate supply chain strategies should be acquired to ensure that FR remains safe and stable when distributed through public welfare Programs like the PDS, mid-day meals, and maternal nutrition schemes. However, the FR value chain faces several challenges, including quality raw materials, optimizing production processes, improving storage infrastructure, addressing distribution gaps, ensuring regulatory compliance, and adopting advanced technologies. Distributing high-quality rice and maintaining quality of micronutrient premixes are essential for effective fortification. Additionally, maintaining quality of FR and FRK requires the use of advanced verification techniques, such as stable isotope ratio analysis. Multi-criteria decision-making (MCDM) approaches are also useful in supplying superior quality FR to the consumers. In the production of FR, extrusion has emerged as the most effective technique for incorporating micronutrients into rice because it offers improved efficiency in nutrient preservation and absorption.

Further, storage conditions play an important role in stability of nutrient content in FR. The exposure of FRK and FR to high temperature and light can be a cause of rapid degradation of vitamin content. Therefore, adequate temperature and relative humidity of the storage place should be maintained to enhance nutrient stability along with minimal light exposure. During storage, another important factor is effective inventory control. Here, various concepts, including Just-In-Time (JIT), EOQ, and real-time tracking can be applied to optimize stock levels and reduce wastage.

Further, these concepts can be merged with cutting-aged technologies (RFID tagging, IoT-based monitoring, and AI-based forecasting) to improve the inventory management. Moreover, during distribution and transportation, there is a chance for FR and FRK to get contaminated and exposed to unwanted environmental conditions, which may negatively affect the supply chain. The use of refrigerated transport systems, smart cushioning materials, blockchain-based tracking technologies, and AI-based route optimization can be implemented to improve the efficiency of FR and FRK distribution without deteriorating the quality of the products. In India, PDS encounters several challenges, including corruption, ration diversion, and inconsistent supply of quality products. In this context, private market linkages, digitization, and blockchain technology can play a crucial role to diminish these challenges along with increasing transparency and accountability.

Maintaining food safety in all stages of the FR supply chain is crucial for acquiring consumer trust. In this context, national and international regulatory bodies such as WHO, FAO, FSSAI, and the FDA have proposed various guidelines to maintain the safety of fortified food during production and distribution. However, there is a lack of strategies to adequately enforce those rules and regulations in supply chain management. Technologies such as electronic documentation, batch level tracking, and blockchain-based verification systems can be utilized in supply chain management by strengthening regulatory oversight and enhancing traceability. In addition, regular testing of quality of FRK and FR has to be reinforced through the application of GMP/GHP and microbiological safety and nutrient content assessments. However, due to high cost of audits, it is difficult for small-scale producers to do audits on a regular basis. The FR supply chain is expected to evolve toward increased digitalization, automation, and sustainability. Real-time monitoring of storage conditions is aided by IoT sensors, while AI is used to manage inventory and predict potential spoilage risks. Besides all these advantages, widespread adoption is still constrained by high implementation costs, improper regulatory frameworks, and lack of skilled labor in FR supply chain. However, through public and private sector collaboration, it is possible to implement AI-powered cutting-edge technologies to improve infrastructure, enhance quality assurance systems, and control storage conditions, which can further develop a robust supply chain for FR aligned with broader goals of public health and food security.

REFERENCES

Anaba, D. C., Agho, M. O., Onukwulu, E. C., and Egbumokei, P. I. 2025. Transforming Vessel and Fleet Operations in Oil and Gas: A Framework for Integrated Operations Planning and Efficiency Optimization. *Gulf Journal of Advance Business Research* 3(1): 262–281.

Arunfred, N., and Bini Marin, V. 2023. Role of Traceability Systems for Food Safety within Post-Harvest Systems: Indian Context. *IntechOpen*. https://doi.org/10.5772/intechopen.110192.

Banerjee, A., Hanna, R., Kyle, J. C., Olken, B. A., and Sumarto, S. 2016. *Contracting Out the Last-Mile of Service Delivery: Subsidized Food Distribution in Indonesia (No. w21837)*. National Bureau of Economic Research.

Chandravarnan, P., Agyei, D., and Ali, A. 2024. The Prevalence and Concentration of Mycotoxins in Rice Sourced from Markets: A Global Description. *Trends in Food Science and Technology* 146: 104394.

Charlebois, S., Latif, N., Ilahi, I., Sarker, B., Music, J., and Vezeau, J. 2024. Digital Traceability in Agri-Food Supply Chains: A Comparative Analysis of OECD Member Countries. *Foods* 13(7): 1075.

Chi, H. Y., Kim, W. R., Kim, J. Y., and Kim, S. H. 2024. Improved Organic and Pesticide-Free Rice (*Oryza sativa L.*) Authentication Based on Multiple Stable Isotope Ratio Analysis and Rice Milling State. *Heliyon* 10(5): e26725.

Cormick, G., Cormick, G., Betrán, A. P., Metz, F., Palacios, C., Beltran-Velazquez, F., García-Casal, M. N., Peña-Rosas, J. P., Hofmeyr, G. J., and Belizán, J. M. 2020. Regulatory and Policy-Related Aspects of Calcium Fortification of Foods: Implications for Implementing National Strategies of Calcium Fortification. *Nutrients* 12(4): 1022.

Dong, Y. 2022. Optimization and Analysis of Raw Material Supply Chain Based on Computational Intelligence. *Mobile Information Systems* 2022(1): 8683598.

Dua, A., and Sahu, S. 2024. Factor Identification for the Procurement of Raw Material in Food Processing Industry. *Management* 1: 13.

Durotoye, T., Yusufali, R., Ajieroh, V., and Ezekannagha, O. 2022. Building the Commitment of the Private Sector and Leveraging Effective Partnerships to Sustain Food Fortification. *Food and Nutrition Bulletin* 44 (1_suppl): S61–S73.

Faiyaz, M. S., Srivastava, G., Singh, S., and Kumar, S. 2024. Emerging Trend in Supply Chain Blockchain and Industry 4.0. In *Supply Chain Management*, edited by Bharti, Shikha Singh, Anand Pandey, and Amit Sachan, 93–118. CRC Press.

Forsman, C., Milani, P., Schondebare, J. A., Matthias, D., and Guyondet, C. 2014. Rice Fortification: A Comparative Analysis in Mandated Settings. *Annals of the New York Academy of Sciences* 1324(1): 67–81.

Granillo-Macías, R. 2020. Inventory Management and Logistics Optimization: A Data Mining Practical Approach. *LogForum* 16(4): 535–547.

Holah, J. 2023. Minimum Hygienic Design Requirements for Food Processing Factories. In *Hygienic Design of Food Factories*, 2nd Ed., edited by John Holah, Huub L.m. Lelieveld and Frank Moerman, 221–240. Woodhead Publishing.

Kumar, A. 2021. Improvement of Public Distribution System Efficiency Applying Blockchain Technology during Pandemic Outbreak (COVID-19). *Journal of Humanitarian Logistics and Supply Chain Management* 11(1): 1–28.

Kumar, P., Choubey, D., Amosu, O. R., and Ogunsuji, Y. M. 2024. Blockchain and Smart Contracts for Supply Chain Transparency and Vendor Management. *World Journal of Advanced Research and Reviews* 23(2): 039–056.

Kuong, K., Laillou, A., Chea, C., Chamnan, C., Berger, J., and Wieringa, F. T. 2016. Stability of Vitamin A, Iron and Zinc in Fortified Rice during Storage and Its Impact on Future National Standards and Programs—Case Study in Cambodia. *Nutrients* 8(1): 51.

Lemos, F. K., and de Paula Castro, L. 2021. Food Retail as the Coordinator Agent of Food Supply Chain: Challenges and Opportunities for Reducing Loss and Waste. *International Journal on Food System Dynamics* 12(3): 246–254.

Lim, J. S., Manan, Z. A., Hashim, H., and Alwi, S. R. W. 2013. Towards an Integrated, Resource-Efficient Rice Mill Complex. *Resources, Conservation and Recycling* 75: 41–51.

Magableh, G. M. 2024. An Integrated Model for Rice Supplier Selection Strategies and a Comparative Analysis of Fuzzy Multicriteria Decision-Making Approaches Based on the Fuzzy Entropy Weight Method for Evaluating Rice Suppliers. *Plos One* 19(4): e0301930.

Malakooti, B. 2013. Inventory Planning and Control. In *Operations and Production Systems with Multiple Objectives*, 313–380. Wiley. https://ssrn.com/abstract=2923492.

Maleš, K. 2020. Karakteristike i uvjeti prijevoza hlađenih tereta. https://repozitorij.pfst.unist.hr/islandora/object/pfst:1033/datastream/PDF/download.

Maulana, B. W., and Arvitrida, N. I. 2022. Design of Risk Mitigation Strategy for Procurement of Raw Materials to Glass Melting Furnace at PT AMFG. *Jurnal Teknobisnis* 8(1): 56–73.

Mejia, L. A., Kuo, W. Y., and Beltran-Velazquez, F. 2019. Provision of Micronutrients in Coexisting Public Health Programs and Risk of Excessive Intake: Regulatory Considerations. *Annals of the New York Academy of Sciences* 1446(1): 66–80.

Mittal, M., and Shah, N. H., eds. 2016. *Optimal Inventory Control and Management Techniques*. IGI Global.

Müller, A., Nunes, M. T., Maldaner, V., Coradi, P. C., de Moraes, R. S., Martens, S., Leal, A. F., Pereira, V. F., and Marin, C. K. 2022. Rice Drying, Storage and Processing: Effects of Post-Harvest Operations on Grain Quality. *Rice Science* 29(1): 16–30.

Mustafa, M. F. M. S., Namasivayam, N., and Demirovic, A. 2024. Food Cold Chain Logistics and Management: A Review of Current Development and Emerging Trends. *Journal of Agriculture and Food Research* 18: 101343.

Nandeep, E. R., Mahajan, H., Mummadi, M. K., Sairam, C., K, V., Kadiyam, J., Meshram, I., Pagidoju, S., Reddy, V. R., Panda, H., and Pullakandham, R. 2024. Implementation, Delivery, and Utilization of Iron Fortified Rice Supplied through Public Distribution System across Different States in India: An Exploratory Mixed-Method Study. *PLOS Global Public Health* 4(8): e0003533.

Nguyen, T. T. L., Pham, T. M. N., Ho, T. B., and Ly-Nguyen, B. 2024. Optimization of Vitamin B_1, B_2, and B_6 Absorption in Nang Tay Dum Floating Rice Grains. *Foods* 13(17): 2650.

Nweje, U., and Taiwo, M. 2025. Leveraging Artificial Intelligence for Predictive Supply Chain Management: Focus on How AI-Driven Tools Are Revolutionizing Demand Forecasting and Inventory Optimization. *International Journal of Science and Research Archive* 14(1): 230–250.

O'Dwyer, D. D., and Vegiraju, S. 2020. Navigating the Maze of Dietary Supplements: Regulation and Safety. *Topics in Clinical Nutrition* 35(3): 248–263.

Papară, C.-M., and Schirliu, Ș.-H. 2024. Logistics Optimization for Resource Allocation and Scheduling Using Time Slots. *International Journal of Advanced Statistics and ITandC for Economics and Life Sciences* 14(1): 131–146.

Parasar, Rohit, and R. V. Bhavani. 2018. "Private Business-Driven Value Chains and Nutrition: Insights from India". The Institute of Development Studies and Partner Organisations. Report. https://hdl.handle.net/20.500.12413/13515.

Pawade, D., Bandiwdekar, C., Kaulgud, P., Bagwe, S., and Kulkarni, A. 2023. Blockchain-Based Public Distribution System. In *International Conference on Recent Developments in Cyber Security*, 605–617. Singapore: Springer Nature Singapore.

Prakash, G. 2018. Managing Welfare Driven Supply Chains: Insights from the Indian PDS. *International Journal of Intelligent Enterprise* 5(1–2): 70–96.

Prasetya, A., and Donoriyanto, D. S. 2024. Optimization of Selection Suppliers for Rice Raw Material Using AHP (Analytical Hierarchy Processes) and TOPSIS (Technique for Order Preference by Similarity to Ideal Solution) Methods. *Indonesian Journals of Industrial Engineering and Management* 5(1): 176.

Quttainah, M. A., Jain, P., Goyal, R., Kanwal, P., and Kumar, A. 2025. Digitalization and Knowledge Management in Entrepreneurial Supply Chains. In *Impacts of Entrepreneurial Orientation on Supply Chain Management*, edited by Muhammad Nawaz Tunio, Aamir Rashid, Muhammad Asif Qureshi, Jawaid Ahmed Qureshi, and Rizwana Rasheed, 237–260. IGI Global.

Reynolds, S. 2024. Unveiling Supply Chain Transparency and Traceability in the Renewable Energy Sector: Challenges and Opportunities. https://doi.org/10.20944/preprints202406.0150.v1.

Roosevelt, S. C., Veemaraj, E., and Kirubakaran, S. 2024. Real Time Stock Inventory Management System. In *2024 8th International Conference on Inventive Systems and Control (ICISC)*, Coimbatore, India, 156–162. IEEE.

Sharma, R., Tripathi, G., Sharma, P., and Tripathi, V. 2023. Development of Architecture for Digital Supply Chain (DSC) Using Blockchain, IoT and AI Technologies. *International Journal of Recent Technology and Engineering* 12(2): 77–83.

Shepherd, A. W., Htenas, A., Alavi, H. R., Clarete, R. L., and Kopicki, R. J. 2011. Trusting Trade and the Private Sector for Food Security in Southeast Asia. https://openknowledge.worldbank.org/bitstream/handle/10986/2384/659560PUB0EPI105737B0Trusting0Trade.pdf?sequence=1.

Shi, K. Y. 2024. Ensuring Food Safety through Effective Supply Chain Management: Challenges and Suggestions. *MedScien* 1(9).

Sinaga, F. H., Nurmalina, R., and Rifin, A. 2023. Efficiency Analysis of Palm Kernel Oil as a Raw Material in Improving Supply Chain Performance: Case Study at PT XYZ. *Jurnal AGRISEP: Kajian Masalah Sosial Ekonomi Pertanian dan Agribisnis* 22 (02): 427–446.

Singh, B., and Singh, S., eds. 2018. *Advances in Postharvest Technologies of Vegetable Crops*. CRC Press.

Singh, S. K., Sahoo, J. P., Saikia, S., Pradhan, P., Mishra, A. P., Behera, L., Lenka, D., and Panda, N. 2023. Biofortification of Rice (*Oryza sativa* L.)." In *Biofortification in Cereals: Progress and Prospects*, edited by Rupesh Deshmukh, Altafhusain Nadaf, Waquar Akhter Ansari, Kashmir Singh and Humira Sonah, 149–172. Singapore: Springer Nature.

Tatuh, D. S., Katiandagho, T. M., Rengkung, L. R., and Lorraine, W. 2015. Analisis Pengelolaan Persediaan Beras di PT. Semarak Kota Bitung. *COCOS* 6(8).

Valarmathi, V. 2024. Design and Implementation of Public Distribution System Using Blockchain. In *2024 IEEE International Conference on Blockchain and Distributed Systems Security (ICBDS)*, 1–5. Pune: IEEE.

Wang, D. 2025. Application of Improved Binary K-Means Algorithm in Time and Cost Optimization for Regional Logistics Distribution Center Location. *Informatica* 49(6), 103–116.

Wang, Junbo, and Xinshi Li. "Dynamic optimization of transportation networks in logistics using long short-term memory neural networks." In *2025 5th International Conference on Neural Networks, Information and Communication Engineering (NNICE)*, Guangzhou, China, 1220–1223. IEEE.

Wu, D. Y., and Zou, F. 2023. Food Processing Ecological Chain Traceability Management System with Decision Support. In *Proceedings of the 2023 4th International Conference on Machine Learning and Computer Application*, Hangzhou China, 466–472.

Zaidi, S. M. R., Alam, A., and Khan, M. Y. 2024. Enhancing Efficiency in Advanced Manufacturing through IoT Integration. *Engineering Headway* 11: 55–61.

14 Micronutrient (Iron, Folic Acid, and Vitamin B_{12}) Analysis and Homogeneity Testing in Fortified Rice Kernels and Fortified Rice

Traditional and Smart Approach

Siddharth Vishwakarma, Nithya A.,
Malla Vandana, Rahul Patel,
Shubham Mandliya,
Chandrakant Genu Dalbhagat,
and Hari Niwas Mishra

14.1 INTRODUCTION

Micronutrient deficiencies, particularly iron, folic acid, and vitamin B_{12}, are a significant public health issue in the majority of low- and middle-income countries. Rice fortification emerged as an evidence-based public health intervention to prevent these deficiencies. Fortified rice kernels (FRK) are extruded rice kernels fortified with iron, folic acid, and vitamin B_{12} as per the fortification guidelines. Fortified rice (FR) consists of a mixture of FRK and normal rice (NR) in the ratio of 1:100 (Dalbhagat, Mahato, and Mishra 2019; Dalbhagat and Mishra 2020). The micronutrients that are added are usually analysed using conventional methods such as atomic absorption spectroscopy (AAS), high-performance liquid chromatography (HPLC), and liquid chromatography-mass spectrometry (LC-MS). Although such analyses are deemed to be precise and are applied on a routine basis in industries, they are most likely to be time-consuming to carry out, require trained technical experts, and are not appropriate for high-throughput or inline industrial quality control analysis. Therefore,

DOI: 10.1201/9781003661245-14

the FRK industry should consider adopting innovative spectroscopic and imaging methods. Especially, hyperspectral imaging (HSI) has been of significant interest due to the fact that it can provide spatial and spectral information simultaneously, making possible reagent-free, nondestructive, and rapid assessment of micronutrient concentration levels (Marín-Méndez et al. 2024; Dalbhagat and Mishra 2021; Fan et al. 2022).

Homogeneity analysis of FR is required to ensure the right mixing of FRK in the FR based on the recommended blending ratio. The analysis ensures that every part of fortified rice has the intended micronutrient composition, and the nutritional quality of the product is, therefore, assured. Traditionally, the homogeneity analysis has been performed using the povidone-iodine test and the iron spot test, which allow differentiation of FRK and NR based on iodine or ferric ion-induced colour reactions. The tests are, however, time-consuming, sample-destructive, and prone to human-factor-induced errors. Rapid, nondestructive alternatives are, therefore, feasible to overcome these shortcomings. Image processing-based methods, both machine learning (ML) and non-ML-based, have been proved to be great tools in the determination of the FRK distribution in the FR. The methods allow automated, objective, and high-throughput determination under controlled imaging conditions and are, therefore, applicable for laboratory as well as field-level quality control (Knez and Šajn 2020). Thus, this chapter addresses the traditional and recent trends in micronutrient determination and homogeneity testing in rice fortification.

14.2 MICRONUTRIENT ANALYSIS

Accurate determination of vitamin and mineral levels in fortificants is a basic requirement in quality assurance (QA) and quality control (QC) systems, particularly in the manufacture of fortified rice. HPLC and AAS are considered traditional but highly effective analytical instruments for micronutrient analysis. They have been used for several decades and are accepted by regulatory bodies and scientific communities for robustness, specificity, and reproducibility. HPLC is preferred for vitamin analysis due to its reliability, speed, and simplified clean-up steps. AAS and inductively coupled mass spectrophotometer (ICP-MS) are widely used for analysing minerals added to FR.

14.2.1 Iron

Mineral analysis of any food samples begins with initial sample preparation, involving dissolution or homogenisation in the laboratory. This includes collecting a representative test portion and decomposing it through hot digestion assisted by acids or alkaline solutions, sometimes under high pressure. In digestion, elements and inorganic species are dissolved by using concentrated acids, organic or alkaline buffers, or techniques like dry ashing, wet acid digestion, or alkaline extraction. These processes may employ traditional conductive heating or microwave techniques (Poitevin 2016). Microwave digestion as a sample preparation technique for iron determination in food has been reported by Jorhem and Engman (2000). The solution obtained upon digestion can be analysed using any one of the methods, such as ICP-MS, AAS, and atomic emission spectrophotometer (AES). AAS is based on the measurement of absorption of electromagnetic radiation by neutral atoms, whereas AES measures the emission of radiation by excited atoms. Since the emission and absorption spectra are unique to an element, these values may be used to estimate the concentration of trace minerals and metals in food samples. ICP-MS is capable of analysing ultra-trace levels of the minerals and can further be used for the determination of various isotopes of the same element (Miller and Rutzke 2020). The method to estimate iron content in the FRK can be found in FSSAI (2023). Kosse et al. (2001) have described a rapid spectrophotometric method, in which the samples were mixed with an iron-extracting solution, heated in a boiling water bath for 15 min, filtered, and mixed with a chromogen reagent. The iron content was determined upon measurement of the absorbance at 535 nm.

14.2.2 Folic Acid (FA)

Folic acid (FA) could be analysed by microbial, biological, chemical, chromatographic, and radiometric methods, depending on the concentration of folic acid, the type of matrix, and the accuracy required (Radhika et al. 2012). HPLC with various detectors, electrochemical methods, electrophoresis, flow injection analysis, and spectrophotometric methods are some of the analytical methods reported in the literature (Singh 2022). The folic acid estimation method for the FRK can be found in FSSAI (2023). Analytical method development and validation studies for folic acid in fortified rice have been carried out by Silveira et al. (2016) and Mahato et al. (2021). Silveira et al. (2016) validated the method for the analysis of folic acid in fortified rice kernels by high-performance liquid chromatography coupled to a diode array detector (HPLC-DAD), while Mahato et al. (2021) used HPLC with a UV detector for the analysis of folic acid in fortified rice and wheat flour.

14.2.3 Cyanocobalamin

Various methods are used for the analysis and quantification of vitamin B_{12} such as microbiological assays, optical biosensor immunoassays, radio protein binding methods, and HPLC methods (Santos et al. 2023). Official AOAC method used for the estimation of vitamin B_{12} uses *Lactobacillus delbrueckii* and is sensitive to detect up to pg/mL level of vitamin B_{12} (Jägerstad and Arkbåge 2003). A chromatographic method for cyanocobalamin analysis in different food matrices uses solid phase extraction (SPE) followed by liquid chromatography with tandem mass spectrophotometry (LC-MS/MS), as reported by Arvapally et al. (2021). Studies on HPLC determination of cyanocobalamin in food matrices with SPE (Heudi et al. 2006) and immunoaffinity columns (IAC) (Nakos et al. 2017) for purification and concentrations have been reported. Ivarsen et al. (2021) reported a reverse HPLC–UV analysis of the vitamin in fortified rice kernels after purification through IAC. The method to estimate vitamin B_{12} content in the FRK can be found in FSSAI (2023).

14.3 ANALYTICAL METHOD DEVELOPMENT FOR MICRONUTRIENT ESTIMATION

14.3.1 Extraction

The extraction procedure aims to release the bonded vitamins to their free forms before the instrumental analysis (Fatima et al. 2019). Samples are subjected to enzymatic digestion or acid treatments such as sulphuric acid and amylase, perchloric acid, trichloroacetic acid, or hydrochloric acid and taka-diastase treatments before the microbiological or HPLC assays (Zafra-Gómez et al. 2006). The enzymatic digestion releases vitamins bound to proteins and sugars (Lebiedzińska et al. 2008). The methods used to extract vitamins from food matrices into the extraction buffer or solvent are homogenisation, vortexing, heating, microwaving, ultrasonication, etc. Ultrasonic-assisted extraction (UAE) is a well-established analytical technique with a wide range of applications from pre-concentration, separation, and determination of chemical species (Gürkan and Altunay 2016). Autoclaving samples at 100°C–121°C for extraction of various water-soluble vitamins, including folic acid and vitamin $B_{12,}$ has been reported in the literature (Ivarsen et al. 2021).

14.3.2 Purification

For sensitive chromatographic techniques, additional cleaning steps such as centrifugation, ultracentrifugation, affinity chromatography, ion-exchange chromatography, and SPE are required (Santos et al. 2023). SPE cartridges, specifically those with C8 and C18 packing and hydrophobic-coated silica beads, have been proven to be effective for sample cleaning. However, there are concerns regarding selectivity SPE columns in this process. IAC has been employed as a process for purification, and it has been proved to be superior in effectiveness compared to SPE, particularly when analysing vitamin B_{12} in food matrices of high complexity. The primary strength of the IAC purification process is specificity for a specific analyte. For instance, the IAC contains antibodies with sole selectivity for vitamin B_{12}, thus excluding extraneous matter with no affinity to the antibodies. However, the primary flaw of using commercial IACs is its high cost, which makes its use in routine vitamin analysis impractical. Moreover, columns containing a single antibody specific to a single vitamin preclude the analysis of multiple vitamins simultaneously (Fatima et al. 2019).

14.3.3 Instrumental Analysis

The majority of official methods available for analysing water-soluble vitamins are spectroscopic, chromatographic, chemical, and microbiological methods (Zafra-Gómez et al. 2006). Among them, HPLC provides improved resolution and sensitivity to most of the B-complex vitamins without any need for sophisticated sample preparation, extraction, or derivatisation procedures (Santos et al. 2023). HPLC and LC methods also exhibit compatibility with mass spectrometry detection, achieve rapid equilibrium, yield superior peaks, and uphold the durability of chromatographic columns. Carbon-packed (C-18) solid phase columns are used most for water-soluble vitamins. Commonly employed mobile phases are a mixture of water and organic solvents like acetonitrile, methanol, ethanol, etc., along with negligible quantity of acidic compounds like trichloroacetic acid (TCA), trifluoroacetic acid (TFA), acetic acid, etc. (Fatima et al. 2019).

14.3.4 Method Validation

Method validation is carried out by the way of laboratory studies to ascertain that the method consistently yields relevant and credible information as per its intended purposes. Validation of any analytical method needs to conform to the International Conference on Harmonisation (ICH) or Eurachem guidelines (Nithya et al. 2023). The general criteria used to verify an analytical method are linearity, accuracy, precision, and sensitivity, including limit of detection (LoD) and limit of quantification (LoQ), range, robustness, and stability (Brouwer et al. 2008).

14.3.5 Linearity and Range

Linearity is determined by the proportional relation between the concentration of the compound of interest and the peak area (Lee et al. 2022). Linearity assesses the degree to which the calibration curve, representing the detector response to concentration, closely resembles a linear relationship within a specified range. Evaluating linearity involves the preparation of standard solutions at six distinct concentrations, typically at 25%, 50%, 75%, 100%, 150%, and 200% of the target concentrations (Djoura 2020).

The range of the method refers to the span of analyte levels, from the lowest to the highest, within which precision, accuracy, and linearity meet acceptable criteria. It is determined based on response curves, linear or nonlinear, and is expressed in the same units as the test results (Radhika et al. 2012).

14.3.6 Accuracy

Accuracy quantifies how closely the results obtained from a method align with the true value, typically expressed as a percentage recovery. Accuracy is determined by spiking a blank sample with three different known concentrations of standard and calculating the percentage recovery (Radhika et al. 2012).

14.3.7 Precision

Precision pertains to the degree of closeness among data values generated through multiple injections under consistent analytical conditions. Precision of a method is expressed in terms of repeatability, intermediate precision, and method reproducibility (Djoura 2020). Precision is determined by assaying a sample by repeated analysis by the method, typically 6–10 times, in order to obtain statistically acceptable results (Radhika et al. 2012).

14.3.8 Limit of Detection and Quantification

LoD refers to the lowest concentration at which the analyte can be reliably detected under the experiment conditions at hand, though not quantitated. Detection limit is most commonly expressed as an analyte concentration (in ppm) in the sample. LoQ refers to the lowest level or concentration of an analyte that is quantitatively measurable with a satisfactory level of precision and accuracy (Ravisankar et al. 2015).

The ICH suggests four calculation methods for LoD and LoQ: visual examination, signal-to-noise ratio, standard response deviation, and standard slope deviation of the linearity plot (Araujo 2009; Ravisankar et al. 2015). LoD is usually defined as the concentration with a signal-to-noise ratio of 3:1, while LoQ is the concentration with a signal-to-noise ratio of 10:1 (Radhika et al. 2012).

14.3.9 Robustness

Robustness indicates a method's ability to withstand minor alterations in its parameters without significant impact. To assess robustness, intentional adjustments are introduced to one or more parameters, which might include variations in solvent strength, wavelength, mobile phase pH, buffer concentration, injection volume, or temperature. Robustness assessment offers insights into the method's dependability under typical operating conditions (Djoura 2020).

14.4 SMART APPROACH USING HYPERSPECTRAL IMAGING

Traditional analysis techniques of iron, folic acid, and vitamin B_{12} content quantification in FRK are AAS, HPLC, and LC-MS. All three are employed in analytical laboratories for micronutrient quantification but have several issues in their application in industries (Cheng et al. 2025). They are time-consuming, involve complex sample preparation, are costly to operate, and require very experienced analysts for appropriate usage and interpretation.

To overcome these limitations, alternative solutions having nondestructive and economical nature are needed that can provide real-time analysis and quality control. Recent advancements in spectroscopic and imaging technology in the form of Raman spectroscopy, near infrared (NIR) and mid-infrared (MIR) spectroscopy, and HSI provide very promising alternatives to the traditional method. Among them, HSI has gained tremendous interest as it involves the integration of imaging and spectroscopy on a single platform and spatial and spectral information acquisition in a broad wavelength range at the same time (Srivastava and Mishra 2022). It allows micronutrient quantification and detection based on their unique absorption and reflectance behaviour without the application of any reagent chemicals and without the need for complex sample preparation. The method allows nondestructive, rapid, and scale-up micronutrient analysis to be applied for industrial purposes (Pu et al. 2019).

14.4.1 Micronutrient Detection Using HSI

HSI provides an innovative, nondestructive technique of micronutrient estimation in FRK. It combines spectroscopy and digital imaging to acquire spatial and spectral information at the same time, obtaining data on the physical, chemical, and biological properties of the sample. In the analysis of FRK, HSI enables fast and accurate quantitation of key micronutrients such as iron, folic acid, and vitamin B_{12} without the need for advanced sample preparation or chemical reagents (Pu, Lin, and Sun 2019).

A typical HSI system used to analyse FRK can include a benchtop visible near infrared (VNIR) imager, such as the one manufactured by Resonon Inc. (Germany). The line-scanning imager records two-dimensional images by scanning the sample stage along the direction of the fixed camera. The four main elements make up the system, which include (a) a Xenon light source, (b) a Pika XC2 hyperspectral camera, (c) an adjustable scanning stage, and (d) a vertical stand used for alignment (Figure 14.1). The process flowchart for the estimation of micronutrients through HSI is illustrated in Figure 14.1.

14.4.2 Preprocessing Methods

Spectral data obtained from HSI are often plagued by intrinsic noise, which negatively impacts the accuracy and reliability of analytic processes. Such noise typically results from instrumentation variability, extraneous environmental conditions, and embedded variability within the sample itself. Therefore, if precision and reliability for spectral analyses are to be enhanced, it is essential that these interferences are removed using preprocessing methods that allow for more effective extraction of relevant chemical information. Several commonly applied preprocessing strategies include multivariate scatter correction (MSC), standard normal variate (SNV), orthogonal signal correction (OSC), detrending, and the Savitzky–Golay derivative filter.

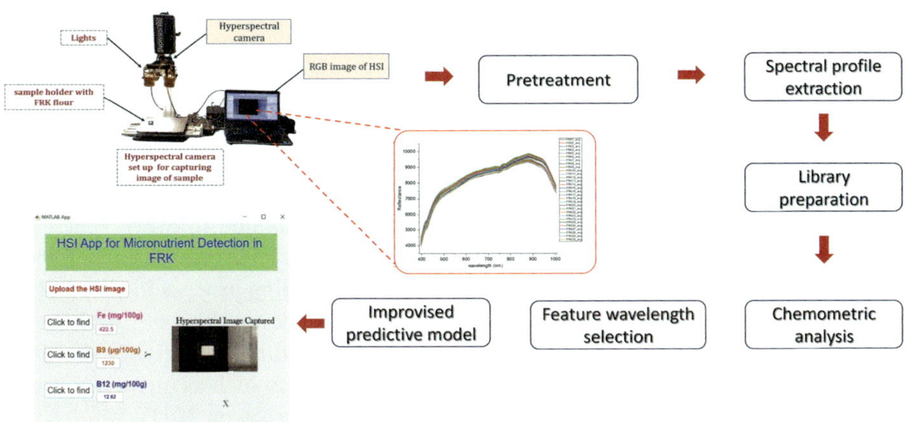

FIGURE 14.1 Process flowchart for micronutrient estimation using hyperspectral imaging system.

Multivariate Scatter Correction (MSC)

MSC is a preprocessing method that has been widely used to counteract the negative impact of light scattering in hyperspectral data, especially for inhomogeneous food constituents. Light scattering results from particle size variation, sample density variations, and surface texture irregularity, all of which negatively impact spectral content but provide no useful spectroscopic information. MSC is effective in removing these multiplicative and additive effects by standardising each spectrum relative to a given baseline, i.e., an average for a population sample. Standardisation increases signal-to-noise and allows spectra to be aligned to facilitate multivariate analysis performed using methods like partial least squares regression (PLSR) or principal component analysis (PCA) (Shao et al. 2020).

Standard Normal Variate (SNV)

SNV is a common method for normalising spectral data in order to minimise multiplicative scatter that exists in spectra. The method normalises every mean value of a spectrum so that all spectra have zero mean and unit standard deviation. The method is especially valuable for eliminating baseline shifts, slope variations, and path length variations, which if eliminated, would mask underlying chemical information. SNV increases spectral homogeneity of samples, which offers greater robustness to calibration models and makes accurate chemical interpretation (Jiang et al. 2023).

Orthogonal Signal Correction (OSC)

OSC forms a unique preprocessing technique aimed at removing orthogonal or extraneous systematics in spectral (*X*) data that are not related to the *Y*-response. OSC pinpoints the dominant non-*Y*-related principle component in the *X*-block and removes this unwanted variation; then, a model is built from the residual data. This is an iterative process that is highly effective in removing unwanted spectral signals and, therefore, improves the predictive accuracy and reproducibility of the model in new datasets. The ultimate model structure, weights, loadings, and scores can be used directly on new datasets, which provides similar predictive results (Sjöblom et al. 1998).

Detrending

Detrending is a preprocessing method employed to eliminate nonlinear trends from spectroscopic data, specifically from hyperspectral data. Detrending is obtained by computing a baseline function for every individual spectrum through the least squares fit of a polynomial. Detrending as a process is unique to every spectrum, unlike transformations done in a uniform manner over wavelengths in a set of spectra. The higher the polynomial order, the better the detrending process in eliminating more baseline effects that can obscure the underlying signal in the data. Addressing such issues facilitates the elucidation of the spectral features underlying the data, and this increases accuracy in tracing analytical procedures such as quantitative and qualitative analysis in hyperspectral imaging (Burger and Gowen 2011).

Savitzky–Golay Derivative

The Savitzky–Golay filter is a digital smoothing filter that is particularly employed to increase spectral resolution in signals without sacrificing valuable features such as peak shape and peak height. It is a local polynomial regression technique applied along moving windows to sample data and then smoothing derivative spectra. Its success has been mainly realised in deconvolution against peak overlapping and sensitivity enhancement for peak detection in complex matrices. Not only does it promote qualitative identification, but quantitative analysis for the targeted analytes is enabled by an increased signal-to-noise ratio (Ruffin et al. 2008).

14.4.3 Model Development

Multivariate calibration modelling is an essential step towards making it possible to quantitate micronutrients like iron, folic acid, and vitamin B_{12} in FRK from spectral data with high precision and speed. Multivariate models define quantitative relationships between the spectral signatures recorded with HSI and reference concentration of target analytes obtained by standard chemical analyses. The chemometric and ML methods can be used for this purpose. Some of the most commonly used methods are PLSR, PCR, and support vector regression (SVR). The methods vary in the way they summarise useful information from high-dimensional spectral data and address multicollinearity, noise, and nonlinearity in the data.

Principal Component Regression (PCR)

PCR is a method of multivariate regression analysis to handle data that is plagued with multicollinearity of the variables. Through the introduction of bias into regression estimates, PCR aims to reduce the standard error and hence increase the precision of the estimate. The standard matrix form of PCR is represented by the Eq. 14.1.

$$Y = \beta X + E \tag{14.1}$$

Where Y is the dependent variable matrix (concentration matrix), X is the independent variable matrix (hyperspectral imaging data), β is the coefficient of regression, and E is the error or the residual matrix.

Partial Least Squares Regression (PLSR)

PLSR is among the widely accepted regression and multivariate analysis methods in spectroscopy, according to Wu et al. (2012). PLSR is run to develop regression models to predict target chemical concentrations (variable matrix Y) from related spectral data (variable matrix X). The PLSR method constructs a linear model by factorising the Y $(N \times M)$ matrix values and X $(N \times K)$ matrix values through the Eq. 14.2–14.3.

$$X = TP^T + E \tag{14.2}$$

$$Y = Tq^T + F \tag{14.3}$$

Where N is the number of observations, M is the number of Y variables, K is the number of X variables, P $(K \times J)$ is the matrix of X-loadings, T $(N \times J)$ is the matrix of X-scores, q $(J \times M)$ is the loading vector of Y, E $(N \times K)$ and F $(N \times M)$ are error matrices, and J is the number of latent variables (LVs).

Both Y and X are expected to be partly modelled by the same LVs. The scores (T) are computed by a linear combination of the X variables and the weight matrix of W^* $(K \times J)$ such that $T = XW^*$, where $W^* = W(P^T W)^{-1}$. The values of W^* are calculated to ensure maximising the covariance between scores and responses. The regression coefficients (β) are calculated from the PLS weights (W) resulting from the model and the optimal number of LVs as $\beta = W \times qT$. The PLSR equation can then be simplified to the Eq. 14.4.

$$\hat{Y} = \beta X + F \tag{14.4}$$

Where $\hat{Y}$ is the predicted Y variable.

The number of latent variables for PLSR was determined based on the minimum predicted residual error sum of square (PRESS) of the validation set (Kamruzzaman et al. 2012). The PRESS was calculated as presented in the Eq. 14.5, as described by Li et al. (2002).

$$\text{PRESS} = \sum_{i=1}^{n} \left(\widehat{y_i} - y_i\right)^2 \tag{14.5}$$

Where $\widehat{y_i}$ and y_i are the predicted and reference values for iron, folic acid, and vitamin B_{12} in the ith sample, respectively.

Support Vector Regression (SVR)

SVR is an important practical subset of support vector machine (SVM), which is a supervised learning technique used for predicting discrete values. The objective of SVR is to create a hyperplane that effectively accommodates the samples, with the aim of minimising the total distance between all samples and the hyperplane. It operates on the assumption that a certain level of error may be accepted between the projected value and the actual output value. In SVR, the loss is only computed when the absolute difference between the predicted and output values exceeds the acceptable error threshold. Thus, it exhibits a certain level of resilience to outliers, robustness, and exceptional capacity for generalisation. It is capable of solving both linear and nonlinear problems. Kernel functions play a significant role in mapping the original data into high-dimensional spaces for nonlinear situations. The kernel function employed in this study was the radial basis function (RBF). A coarse grid of SVM parameter values is used to search for optimal values like the cost function, gamma value, epsilon value, and number of support vector components (Huang et al. 2021).

14.4.4 Model Evaluation

The models developed using the spectral reflectance of the samples in the calibration sets were tested using the test sets. The accuracy of the iron, folic acid, and vitamin B_{12} concentrations predicted using the HSI cameras was evaluated and compared against their corresponding reference concentrations. The best models were selected based on having the highest coefficient of determination (R^2) and the lowest root mean squared error (RMSE) represented in Eq. 14.6 and 14.7 for the calibration R_{cal}^2 and RMSE_{cal} in the cross-validation R_{CV}^2 and RMSE_{CV} and test sets (R_P^2 and RMSE_P) (Marín-Méndez et al. 2024; Cheng et al. 2025).

$$R^2 = 1 - \frac{\sum_{i=1}^{n} \left(Y_i - \widehat{Y}_i\right)^2}{\sum_{i=1}^{n} \left(Y_i - \hat{Y}\right)^2} \tag{14.6}$$

$$\mathrm{RMSE} = \sqrt{\frac{\sum_{i=1}^{n} \left(Y_i - \widehat{Y}_i\right)^2}{N}} \tag{14.7}$$

Where Y_i and $\widehat{Y}_i$ are the reference and predicted values of target variables (iron, folic acid, and vitamin B_{12}) in the ith sample, respectively. $\hat{Y}$ is the mean of reference values, and N is the number of samples.

The ratio of performance to deviation (RPD) was also calculated using Eq. (14.8) to further evaluate the reliability of the predictions (Qiao et al. 2022). Zhao et al. (2022) suggested that RPD values greater than 3 are useful for screening, values greater than 5 can be used for quality control, and values greater than 8 for any application.

$$\mathrm{RPD} = \frac{SD_t}{\mathrm{RMSE}_t} \tag{14.8}$$

Where SD_t and RMSE_t are the standard deviation of the reference and root mean square error of prediction in the test set, respectively.

In order to evaluate the accuracy of the results even more, the range error ratio (RER) was also computed using Eq. 14.9 (Yasmin et al., 2019). In general, RER values around or above ten indications of a good model, as demonstrated by Williams and Norris (1987).

$$\mathrm{RER} = \frac{y_{\max} - y_{\min}}{\mathrm{RMSE}_t} \tag{14.9}$$

Where $y_{\max}$ and $y_{\min}$ are maximum and minimum reference values, respectively.

The difference between $\mathrm{RMSE}_{\mathrm{cal}}$ and RMSE_{CV} increases when a model is overfitted, indicating that it is overly fitted to the calibration data and lacks accuracy in predicting new samples. The error ratio was also computed using Eq. (14.10), as stated by Ezenarro et al. (2023). The error ratio may be used to determine how much overfitting is in a regression model. A value of 0 indicates that the errors for both sets of samples are identical, while a value of 1 suggests that the calibration error is zero or that the cross-validation error is significantly larger.

$$\text{Error ratio} = 1 - \frac{\text{RMSE}_{\text{cal}}}{\text{RMSE}_{CV}} \tag{14.10}$$

Selection of Optimum Wavelength

Contiguous spectral bands may include duplicate information because of the strong correlation between the spectral bands in hyperspectral images. Analysing spectral data is computationally difficult and perhaps ineffective because of the strong connection. Therefore, extracting uncorrelated components is crucial for improving efficiency and effectiveness as well as solving collinearity issues. Multiple feature wavelengths, also known as optimal or significant wavelengths, may be chosen to improve the prediction accuracy. After selecting the appropriate wavelengths, a model can be constructed based on only such particular wavelengths instead of the range of spectrum. The objective of wavelength selection techniques is to determine the best wavelengths having significant information regarding such micronutrients. Removing the irrelevant variables would simplify model calibration and result in more precise and stable outcomes (Wu et al. 2012). Recursive partial least squares (rPLS) and regression coefficients (RC) are two common techniques that can be utilised for the selection of the best wavelengths.

Regression Coefficients (RC)

RC is a method that has been used to choose hyperspectral wavelengths by excluding wavelengths with low weights in the model. In the present study, wavelengths with low weights (i.e., the lowest RC coefficients) were excluded from the full spectral models. The models were re-evaluated using the remaining wavelengths, as outlined in Tahmasbian et al. (2017). Simply said, if the R^2 value of the calibration set from the rebuilt model was lower than that of the original model, the redeveloped model was deemed unacceptable and discarded. The wavelengths that were previously excluded were included back into the model. The procedure was iterated until the chosen wavelengths yielded an R^2 value for the validation set in the re-developed model that was equal to or greater than the R^2 value of the validation set in the original model, which used all wavelengths.

Recursive Partial Least Square (rPLS)

In the rPLS method, the regression coefficients, which are treated as variable weights, are continuously modified using the response vector (Y) to identify the most significant variables. During each iteration, the updated weighted X, ($X_R - 1$), and Y (the response vector) are utilised to construct a new PLS model (Rinnan et al. 2014). The regression coefficients of this model, saved in a vector (b), are then used to reweight X_{R-1} using the Eq. 14.11.

$$X_R = X_{R-1} \times \text{diag}\ (b) \tag{14.11}$$

Where X_R represents the new weighted data matrix. The significant factors progressively acquire greater weights, while the irrelevant variables to Y are reduced in weight. The technique reaches convergence with a small number of variables. However, the resulting model's prediction performance is not always optimal. A simple segmented cross-validation approach has been suggested to identify the ideal predictive model. In this approach, variables with weights close to zero are deemed insignificant, whereas variables with high absolute weight values are considered important.

14.4.5 App Interface Development

A graphical user interface (GUI) can be developed to enhance the functionality and efficacy of hyperspectral imaging analysis for estimating levels of iron, folic acid, and vitamin B_{12} in FRK in an attempt to optimise user interaction with predictive chemometric models for fast analysis without requiring expert knowledge for spectral data interpretation.

The app can be used to upload a hyperspectral image of the FRK sample in '.bil' (Band Interleaved by Line) file format through an easy-to-use front-end interface. After an image has been submitted, the app initiates a backend processing pipeline that employs optimised multivariate models. Based upon parameters defined within the system, the submitted image will either be preprocessed automatically or will be evaluated in its raw, unprocessed form, depending upon which processing pipeline has been chosen for use in the app's backend.

The GUI layout is depicted in Figure 14.1 and has elements for image uploading, selecting preprocessing options, running models, and presenting outputs. In addition, instantaneous feedback is provided through a display of the hyperspectral image of the sample loaded and corresponding calculated concentrations of iron, folic acid, and vitamin B_{12} from calibrated models.

14.4.6 Challenges

Despite the significant potential for HSI and chemometric modeling for micronutrient quantification in FRK, many challenges need to be overcome if broader implementation and higher efficacy can be achieved. Natural variability in FRK samples, defined in terms of particle size, surface texture, and moisture content, results in high spectral variability, complicating the extraction of chemically relevant information. The use of effective forms of preprocessing, including SNV, MSC, and Savitzky–Golay derivatives, is critical and requires careful parameter optimisation. The choice of optimal wavelength selection for balancing complexity in modelling versus accuracy poses a notable challenge, particularly when applied in high-dimensional hyperspectral datasets. Additionally, predictive models often suffer from limits to robustness and transferability, particularly when applied across different imaging configurations or under different environmental conditions. Industrial implementation raises further challenges in terms of variability in light conditions, vibrations, and temperature gradients along manufacturing lines, which can warp spectral readings. Finally, high capital investment for HSI machines and the requirement for specialist training in spectral handling curtail the availability for small- and medium-sized enterprises.

14.5 HOMOGENEITY TESTING

Homogeneity is the uniform distribution of FRK in NR during FR production. Homogeneity tests are conducted after blending to evaluate blending efficiency or blending homogeneity or following packaging, transit, and storage for the determination of any potential FRK segregation (BIS 2021a). A homogeneous mixture ensures that every part of FR has an equal proportion of FRK-to-NR and, hence, an equal proportion of nutrients. Issues with an even FRK distribution are encountered due to technical and operational limitations or issues in FRK production. Different sizes, weights, and shapes of FRK than the regular rice grains can cause segregation during mixing, packaging, storage, or distribution. Homogeneity is an important quality control parameter in the production of FR and helps in the success of rice fortification programmes (DFPD 2022). The homogeneity testing can be done using both the traditional and modern approaches. Both the methods are discussed in the following sections.

14.5.1 Sampling Procedures

Sampling is the process of taking samples that represent the batch of goods (Makwana et al. 2023). It is done after blending to check blending efficiency and after packaging and transportation to check for segregation. It is a crucial step in the context of homogeneity, as the sample taken represents the entire lot. It involves careful selection of the number of bags from which samples are to be taken, collection of increments, pooling together of these increments to get a bulk sample, and then dividing this bulk sample into laboratory samples for analysis (FSSAI 2023; Jideani and Jideani 2024).

Sample Collection after Blending

After blending, samples are collected to check the blending efficiency. Samples are collected systematically from the multiple points within the blender, including the top, middle, bottom, and edges. It is performed at regular intervals to assess the progression and completion of the blending process. Sampling probes such as thief samplers or scoop probes are used to effectively collect the samples from specific depths and locations. The collected samples are combined to form a composite for an overall assessment of the blend or analysed separately to detect any localised inconsistencies (FSSAI 2023).

Sample Collection after Packaging, Storage, and Transport

The blended FR is packed and stored in bags or sacks for transportation. At this stage, homogeneity is assessed to check for any potential segregation during transport. It starts by identifying the number of bags in the lot. Then, determine the number of bags to be selected and increments to be collected from each selected bag. From lots consisting of up to 10 bags, a sample must be taken from every bag. From lots containing between 10 and 100 bags, samples should be randomly taken from

10 bags. From lots with more than 100 bags, the number of bags (n) to be sampled should approximately equal the square root of the number of bags (x), i.e., $n = \sqrt{x}$. The bags are selected based on a suitable sampling plan. Sampling probes such as bag trier and grain trier are used to collect the samples from different sections of each bag. Samples are collected from the top, middle, and bottom of the bag to ensure representative sampling. The collected increments are combined and mixed thoroughly to form a homogeneous composite bulk sample (FSSAI, 2023).

Preparation of Laboratory Samples

The obtained bulk sample is then divided to obtain the required number of laboratory samples, either by the coning and quartering method or by using a mechanical sample divider. In the coning and quartering method, samples are spread on a clean, non-absorbent surface in the form of a conical heap, and the top is flattened. A sample is then divided into four equal quarters. Two diagonally opposite quarters are discarded, and the remaining two samples are mixed. The process is repeated until the desired sample size is achieved. Alternatively, a mechanical sample divider can be used to split the bulk sample evenly. Mechanical dividers such as riffle, Boerner, Gamet, and rotary dividers can also be used. They operate on different principles, such as gravity, centrifugal force, and mechanical rotation, to make subsamples (FSSAI 2023).

14.6 TRADITIONAL METHODS (CHEMICAL METHODS) FOR HOMOGENEITY TESTING

Traditionally, visual inspection and counting are used to identify FRK and assess homogeneity. Since FRK looks like regular rice, differentiating them can be quite challenging. In overcoming this limitation, colour-based chemical reactions are utilised to differentiate FRK from NR (DFPD 2022).

14.6.1 Povidone-Iodine Test

The starch-povidone iodine test is a cost-effective and rapid technique to visually detect the presence of FRK in FR. The test is based on the interaction between iodine and exposed amylose molecules in extruded FRK. In NR kernels, amylose is tightly packed within the starch granule, and the exposure is limited. Upon extrusion, amylose molecules are more accessible for iodine binding (Huang et al. 2022). The iodine molecule inserts into the helical cavity of amylose molecules and produces a blue-black colour amylose–iodine complex that remains stable or intact even after washing. On the contrary, NR kernels initially turn blue but lose their colour upon washing, as iodine cannot strongly bind to intact granules.

The following are the procedures:

- Take 50 g of sample in a plastic tray or cup and spread evenly.
- Pour 1% povidone-iodine solution over the kernels and mix well to provide even contact.

- Observe the kernels for colour change.
- After the kernels turn blue-black, drain the excess iodine solution and wash the sample gently with water till the coloured FRK is differentiated from the NR.
- Separate and count the blue-coloured kernels.

14.6.2 Iron Spot Test

The iron spot test, also known as the HCl-KSCN test, is a simple and rapid method for visually identifying the presence of FRK in FR. This method is based on forming a red-coloured, ferric thiocyanate complex (Nichols et al. 2012). When the sample is treated with hydrochloric acid (HCl), the iron (Fe^{2+} or Fe^{3+}) present in the FRK converts to a ferric (Fe^{3+}) form. These ions react with potassium thiocyanate (KSCN) to form a ferric thiocyanate complex, which imparts a red to dark red colour to the iron-fortified kernels. NR kernels do not contain iron and therefore do not undergo this colour change.

The reaction mechanism can be outlined as:

$$Fe^{3+} + 3KSCN \rightarrow Fe(SCN)_3 \tag{14.12}$$

The following is the procedure for performing iron spot test:

- Take 50 g of laboratory samples and spread them on a tray.
- Add enough 2N hydrochloric acid (HCl) to wet the kernels evenly.
- Add an equal volume of 10% potassium thiocyanate (KSCN) solution and gently mix to provide complete contact, and leave it for 2–3 min.
- Iron-fortified rice kernels will turn red to dark red, whereas NR kernels will remain unchanged in colour.
- Separate and count the red-coloured kernels (FRK).

14.6.3 Calculation of Homogeneity

After performing either the povidone-iodine test or the iron spot test, considering m is the weight of FRK in the drawn sample and w is the total weight of the sample drawn FRK concentration can be calculated by using Eq. 14.13. The weight of FRK (m) can be calculated by multiplying the number of coloured FRK kernels in the sample by the average mass of a single FRK (measured before chemical testing).

$$\text{FRK concentration }(\%) = \frac{m}{w} \times 100 \tag{14.13}$$

The homogeneity of FRK in FR can be assessed using the statistical variance (σ^2) and RSD of sample concentrations.

If N is the number of random samples collected for checking the homogeneity in a specific batch of FR, the variance (σ^2) can be calculated using the Eq. 14.14.

$$\sigma^2 = \frac{1}{N-1} \sum_{i=1}^{N} (x_i - \bar{x})^2 \tag{14.14}$$

Where $\bar{x}$ and x_i are the intended (target) concentration of FRK and the concentration of FRK in the ith random sample, respectively.

The RSD, which indicates the variability of FRK concentration relative to the target concentration (x), is calculated as shown in the Eq. 14.15.

$$\text{RSD}\ (\%) = \frac{\sqrt{\sigma^2}}{x} \times 100 \tag{14.15}$$

An ideal (perfectly uniform) mixture would have RSD = 0%, meaning that the concentration of FRK in any sample equals the target blend concentration. The lower the RSD (%), the better the homogeneity of blend (FR). Currently, the FRK concentration variation is allowed up to ± 10% (DFPD 2023; IS 17782 BIS 2021).

14.7 SMART APPROACH FOR HOMOGENEITY TESTING

The chemical methods used for detecting FRK possess several inherent drawbacks, including being labor-intensive, costly, time-consuming, and susceptible to human error. Additionally, their destructive nature inevitably leads to the wastage of samples after analysis. Consequently, there is a significant demand for more efficient, rapid, user-friendly, and nondestructive alternatives (Aher 2024). Image processing-based techniques have emerged as effective nondestructive methods for identifying FRK within FR, enabling accurate evaluation of their proportion in mixtures. Broadly, these image-based methods can be categorised into two primary approaches: ML-based and non-ML-based methods. The ML-based approach leverages algorithms and computational models to automatically learn distinguishing features from image data, enhancing the speed and accuracy of FRK detection. Conversely, the non-ML-based methods utilise classical image processing techniques, relying on predefined feature extraction and thresholding criteria to identify and quantify FRK. Both methodologies are explored in detail in subsequent sections.

14.7.1 Development of Controlled Lighting Condition

Before implementing ML and non-ML approaches, it is crucial to establish standardised lighting conditions for capturing images and fixing the image-capturing device. Initially, a smartphone camera with good resolution may be a suitable starting point; however, an Arduino-based camera system can subsequently be employed for consistency and improved functionality. It should be kept in mind while regulating the lighting conditions that no reflections from LED illuminations are captured as well as the rice grains should have proper uniform lighting without any shadows or highlights (Aher 2024).

14.7.2 Machine Learning Approach

The ML approach involves developing a big database of images of various FRK and NR kernels, which are classified based on their sizes, shapes, colours, orientations, and cracked kernel frequency. Individual images in the database are tagged as FRK or NR. Multiple images must be captured to develop a strong and representative image database. The collected data is split into training, testing, and validation sets in a typical ratio of 70:20:10, respectively. This structured split allows for accurate and unbiased model training and testing (Vishwakarma et al. 2023).

Convolution neural network (CNN) models are widely applied in ML-based approaches due to their exceptional performance in image recognition and classification tasks. They operate by applying convolutional filters to input images, which can automatically learn hierarchical features such as edges, textures, and shapes. Hierarchical feature extraction renders CNNs highly efficient in distinguishing FRK and NR kernels based on subtle differences in shape, size, colour, and texture (Medus et al. 2021).

In addition to CNNs, the YOLO model family, including YOLOv5 and YOLOv8, can be employed in real-time object detection tasks. YOLO models divide an image into a grid and predict class probability and bounding boxes for every grid cell at the same time. This makes it possible to detect more than one object in a given image to be performed quickly and accurately. This is beneficial in FRK detection, where more than one kernel needs to be detected and classified in real time (Aher 2024).

To quantify the performance of such ML models, a number of metrics are employed. A confusion matrix is typically employed to provide the complete breakdown of model predictions, providing true positives, false positives, true negatives, and false negatives. Metrics such as accuracy, precision, recall, and F1-score can be derived from the confusion matrix, offering insights into the model's reliability and effectiveness. Additionally, for object detection models like YOLO, metrics such as mean Average Precision (mAP) are utilised to assess the precision and recall across different Intersection over Union (IoU) thresholds, providing a comprehensive evaluation of the model's detection capabilities (Aher 2024).

14.7.3 Non-Machine Learning Approach

The non-ML approach involves detecting specific functional differences between FRK and NR based on characteristics such as size, shape, and colour. The primary objective of this approach is to amplify these differences through various image processing techniques, including image segmentation, thresholding, edge detection, morphological operations, feature extraction, and colour space analysis. This methodology does not require a large image library but must be validated using diverse images to enhance robustness and reliability. Consequently, comprehensive testing across multiple image scenarios is essential to ensure the non-ML method's consistency and accuracy in identifying and quantifying FRK. For example, a Raspberry Pi-based setup can be employed, positioned above the sample holder equipment with an attached camera aligned through a specific opening. This setup ensures consistent image capture in terms of illumination, texture, sharpness, and other visual attributes across various samples. Multiple images obtained from this configuration demonstrate consistent quality and facilitate reliable differentiation between FRK and NR. Subsequently, custom code integrating observed fundamental differences can be developed to automate the identification of FRK within the rice samples (Aher 2024).

14.7.4 App Interface Development

The app interface is essential for facilitating user-friendly interaction with the developed detection methods. Python, MATLAB, or Android platforms can be utilised to design and develop intuitive graphical interfaces that streamline the image capturing and processing workflow. For instance, a Python-based GUI created using Tkinter (Figure 14.2) could effectively integrate image acquisition, processing, and result display. Users could capture images seamlessly through this interface by placing evenly distributed FR samples within the controlled environment. The images would then be processed through ML or non-ML algorithms on the backend, displaying processed results that distinguish FRK from NR promptly on the same interface, thereby enabling efficient and intuitive user interaction. A process flowchart for homogeneity testing using image analysis through the GUI interface is shown in Figure 14.2.

14.7.5 Challenges

Despite the advantages of nondestructive techniques for FRK detection, several challenges persist. First, maintaining consistent lighting and image quality is critical for accurate analysis; variations in ambient conditions can significantly affect results. Second, differentiating between FRK and NR becomes complex in cases where visual characteristics (e.g., shape, colour, or size) are not distinct enough, particularly for poorly fortified or blended kernels. Third, non-ML approaches depend heavily on manually defined thresholds and parameters, which may limit adaptability and generalisability across diverse samples. Nevertheless, despite these challenges, nondestructive techniques hold substantial promise for rapid and effective homogeneity analysis. With continued improvements in image acquisition, algorithm design, and system integration, these methods are well-positioned to become practical tools for on-site and industrial applications of FRK detection.

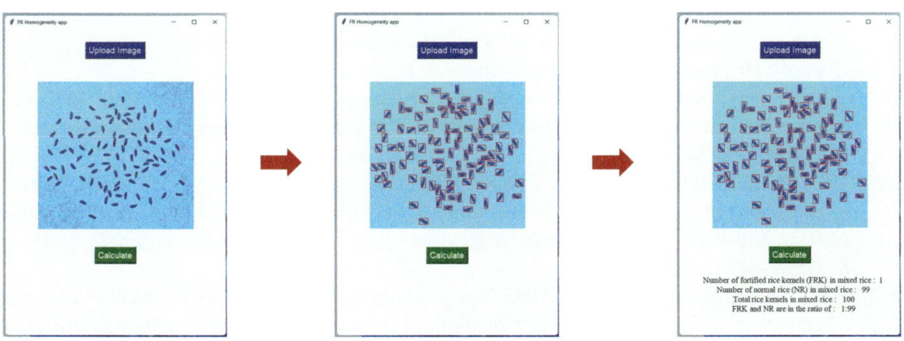

FIGURE 14.2 Homogeneity testing with GUI interface.

14.8 CONCLUSIONS

The whole analysis of iron, folic acid, and vitamin B_{12} in FRK and homogeneity testing in FR are vital to the maintenance of the nutritional integrity and quality of rice fortification programmes. Though conventional analytical techniques like AAS, HPLC, and LC-MS are very accurate and precise, they are not suitable for application in the large-scale industry because they are resource-demanding. On the other hand, HSI, in combination with ML and chemometrics, is a promising nondestructive and faster alternative to micronutrient quantification that enables real-time quality control. Similarly, substituting conventional manual, reagent-based homogeneity testing with image-based ML and non-ML methods enhances detection accuracy, reduces sample loss, and provides scalability. In spite of implementation and technical hurdles, the use of digital imaging and intelligent sensing devices can drastically improve the reliability, effectiveness, and ease of quality control of rice fortification processes.

REFERENCES

Aher, Pranali Jaywant. 2024. "Development of Homogeneity Testing App for Fortified Rice". Master's Disseratation, Indian Institute of Technology Kharagpur.

Araujo, Pedro. 2009. "Key Aspects of Analytical Method Validation and Linearity Evaluation." *Journal of Chromatography B: Analytical Technologies in the Biomedical and Life Sciences* 877(23): 2224–34.

Arvapally, Manasa, Ankita Asati, Narendra Kumar Nagendla, and Mohana Krishna Reddy Mudiam. 2021. "Development of an Analytical Method for the Quantitative Determination of Multi-Class Nutrients in Different Food Matrices by Solid-Phase Extraction and Liquid Chromatography-Tandem Mass Spectrometry Using Design of Experiments." *Food Chemistry* 341: 128173.

Brouwer, Veerle De, Sergei Storozhenko, Jet C. Van De Steene, Sarah M.R. Wille, Christophe P. Stove, Dominique Van Der Straeten, and Willy E. Lambert. 2008. "Optimisation and Validation of a Liquid Chromatography-Tandem Mass Spectrometry Method for Folates in Rice." *Journal of Chromatography A* 1215(1–2): 125–32.

Bureau of Indian Standards (BIS). 2021a. "Fortified Rice-Specification (IS 17780: 2021). https://standardsbis.bsbedge.com/BIS_SearchStandard.aspx?Standard_Number=17780&id=0

Bureau of Indian Standards (BIS). 2021b. "Fortified Rice Kernels — Specifications (IS 17782: 2021)." https://standardsbis.bsbedge.com/BIS_searchstandard.aspx?Standard_Number=IS+17782&id=34819

Burger, James, and Aoife Gowen. 2011. "Data Handling in Hyperspectral Image Analysis." *Chemometrics and Intelligent Laboratory Systems* 108(1): 13–22.

Cheng, Ming-fang, Arvind Mukundan, Riya Karmakar, Muhamed Adil Edavana Valappil, Jumana Jouhar, and Hsiang-chen Wang. 2025. "Modern Trends and Recent Applications of Hyperspectral Imaging: A Review." *Technologies* 13(5): 170.

Dalbhagat, Chandrakant Genu, Dipendra Kumar Mahato, and Hari Niwas Mishra. 2019. "Effect of Extrusion Processing on Physicochemical, Functional and Nutritional Characteristics of Rice and Rice-Based Products: A Review." *Trends in Food Science and Technology* 85: 226–40.

Dalbhagat, Chandrakant Genu, and Hari Niwas Mishra. 2020. "Studies on Manufacture of Micronutrient Fortified Rice Kernels Using Extrusion Technology." Indian Institute of Technology Kharagpur.

Dalbhagat, Chandrakant Genu, and Hari Niwas Mishra. 2021. "Drying Modeling, Cooking Characteristics, Pasting Properties, and Crystallinity of Fortified Rice Kernels." *Journal of Food Processing and Preservation* 45(6): 1–9.

Department of Food and Public Distribution (DFPD). 2023. "Fortification of Rice and Its Distribution under Public Distribution System: Operational Guidelines." Retrieved May 16, 2025, from https://cdnbbsr.s3waas.gov.in/s3e21e4e58ad9ab56e8a4634046da90113/-uploads/2023/06/2023060963.pdf

Djoura, Volodymyr. 2020. "Method Development and Validation for Separation of Three Fat Soluble Vitamins Using Reversed-Phase Liquid Chromatography and Solid Phase Extraction." *ProQuest LLC*, Northeastern Illinois University.

Ezenarro, Jokin, Daniel Schorn-García, Laura Aceña, Montserrat Mestres, Olga Busto, and Ricard Boqué. 2023. "J-Score: A New Joint Parameter for PLSR Model Performance Evaluation of Spectroscopic Data." *Chemometrics and Intelligent Laboratory Systems* 240: 104883.

Fan, Shuang, Zhuopin Xu, Weimin Cheng, Qi Wang, Yang, Junyao Guo, Pengfei Zhang, and Yuejin Wu. 2022. "Establishment of Non-Destructive Methods for the Detection of Amylose and Fat Content in Single Rice Kernels Using Near-Infrared Spectroscopy." *Agriculture* 12(8): 1258.

Fatima, Zakia, Xiangzi Jin, Yilin Zou, Han Yeong Kaw, Maurizio Quinto, and Donghao Li. 2019. "Recent Trends in Analytical Methods for Water-Soluble Vitamins." *Journal of Chromatography A* 1606: 360245.

Food Safety and Standards Authority of India (FSSAI). 2023. "Guidelines on Sampling of Fortified Rice (FR), Fortified Rice Kernels (FRK) and Vitamin Mineral Premix for FRK." Retrieved May 16, 2025, fromhttps://www.fssai.gov.in/upload/uploadfiles/files/Order%20on%20Guidelines%20on%20sampling%20of%20Fortified%20Rice_FR%2C%20Fortified%20Rice%20Kernels-17-11-2023.pdf

Gürkan, Ramazan, and Nail Altunay. 2016. "A Simple and Efficient Ultrasonic-Assisted Extraction Procedure Combined with UV-Vis Spectrophotometry for the Pre-Concentration and Determination of Folic Acid (Vitamin B9) in Various Sample Matrices." *Food Additives and Contaminants — Part A: Chemistry, Analysis, Control, Exposure and Risk Assessment* 33(7): 1127–38.

Heudi, Olivier, Tamara Kilinç, Patric Fontannaz, and Elaine C. Marley. 2006. "Determination of Vitamin B12 in Food Products and in Premixes by Reversed-Phase High Performance Liquid Chromatography and Immunoaffinity Extraction." *Journal of Chromatography A* 1101(1–2): 63–68.

Huang, Feng Hua, Yan Hong Liu, Xin Yi Sun, and Hua Yang. 2021. "Quality Inspection of Nectarine Based on Hyperspectral Imaging Technology." *Systems Science and Control Engineering* 9(1): 350–57.

Huang, Xiaoyue, Hongsheng Liu, Yue Ma, Shihua Mai, and Cheng Li. 2022. "Effects of Extrusion on Starch Molecular Degradation, Order–Disorder Structural Transition and Digestibility—A Review." *Foods* 11(16): 2538.

Ivarsen, Elise, Christoffer P. Andersen, Sabine M. Jensen, Carsten T. Pedersen, and Anders K. Svaneborg. 2021. "Quantification of Retinyl Palmitate, Thiamine, Niacin, Pyridoxine, Folic Acid, Cyanocobalamin, Zinc, and Iron by Chromatographic Methods in Fortified Kernels and Fortified Rice." *Acta Chromatographica* 33(2): 145–52.

Jägerstad, Margaretha, and Karin Arkbåge. 2003. "Cobalamins – Properties and Determination." In *Encyclopedia of Food Sciences and Nutrition*, edited by Benjamin Caballero, Luiz Trugo, and Paul Finglas, 1419–27. Amsterdam: Academic Press, Elsevier.

Jiang, Xinna, Youhua Bu, Lipeng Han, Jianping Tian, Xinjun Hu, Xiaobing Zhang, Dan Huang, and Huibo Luo. 2023. "Rapid Nondestructive Detecting of Wheat Varieties and Mixing Ratio by Combining Hyperspectral Imaging and Ensemble Learning." *Food Control* 150: 109740.

Jideani, Victoria, and Afam I. O. Jideani. 2024. "Sampling." In *Statistics in Food and Biotechnology: From Theory to Practical Applications*, edited by Gerhard Schleining, Saverio Mannino, and Prisana Suwannaporn, 155–90. Cham: Springer International Publishing.

Jorhem, Lars, and Joakim Engman. 2000. "Determination of Lead, Cadmium, Zinc, Copper, and Iron in Foods by Atomic Absorption Spectrometry after Microwave Digestion: NMKL Collaborative Study." *Journal of AOAC International* 83(5): 1189–1203.

Kamruzzaman, Mohammed, Gamal Elmasry, Da Wen Sun, and Paul Allen. 2012. "Non-Destructive Prediction and Visualization of Chemical Composition in Lamb Meat Using NIR Hyperspectral Imaging and Multivariate Regression." *Innovative Food Science and Emerging Technologies* 16: 218–26.

Knez, Simon, and Luka Šajn. 2020. "Food Object Recognition Using a Mobile Device: Evaluation of Currently Implemented Systems." *Trends in Food Science & Technology* 99: 460–71.

Kosse, Jennifer S., Andrew C. Yeung, Ana I. Gil, and Dennis D. Miller. 2001. "A Rapid Method for Iron Determination in Fortified Foods." *Food Chemistry* 75(3): 371–76.

Lebiedzińska, A., M. Dąbrowska, P. Szefer, and M. Marszałł. 2008. "High-Performance Liquid Chromatography Method for the Determination of Folic Acid in Fortified Food Products." *Toxicology Mechanisms and Methods* 18(6): 463–67.

Lee, Wonjong, Yoon Bok Lee, Moon Haeng Huh, Jae Kwon Choi, and Muhammad Faisal Manzoor. 2022. "Determination of the Chemical Stability of Cyanocobalamin in Medical Food by a Validated Immunoaffinity Column-Linked HPLC Method." *Journal of Food Quality* 2022: 1619936.

Li, Baibing, Julian Morris, and Elaine B. Martin. 2002. "Model Selection for Partial Least Squares Regression." *Chemometrics and Intelligent Laboratory Systems* 64(1): 79–89

Mahato, Ansuman, Samir Vyas, and Niladri Sekhar Chatterjee. 2021. "HPLC-UV Estimation of Folic Acid in Fortified Rice and Wheat Flour Using Enzymatic Extraction and Immunoaffinity Chromatography Enrichment: An Interlaboratory Validation Study." *Journal of AOAC International* 103(1): 73–77.

Makwana, Dhaval, Priti Engineer, Amisha Dabhi, and Hardik Chudasama. 2023. "Sampling Methods in Research: A Review." *International Journal of Trend in Scientific Research and Development* 7(June): 762–68.

Marín-Méndez, Juan-Jesús, Paula Luri Esplandiú, Miriam Alonso-Santamaría, Berta Remirez-Moreno, Leyre Urtasun Del Castillo, Jaione Echavarri Dublán, Eva Almiron-Roig, and María-José Sáiz-Abajo. 2024. "Hyperspectral Imaging as a Non-Destructive Technique for Estimating the Nutritional Value of Food." *Current Research in Food Science* 9(May): 100799.

Medus, Leandro D., Mohamed Saban, Jose V. Francés-Víllora, Manuel Bataller-Mompeán, and Alfredo Rosado-Muñoz. 2021. "Hyperspectral Image Classification Using CNN: Application to Industrial Food Packaging." *Food Control* 125(July): 107962.

Miller, Dennis D., and Michael A. Rutzke. 2020. "Atomic Absorption Spectroscopy, Atomic Emission Spectroscopy, and Inductively Coupled Plasma-Mass Spectrometry." In *Food Analysis*, 4th ed., edited by S. Suzanne Nielsen, 419–438. Springer Cham.

Ministry of Consumer Affairs: Department of Food & Public Distribution (DFPD). 2022. "Standard Operating Procedure (S.O.P) for Quality Management Protocols for Fortified Rice Kernels (FRK) and Fortified Rice (FR)." Retrieved May 16, 2025, from https://anaajkharid.in/Images/FRK/SoP%20for%20FRK%20and%20FR%2015.03.2022.pdf

Nakos, M., Iliyana Pepelanova, Sascha Beutel, Ulrich Krings, Ralf Günter Berger, and Thomas Helmut Scheper. 2017. "Isolation and Analysis of Vitamin B_{12} from Plant Samples." *Food Chemistry* 216: 301–8.

Nichols, Erin K., Nancy Aburto, Hanan Masa' D, James Wirth, Kevin Sullivan, and Mary Serdula. 2012. "Performance of Iron Spot Test with Arabic Bread Made from Fortified White Wheat Flour." *Food and Nutrition Bulletin* 33(3): 202.

Nithya, A., Chandrakant Genu Dalbhagat, and Hari Niwas Mishra. 2023. "Development of a Simple Method for Extraction and Analysis of Folic Acid in Fortified Rice Kernels by HPLC and Its Application in Vitamin Retention Studies." *Journal of Food Measurement and Characterization* 17: 6440–6451.

Poitevin, Eric. 2016. "Official Methods for the Determination of Minerals and Trace Elements in Infant Formula and Milk Products: A Review." *Journal of AOAC International* 99(1): 42–52.

Pu, Hongbin, Lian Lin, and Da-Wen Sun. 2019. "Principles of Hyperspectral Microscope Imaging Techniques and Their Applications in Food Quality and Safety Detection: A Review." *Comprehensive Reviews in Food Science and Food Safety* 18(4): 853–66.

Qiao, Mengmeng, Yang Xu, Guoyi Xia, Yuan Su, Bing Lu, Xiaojun Gao, and Hongfei Fan. 2022. "Determination of Hardness for Maize Kernels Based on Hyperspectral Imaging." *Food Chemistry* 366: 130559.

Radhika, Kommu, Srinath Nissankararao, K. S. Sumanth, S. Neelima, and V. Krishna Prasanna. 2012. "Method Development and Validation for the Simultaneous Estimation of Nicotinic Acid, Folic Acid and Cyanocobalamin Using RP-HPLC." *International Journal of Research in Ayurveda and Pharmacy* 3(5): 701–5.

Ravisankar, Panchumarthy, C. Naga Navya, D. Pravallika, and D. Navya Sri. 2015. "A Review on Step-by-Step Analytical Method Validation." *IOSR Journal of Pharmacy* 5(10): 2250–3013.

Rinnan, Åsmund, Martin Andersson, Carsten Ridder, and Søren Balling Engelsen. 2014. "Recursive Weighted Partial Least Squares (RPLS): An Efficient Variable Selection Method Using PLS." *Journal of Chemometrics* 28(5): 439–47.

Ruffin, Chris, Roger L. King, and Nicolas H. Younan. 2008. "A Combined Derivative Spectroscopy and Savitzky-Golay Filtering Method for the Analysis of Hyperspectral Data." *GIScience & Remote Sensing* 45(1): 1–15.

Santos, Albano Joel Moreira, Sheyma Khemiri, Sara Simões, Catarina Prista, Isabel Sousa, and Anabela Raymundo 2023. "The Importance, Prevalence and Determination of Vitamins B6 and B12 in Food Matrices: A Review." *Food Chemistry* 426.

Shao, Yuanyuan, Yongxian Wang, Guantao Xuan, Zongmei Gao, Zhichao Hu, Chong Gao, and Kaili Wang. 2020. "Assessment of Strawberry Ripeness Using Hyperspectral Imaging." *Analytical Letters* 54(10): 1547–60.

Silveira, C. M. M., C. M. Della Lucia, M. R. Pirozi, T. A. Montini, and H. M. Pinheiro-Sant'Ana. 2016. "Methods for Analysis of Thiamin and Folic Acid by HPLC-DAD in Fortified Rice Pure and Mixed to Milled Rice before and after Different Cooking Techniques." *Acta Chromatographica* 28(4): 455–72.

Singh, Jagdish. 2022. "Vitamin B9 in Dark Green Vegetables: Deficiency Disorders, Bio-Availability, and Fortification Issues." In *B-Complex Vitamins – Sources, Intakes and Novel Applications*, edited by Jean Guy LeBlanc, 11:13. IntechOpen.

Sjöblom, Jonas, Olof Svensson, Mats Josefson, Hans Kullberg, and Svante Wold. 1998. "An Evaluation of Orthogonal Signal Correction Applied to Calibration Transfer of Near Infrared Spectra." *Chemometrics and Intelligent Laboratory Systems* 44(1–2): 229–44.

Srivastava, Shubhangi, and Hari Niwas Mishra. 2022. "Detection of Insect Damaged Rice Grains Using Visible and Near Infrared Hyperspectral Imaging Technique." *Chemometrics and Intelligent Laboratory Systems* 221: 104489.

Tahmasbian, Iman, Zhihong Xu, Kadum Abdullah, Jun Zhou, Roya Esmaeilani, Thi Thu Nhan Nguyen, and Shahla Hosseini Bai. 2017. "The Potential of Hyperspectral Images and Partial Least Square Regression for Predicting Total Carbon, Total Nitrogen and Their Isotope Composition in Forest Litterfall Samples." *Journal of Soils and Sediments* 17: 2091–2103.

Vishwakarma, Siddharth, Shubham Mandliya, Chandrakant Genu Dalbhagat, Pallavi Kumari Singh, and Hari Niwas Mishra. 2023. "Rheological Modeling of Marjoram Fortified Rice Dough: Empirical and Machine Learning Approach." *Journal of Food Process Engineering* 46(8): 1–15.

Williams, Phil, and Karl Norris. 1987. *Near-Infrared Technology in the Agricultural and Food Industries*. American Association of Cereal Chemists, Inc.

Wu, Di, Hui Shi, Songjing Wang, Yong He, Yidan Bao, and Kangsheng Liu. 2012. "Rapid Prediction of Moisture Content of Dehydrated Prawns Using Online Hyperspectral Imaging System." *Analytica Chimica Acta* 726: 57–66.

Yasmin, Jannat, Mohammed Raju Ahmed, Santosh Lohumi, Collins Wakholi, Hoonsoo Lee, C. Mo, Byoung-Kwan Cho. 2019. "Rapid Authentication Measurement of Cinnamon Powder Using FT-NIR and FT-IR Spectroscopic Techniques." *Quality Assurance and Safety of Crops & Foods* 11(3): 257–67.

Zafra-Gómez, Alberto, Antonio Garballo, Juan C. Morales, and Luis E. García-Ayuso. 2006. "Simultaneous Determination of Eight Water-Soluble Vitamins in Supplemented Foods by Liquid Chromatography." *Journal of Agricultural and Food Chemistry* 54(13): 4531–36.

Zhao, Zhen, Liang Xu, Jianmin Gao, Lei Xi, Qicheng Ruan, and Yunlong Li. 2022. "Multi-Objective Optimization of Parameters of Channels with Staggered Frustum of a Cone Based on Response Surface Methodology." *Energies* 15(3): 1240.

15 Consumer Acceptance and Sensory Evaluation of Fortified Rice Kernels and Fortified Rice

Yuvraj Bhosale, Ghumika Pandita, Chandrakant Genu Dalbhagat, V. R. Sinija, and Hari Niwas Mishra

15.1 INTRODUCTION

Major imbalance in the intake, absorption or utilisation of various micronutrients leading to multiple medical and nutritional conditions that measurably affect structure and function of body is known as malnourishment. Among these deficiencies, iron-deficiency anaemia (IDA) is caused due to insufficient iron levels in the body leading to the decreased synthesis of RBC. IDA is one of the most prevalent disorders that are observed in large proportion of population and can be identified by microcytic (small) and hypochromic (pale) red blood cells on laboratory evaluation (Kaur 2016). Around one billion individuals mainly women and children are affected by IDA predominantly in developing countries. Apart from IDA (microcytic anaemia), lower intakes of folic acid (vitamin B_9) and cyanocobalamin (vitamin B_{12}) interfere with maturity and physiology of RBCs giving rise to megaloblastic anaemia. Together, these disorders can cause the visible symptoms, viz., impaired neurodevelopment, lower work productivity, increased maternal and perinatal morbidity, weakened immunity and chronic cardiac diseases.

Considering prevalence of anaemia, various dietary and medicinal interventions were undertaken for its eradication. However, these interventions do not consider the deficiency of cyanocobalamin and their effective implementation, which is difficult to monitor. Taking these rationales into consideration, food fortification was suggested as a more effective and implementable intervention to tackle IDA (Mishra, Mishra, and Rao 2012). To initiate with, the fortification of rice was done by the techniques such as dusting and coating. However, the techniques were least feasible due to nutrient losses during washing, cooking and draining of excess water after cooking.

Progressing further, Mishra, Mishra, and Rao (2012) complied the technological advances in the field of extrusion for the development of fortified rice (FR) analogue

DOI: 10.1201/9781003661245-15

to deliver the effective technology by implementation of extrusion for fortification of rice with iron, folic acid and cyanocobalamin.

The technological advancement and government policies led to the implementation of a fortification programme in India for the fortification of staple foods with various micronutrients ("India" 2025). However, failure to maintain key sensory attributes such as colour, appearance, aroma and taste in the rice distributed through public distribution system (PDS) leads to consumer speculation of adulteration or other forms. These deviations from the expected sensory attributes can undermine consumer trust, raising concerns about product authenticity, safety and overall quality.

Sensory attributes are crucial for product success, influencing consumer acceptability and preference. High sensory quality increasingly drives success across various food categories, particularly in the fortified food market. Consumers differ in their sensory preferences based on biological, cultural, psychological, social, economic, environmental and political background. Understanding these aspects for product development leads to the success of the product in the market. Although sensory attributes such as colour, aroma, texture and taste play a crucial role in determining consumer acceptance and overall food quality, they are often overlooked in practice due to various factors. These include limited resources, lack of technical expertise, prioritisation of economic considerations over sensory quality and inadequate integration of sensory evaluation in food product development and distribution systems.

Considering the above rationales, this chapter explores the importance of assessing consumer acceptance of FR by examining key sensory attributes. It discusses how raw materials and processing conditions affect the sensory quality attributes and outlines evaluation methods for sensory parameters, including analytical tools and panel requirements. This chapter also highlights factors influencing consumer preferences, such as visual appeal and cooking behaviour and identifies future research needs, including standardised sensory tools and long-term acceptability studies.

15.2 SENSORY ATTRIBUTES OF FORTIFIED RICE KERNEL AND FORTIFIED RICE

Developed fortified rice kernel (FRK) and FR are specifically manufactured to closely resemble unfortified rice in terms of sensory characters, with the primary objective of ensuring nutritional delivery with high consumer acceptance. This sensory evaluation is crucial, as observable differences in attributes such as colour, aroma, taste and texture may lead to consumer rejection regardless of the nutritional benefits served through fortification (Standard 2021; Dalbhagat and Mishra 2021a). The studies indicate that with proper selection of raw materials and process parameters optimisation, FR can maintain a natural-like appearance, pleasant aroma, desirable taste and appropriate texture (Dalbhagat and Mishra 2021b; Nithya, Dalbhagat, and Mishra 2022; Beinner et al. 2010a). The following are the major sensory parameters that need to be considered for unidentifiable blending of FRK in FR.

15.2.1 Appearance

Appearance is the primary sensory parameter that consumers experience while purchasing the food giving a rapid judgement of freshness, ripeness and spoilage. It plays an important role in shaping consumer expectations in terms of quality, flavour and safety and determines the marketability of the product. The concept of appearance includes several visual characteristics, viz., colour, size, shape, surface texture, glossiness, clarity and visual uniformity. FRK is typically manufactured by mixing a prescribed quantity of vitamin-mineral premix (VMP) into rice flour followed by processing through extrusion. For effective consumer adoption, the physical characteristics of FRK, including shape, size, translucency and surface texture, must closely mimic the raw/parboiled rice. Visually differentiable attributes can result in the FRK standing out when blended with raw/parboiled rice, forcing the impression of malpractices such as adulteration. Contrarily, when crafted with utmost precision, the FRK showed elevated acceptability even in comparison with biofortified rice (Mamoriska et al. 2022).

15.2.2 Colour

Colour is one of the important sensory attributes of food that significantly influences consumer perception often determining the acceptability of food. It serves as an immediate indicator of quality. In the case of common white rice, polishing removes the outer bran and starch, and protein present in the endosperm gives rice its characteristic translucency or white colour. Starch and protein undergo gelatinisation, retrogradation and denaturation during various unit operations of FRK manufacturing, altering their natural translucency. Moreover, the sources of iron, folic acid and cyanocobalamin further contribute to the colour alteration of FRK. In many cases, the FRK resemble chalky grain leading to the removal of FRK from FR and reduced micronutrient availability. This makes it very crucial to optimise FRK manufacturing unit operation for an acceptable colour of FRK and FR.

15.2.3 Shape and Physical Dimensions

The shape, density and physical dimensions of raw/parboiled rice grains must be resembled by FRK in FR to ensure consumer acceptance and facilitate their integration into the regular diet. This resemblance is essential as it eliminates the possibility of consumers distinguishing between fortified grains and unfortified grains (Nithya, Dalbhagat, and Mishra 2022). The rice-like shape of FRK is typically achieved by passing thermally processed rice flour and VMP mixture through a dedicated rice-shaped die. Naturally, rice varieties can be found in a wide range of sizes and shapes and are usually categorised by their length or length-to-breadth (L/b) ratio. In the Indian scenario, the agencies such as the Bureau of Indian Standards (BIS) and the Food Corporation of India (FCI) prescribed rice varieties with L/b ratio $\geq$2.5 as grade A (long varieties) and $<$2.5 as common varieties (short varieties). High yielding Indian varieties are typically ranging from 5.2 to 6.8 mm long and 1.9 to 2.5 mm thick, while paddy grains can reach up to 14 mm in length. Though, matching the physical dimensions and characteristics of FRK with widely consumed local varieties is critical. The physical dimensions of it must be carefully measured using grain analyser or image processing software and controlled to ensure uniform blending of FRK. It ensures not only sensorial acceptability of consumer but also maintains economic value, cooking properties and effectiveness and sustainability of the fortification programme (Dalbhagat and Mishra 2021b; Nithya, Dalbhagat, and Mishra 2022).

15.2.4 Texture

The texture of the food material plays an important role in consumer acceptance. In the case of FRK, when conditioned rice flour passes through the extruder, the starch gets gelatinised forming a viscoelastic melt, which further compresses due to screw. This compaction of gelatinised rice starch forms intramolecular hydrogen bonds. These hydrogen bonds get stronger during the drying of FRK as the moisture is removed and retrogradation of starch occurs. This mechanism is responsible for providing FRK its distinguishing texture after drying, also known as the pre-cooked texture of FRK. Furthermore, as the FRK is cooked, re-gelatinisation of the starch occurs, which eventually leads to softening of the pre-cooking texture. The texture of FR is dependent on the formulation, fortification level and processing techniques. Several studies emphasise that achieving sensory attributes close to raw/parboiled rice is essential for consumer satisfaction, and some such studies are tabulated in Table 15.1. Maintaining such quality parameters ensures the success of the fortification programme without compromising sensory attributes.

15.2.5 Taste and Aroma

The compositional components, such as starch, protein and sugar, present in rice provide the distinctive taste. When it comes to aroma, the volatile alcohols, aldehydes, ketones and hydrocarbons are responsible for the aroma of various rice varieties. N-hexanol, hexanal, 6,10-dimethyl-2-undecanone and 6,10,14-trimethyl -2-pentadecanone are some of the volatile compounds that are identified in different rice varieties (Lin et al. 2010). However, changes in soil condition, cultivating environment, geographical condition, etc., alter the composition of macronutrients as well as volatile compounds, hence changing the taste and aroma of rice. As the FRK is made with rice flour as base material, it is supposed to contain the macromolecules and volatile compounds that are responsible for the taste and aroma of the particular rice variety. Hence, it is suggested to use rice broken or rice flour of the same variety in which the FRK will be blended further. Studies have demonstrated that excluded FRK can achieve good palatability with the fortification of valuable micronutrients to desirable limits. In the case of aroma, rice fortified with iron at moderate levels does not affect the aroma. Some studies revealed that FR was preferred over raw/parboiled rice for its smell. However, very high levels of iron fortification (e.g. 450 mg/kg rice) may produce off flavour leading to reduced acceptability (Prom-u-thai et al. 2009).

TABLE 15.1
Studies Presenting Need for Achieving Sensory Attributes Similar to raw/parboiled Rice

Aspect	Observation	Source
Texture similarity	Cooked extruded fortified grains show texture comparable to rice, ensuring consumer acceptance	Moretti (2006)
Starch gelatinisation	High-quality extruded fortified rice requires a high degree of starch gelatinisation without excessive breakdown. This balance is crucial for maintaining desirable texture	Ayoub et al. (2013)
Sensory characteristics	Fortified rice, when properly processed, can match the texture and sensory qualities of commercial cooked rice	Ayoub et al. (2013)
Chewiness and flakiness	Iron fortification, especially at high levels, can affect texture (e.g., increased chewiness or altered flakiness), potentially reducing acceptability if not controlled	Prom-u-thai et al. (2009)
Impact of fortification level	At optimised iron levels (e.g., 250 mg Fe/kg paddy rice), no adverse effects on texture were observed. This suggests the importance of appropriate dosage and method	Prom-u-thai et al. (2009)
Preparation method influence	The processing method (e.g., extrusion, parboiling) and type of fortificant (e.g., iron salt form) significantly impact the final texture	Moretti (2006)

Source: Dalbhagat (2020).

15.3 SENSORY EVALUATION TECHNIQUES

Sensory parameters prescribed in previously are essential to evaluate for manufacturing superior quality FRK. Considering that there are several sensory evaluation protocols that can be followed to evaluate the sensory quality of the product. Some protocols are for understanding the threshold of the sensory panel and panel selection, whereas other parameters for describing the quality of products. The following section describes the various protocols that are followed for sensory evaluation of various products followed by special emphasis on protocols that are used for evaluation of sensory attributes of FRK and its products.

15.3.1 Use of Trained Panels vs Consumer Panels

Trained sensory panels are framed with individuals who have undergone specific training to develop elevated sensitivity and consistency in analysing responses of various sensory organs towards a food product. The primary aim of the trained panel is to provide an objective and detailed analysis of sensory attributes. These panels use structured methods like quantitative descriptive analysis, triangle test, duo-trio test and ranking test (Pande et al. 2025). Trained analysts are capable of detecting minute differences in sensory attributes of the products, making them especially valuable in product formulation, quality control and consumer acceptance. The utilisation of trained panels for sensory evaluation shares the key advantage of better discrimination ability, higher prediction consistency and the ability to identify specific changes in flavour and texture. These trained panels play a vital role in investigating the effect of ingredients or fortification on specific sensory attributes (Prom-u-thai et al. 2009). In a study evaluating the sensory attributes of FR, the trained panel was able to identify the alteration in chewiness and flakiness of rice due to fortification of iron. However, trained panels also have some limitations. The process is resource-intensive, and the panel needs expensive training to attain the sensory sensitivity for a particular attribute. Moreover, the data derived from descriptive analysis of fortified foods with trained panels do not coincide with the findings of consumer preferences, as the panel may overlook everyday consumer responses (Beinner et al. 2010b).

In contrast, consumer panels consist of randomly selected semi-trained or untrained panellists from the target population. These panels are used to evaluate the overall acceptability of food products, along with their likeness and market potential (Woods and Gallego-Castillo 2020; Beinner et al. 2010b). The methods employed in consumer panels are simple and preference-oriented. They mainly consist of 5-point, 7-point or 9-point hedonic scale, paired comparison and double-blind acceptability tests (Hof 2006; Rai et al. 2019). Though consumer panels are scientifically non-systematic, they provide direct insights into product's market performance and relatable consumer acceptance for a specific population. This advantage of consumer panel-based sensory analysis is an indispensable tool to predict product launch, market studies and public health investigations such as the rice fortification programme. Despite having an in-depth impact on product acceptability, these evaluations have very low discrimination ability, higher variability and minimum repeatability between samples and populations.

15.3.2 Sensory Methods

Sensory evaluation is a science used to evaluate, measure, analyse and interpret human sensory responses (vision, smell, touch and taste) to the properties of food. There are several sensory procedures that include sample identification, description of sensory attributes and analyse consumer preferences (Varela and Ares 2012). Broadly, the sensory methods are classified into three categories that are described as follows.

15.3.3 Discrimination Test

Discrimination tests are usually used to identify observable sensory differences between fortified and non-FR. These tests often involve trained or untrained judges who are asked to identify odd samples among the set considered for analysis. Among several available protocols, the triangle test and duo-trio test are most often considered to analyse the FR samples.

Triangle Test

The triangle test is mostly used as the discrimination test to identify the perceptible sensory difference based on the specific quality attributes between the two products. The panelists are asked to evaluate and find out the odd sample, followed by data analysis helping to draw a firm conclusion. This method is applicable only when the perceptible difference in the product is small. The easily noticeable or large difference may add constraints in the use of this method (Sinkinson 2017).

Duo-Trio Test

This test considers three samples in total: one is a reference sample and two are separate samples. Among these two samples, one exactly matches the reference samples. In the case of FR, the reference can be either fortified or non-FR. Beinner et al. (2010a) reported on the use of Duo-Trio triangle test to distinguish iron FR and unfortified rice. Initially, panelists could not differentiate the FR on the basis of colour, odour, and taste. However, during further analysis, out of 38 panelists, 17 could identify the FR sample. A similar finding was also presented by Moretti et al. (2006) wherein 18 panelists conducted the triangle test with rice fortified with ferric pyrophosphate and found that the extruded rice containing ferric pyrophosphate 5 mg/100 g closely resembled unfortified rice. Rai et al. (2019) found that 37.5% of respondents identified uncooked FR, whereas 39.3% of respondents identified cooked FR. If less than half of the panellist could detect the difference, it is assumed that there is no perceptible differences or minimum difference between the conventional and fortified products.

15.3.4 Descriptive Tests

Descriptive tests give a detailed analysis of various sensory attributes of a product, enabling the quantification and characterisation of each sensory response. As the descriptive analysis demands utmost precision in giving responses, a trained panellist must be involved in conducting these tests. Studies have followed the protocols of descriptive analysis to understand the effect of fortification on the sensory attributes such as appearance, texture and flavour along with physical properties such as length, width, weight, shape, chalkiness and appearance. Though descriptive analysis is capable of detailing the exact responses for sensory attributes, there are very few studies of FR.

15.3.5 Affective Tests

Consumer acceptance tests, also known as affective tests, are used to evaluate whether consumers accept or reject the fortified food product. These tests are especially valuable in evaluating the acceptability of FR as they play a critical part in determining the role of fortification on consumer perception. An effective test is mostly used with consumer panels that are generally of an untrained nature. These untrained panels usually evaluate the product based on basic sensory attributes like colour, aroma, taste, appearance, and overall acceptability.

Hedonic Scale

The most commonly used sensory evaluation methodology is the hedonic scale. It assumes the preference of panellists on the basis of "degree of like" from extremely like to extremely dislike. The hedonic scale can be categorised as 5-, 7- or 9-point hedonic scale. Sereewat et al. (2015) used a 5-point hedonic scale to judge the colour and texture of rice-based spaghetti. While the texture of cooked rice spaghetti was acceptable, the colour of cooked spaghetti received a lower score. Rai et al. (2019) conducted an acceptability study of the FR among 195 households in Nepal and reported that the sensory qualities of the FR were acceptable with 3.9 scores out of 5.

The study on iron FR found that a mean score for appearance, aroma, taste and colour ranged from 5.6 to 5.9 on a 7-point hedonic scale. The data was interpreted as "like" to "like moderately", which shows that iron fortification did not impact the sensory attributes of FR (Beinner et al. 2010b). Woods and Gallego-Castillo (2020) and Rai et al. (2019) also followed the same approach of utilising a 7-point and 5-point hedonic scale for sensory evaluation of zinc FR in Bolivia and Colombia and Iron FR in Nepal, respectively. The Bolivian and Colombian varieties showed a significant effect of zinc fortification on physical properties such as colour, texture and cooking behaviour. In case of the rice fortification programme in Nepal, FR scored 3.9 on a 5-point hedonic scale for overall acceptability, with no statistical significance between fortified and non-FR, revealing the consumer acceptance of fortified products. Though studies have shown that fortification does not negatively affect the sensory properties of rice, the level of fortification especially needs to be considered to evaluate the alteration of sensory attributes. Additionally, the properties such as hue, brightness and grain weight influence the consumer acceptability in the markets where the visual appeal of the rice is given more importance than its nutritional value. Moreover, the methods of cooking also need to be considered for their effectiveness on sensory attributes and consumer acceptability.

15.4 STATISTICAL METHODS USED FOR ANALYSIS

The data obtained from sensory analysis must be evaluated to analyse its statistical significance. The common statistical tools that can be used for analysing statistical data are univariate and multivariate statistical analysis. Apart from these tools, preference mapping and regression analysis can also be used to evaluate the data obtained from the evaluation of sensory attributes.

15.4.1 Univariate Analysis

Univariate methods are fundamental for sensory data analysis because they allow the evaluation of a single sensory variable at a time. The most commonly applied method is analysis of variance (ANOVA), which identifies statistically significant differences between products or populations. ANOVA can be used either with fixed models, where all factors are fixed, or with mixed models, which include both fixed and random effects accounting for variance in fortification. However, while applying ANOVA, it is very important to consider the presence of replications and potential sources of error that can affect the reliability of the conclusion. There are several studies that reported the effectiveness of ANOVA in analysing the data of sensory attributes of FR products (Adanse et al. 2022; Beinner et al. 2010b). These studies evaluate overall acceptability, appearance, colour, aroma, taste, and texture and found that there is very least significant effect of fortification on rice products.

15.4.2 Multivariate Analysis

Sensory data represents various attributes that are evaluated simultaneously, making multivariate statistical analysis a significant tool to draw meaningful insights. Principal component analysis (PCA) and Generalised Procrustes analysis (GPA) are the most commonly used multivariate analysis techniques of sensory attributes. PCA is used for uncovering latent structures in data and reducing dimensionality. It provides visual plots that help interpret how sensory variables contribute to product differentiation and clustering (Mandliya et al. 2023; Pande et al. 2025). GPA is another multivariate method applied to assess differences in vocabulary and scaling of sensory attributes. GPA aligns matrices of individual panellists to a consensus configuration, effectively minimising individual bias and enhancing agreement visualisation. Statistical significance in GPA can be assessed using Monte Carlo simulation. Despite being effective in evaluating differences between statistical data, FR has not been exclusively studied.

15.4.3 Other Analytical Tools

Partial least squares regression (PLSR), Principal component regression (PCR), conjoint analysis, multidimensional scaling (MDS) and cluster analysis are some of the other multivariate analysis techniques that can be implemented for statistical analysis of data obtained from the evaluation of sensory attributes. Modern sensory analysis relies on statistical software that simplifies data handling and interpretation. Consumer check is one-of-a-kind, user-friendly, open-source platform specifically designed for sensory and consumer science, offering tools like PCA, PLSR, PCR, Preference mapping and Conjoint analysis. Other platforms like R-package provide a wide range of tools that can be utilised for sensory data statistical analysis. Furthermore, Python, MATLAB and SAS software can also be implemented for sensory data evaluation.

15.5 INFLUENCE OF RAW MATERIALS AND PROCESS PARAMETERS ON SENSORY ATTRIBUTES

15.5.1 Influence of Raw Materials

The quality of raw material significantly impacts final quality of finished food products altering nutritional, sensory and technological aspects. High-quality food products can only be produced from high-quality raw materials as raw materials influence the end products quality to a significant level. These quality parameters include physical quality parameters, chemical composition and sensory quality. The quality of FRK is very highly dependent on its raw material, which includes rice flour and VMP. This section elaborates the effect of raw material changes on the quality of FRK.

Rice Broken/Flour

Rice broken or rice flour is utilised as a basic ingredient for the manufacturing of FRK. The quality parameters such as impurities/dockage, particle size of rice flour, amylose and amylopectin content significantly affect the quality of FRK. Impurities and dockage in raw rice can negatively affect the quality of FRK. During milling, dockage like husk gets reduced into fine particles and gets mixed with rice flour, which consequently becomes part of FRK. These fine particle impurities negatively affect colour, density, water absorption capacity, degree of gelatinisation, texture and structural integrity of FRK. Maintaining the structural integrity of the rice kernel is crucial for the nutrient delivery through FRK. This makes it important to remove the extraneous matters from the raw material.

Furthermore, the particle size of rice flour plays a critical role in physical, textural and techno-functional properties of FRK. Fine particles are usually associated with increased surface area, high water absorption capacity and enhanced solubility. Physically, smaller rice flour particles tend to absorb more water, subsequently increasing the viscosity at the time of extrusion allowing greater breakdown and gel hardness. This in turn increases the hardness of the kernel and affects opacity. Flours with smaller particle size also tend to gelatinise at lower temperature influencing cooking behaviour of FRK. In terms of texture, smaller particles improve the quality of rice-based products. Considering these aspects, the advised particle size of rice flour is 150–250 μm (Standard 2021).

The amylose and amylopectin content in rice starch significantly influence textual, cooking and organoleptic properties depending on expansion, hardness and gelatinisation of FRK (Nithya, Dalbhagat, and Mishra 2022; Nithya et al. 2024). Rice starches with 20%–32% amylose exhibited the highest expansion ratio, which provide dry and fluffy texture during consumption. In contrast, starch with 55% amylose had lower expansion ratio, which gives hard texture. Amylopectin, on the other hand, retains moisture during extrusion by gelatinisation, which in turn expands and affects the density and volume of FRK. This revealed that the alteration of amylose-to-amylopectin ratio shows significant effect on texture and taste of FRK. Lower amylose content in rice with lower amylose/amylopectin ratio has an adverse effect on taste of rice (Liu et al. 2025). A higher degree of starch gelatinisation with elevated degree of starch damage in FRK containing higher amylose gives positive effect on its taste.

Vitamin-Mineral Premix

Incorporating VMP in FRK can alter sensory attributes, particularly colour, appearance and overall acceptability. The specific type of iron and vitamins used in VMP manufacturing plays a critical role, as VMP causes noticeable changes in FRK and alters its natural rice-like resemblance (Hof 2006). The source and concentration of iron in FRK significantly influence its sensory qualities. Iron compounds like ferrous sulphate, ferric pyrophosphate and sodium-EDTA-iron are the prescribed sources of iron. Sources like ferrous sulphate and sodium-EDTA-iron have high bioavailability; however, their distinct colour often causes discolouration reducing the acceptability of FRK (Hof 2006). To tackle this issue, micronised ferric pyrophosphate is preferred due to its lower interference with colour. However, the micronised FePP with particle size of D-90, 1–3 μm, minimally alters sensory attributes with superior bioavailability (Standard 2021).

Vitamin sources in FRK can subtly affect sensory quality attributes, particularly colour and appearance. Though less impactful than iron when used in small amounts (Hof 2006; Dalbhagat and Mishra 2019).

15.5.2 Influence of Process Parameters

The sensory quality of FRK is strongly influenced by extrusion parameters, viz., moisture content, temperature and screw speed. Moisture content of conditioned rice flour plays a crucial role in physical and sensory quality attributes. An optimal moisture content, nearly 30%–32%, helps to achieve a desirable degree of gelatinisation improving softness and chewiness. However, too much moisture can lower the expansion ratio, alter the density and negatively impact brightness, hue and visual appeal.

Extrusion temperature, particularly barrel and die temperatures, significantly affects the performance of an extruder. Extrusion temperatures higher than optimised levels can negatively alter colour density and eating texture of FRK. Contradictory, lower extrusion temperature reduces the degree of gelatinisation, resulting in brittle grains that disintegrate during cooking.

Screw speed directly affects gelatinisation and the conversion of rice matrix to a viscoelastic melt. Optimised screw speed generates effective compaction providing FRK its structural integrity and colour. Optimal screw speed led to improved product expansion and a more desirable chewy and firm texture. In a nutshell, careful optimisation of the parameters ensures high-quality FRK and FR manufacturing (Dalbhagat and Mishra 2019).

15.6 FACTORS INFLUENCING CONSUMER ACCEPTANCE

There are numerous factors that influence the consumer acceptance of the FR due to various facts including education, awareness and acceptance, which play a major role in its market establishment. Furthermore, different other characteristics exhibited by the FR leading to the change in its consumer acceptance are discussed.

15.6.1 Visual Appeal and Grain Uniformity

Visual appeal and uniformity of grains play a significant role in the victorious market adoption of the FR with improving its economic value (Prom-u-thai and Rerkasem 2020). Uniformity, appearance and shape of the grain establish the blending of the FR with the regular rice. The problems arise due to consumer expectation of rice grain being translucent, grain uniformity and density difference causing FR to float on the water as well as whiteness, which is not completely achieved by FR due to browning reactions occurred during extrusion (Scheuchzer et al. 2023). Henceforth, the process of rice fortification is optimised in such a way that the maximum FR produced corresponds to conventional rice closely. This leads to the enhanced consumer satisfaction.

15.6.2 Cooking Behaviour and Eating Quality

FR is enriched with micronutrients, viz., iron, folic acid and cyanocobalamin, which might cause any culinary change along with sensory aspects of the rice. Furthermore,the change in the cooking time of FR due to rice gelatinisation, protein bond disruption causing enhanced water uptake. This factor also leads to sticky rice unlike other rice varieties produced in India (Wahengbam, Green, and Hazarika 2019). Henceforth, these changes in the cooking behaviour may also cause changes in the eating quality such as hardness and gumminess. However, these changes stand to be acceptable till the required level of the fortification is achieved. Thus, the application of numerous optimisation techniques in the process control would result in the similarity of FR with the regular rice varieties (Dalbhagat and Mishra 2019).

15.6.3 Perception of Fortified vs Non-fortified Rice

The lack of awareness about fortification leads to misconception of it merely being a low-quality rice due to lack of translucency and non-uniform kernel size (Scheuchzer et al. 2023). Additionally, the misconception also arises from the fact that during washing of rice, if the rice is not well processed, it leads to its density difference from the regular rice and ends up being removed due to the perception of it being a low-quality rice (Scheuchzer et al. 2023). Besides that, the unfortified rice has grain size uniformity and translucency. Additionally, the addition of micronutrients as well may change the characteristics of the kernel such as dark colour and non-uniform size (Scheuchzer et al. 2023). Henceforth, the resemblance of the FR with the non-FR plays an important role in overall acceptability, followed by the awareness regarding the same within the people.

15.7 STRATEGIES TO IMPROVE CONSUMER ACCEPTABILITY

15.7.1 Optimisation of Extrusion and Blending

The optimisation of the extrusion and blending processes of the FR is pivotal in the production of supreme quality FRK. The main quality disputes are reading its physical appearance and sensory attributes. In the extrusion process, temperature, moisture content, screw speed, particle size and composition are necessary for the optimisation and production of high-quality rice analogue (Dalbhagat and Mishra 2019). Temperature plays an important role in achievement of optimum gelatinisation to form stable conventional rice like kernels. Furthermore, the speed of extruder determines the residence time of the product, which influences the gelatinisation degree along with the overall quality of the product. This factors also influences the final shape and texture. The blending ratio is optimised to 1:100 for FRK to unfortified rice kernels. This helps in achieving the overall nutritional requirement as per RDA. Furthermore, the FRK is required to mimic the conventional rice in its appearance, colour, taste and aroma such that the batch homogeneity is maintained. This is followed by the nutrient retention during processing to ensure that, the micronutrients required by the RDA reach the consumers.

15.7.2 Use of Food-Grade Additives to Improve Appearance

Fortification although beneficial nutritionally causes challenges in the sensory attributes. These factors cause negative depiction in the consumers perception. The addition of iron along with the browning reactions leads to darker colour than the regular rice, which can be curbed down by utilisation of the stabilised, encapsulated VMP and reduced browning. Furthermore, the texture deficiencies such as stickiness, flakiness and chewiness reduce its overall acceptability, which can be reduced by utilisation of starches acting as texturisers along with the food-grade hydrocolloids giving the desired mouthfeel. Additionally, the addition of the micronutrients may also lead to off flavour and undesirable aroma production, which can be reduced by the utilisation of the food-grade flavour enhancements that may mimic the regular rice aroma such that successful masking could be achieved (Miller and Welch 2013).

15.8 SENSORY EVALUATION OF FR

The sensory study of FR can be done in three ways, viz., discrimination, descriptive and affective test. It is desired that the FRK should not be distinguishable when blended with raw/parboiled rice. Therefore, the sensory evaluation of FR employs protocols similar to those used for FRK, assessing attributes including appearance, colour, aroma, texture and overall acceptability. For sensory evaluation, the FR is prepared by blending the FRK in raw/parboiled rice in the ratio of 1:100 (FSSAI guideline 2021). The triangle test was performed to assess the distinguishable difference between the raw and cooked unfortified and FR samples (Figure 15.1).

The number of participants (*N*), the maximum number of correct answers and the maximum allowable proportion of distinguisher (P_d) were taken from the standard table of triangle test (Meilgaard, Carr, and Civille 1999). All the participants said that one of the samples is different from the given triangle of uncooked and cooked rice. Around 20.8% (5 panelists) and 29.16% (7 panelists) of participants were able to correctly find the raw and cooked FR. The result of triangle test indicated that the number of correct answers was lower than the maximum allowable correct answer (Table 15.2). This indicated that there was no significant difference between the raw and cooked unfortified rice and FR.

The acceptability study was conducted using the 9-point hedonic test in which cooked unfortified rice and FR were evaluated based on liking score for appearance, colour, aroma, texture, taste and overall acceptability (Figure 15.2). A total of 29 panelists participated in the study. The result indicated that the mean hedonic score of quality attributes of FR was close to that of the unfortified rice (Table 15.3).

(a) (b)

FIGURE 15.1 Photographs of (a) uncooked and (b) cooked unfortified rice and FR samples prepared for the triangle test.

TABLE 15.2
Result of the Triangle Test for FR

Particulars	Uncooked Rice	Cooked Rice
Number of panelists participated	24	24
Observed % of respondents identifying FR	20.8	29.16
Respondents who correctly identified FR	5	7
Maximum allowable correct answers	13	13
*p-value	0.10	0.73

Source: *Dalbhagat (2020).*
The statistical analysis is based on P_d=50%, α=0.05 and β=0.05; *p-value is calculated from the χ^2 test.

The statistical analysis showed that there was no significant difference found between the quality attributes of FR and unfortified rice (Table 15.3). The radar plot indicating the quality attributes of FR and unfortified rice is shown in Figure 15.3. These studies indicated that the blending of FRK in raw rice in 1:100 proportion did not alter the quality attributes in FR, and the FR was accepted by all panelists.

15.9 REGULATORY AND QUALITY CONTROL ASPECTS

The maintenance of the regulatory framework of the FR is pivotal in order to ensure public health impact and compliance of the product. However, there are no regulatory protocols for sensory evaluation of fortified products. The rice fortification guidelines

FIGURE 15.2 Photographs of cooked unfortified rice (a) and FR (b) prepared for the 9-point hedonic test.

TABLE 15.3
Acceptability of FR and Unfortified Rice

Parameters	FR	Unfortified Rice	t-value	p-value
Appearance	7.37 ± 0.82^a	7.62 ± 0.97^a	0.98	0.33
Colour	8.00 ± 0.92^a	7.75 ± 1.24^a	–0.84	0.40
Aroma	7.37 ± 1.14^a	7.20 ± 1.04^a	–0.60	0.55
Texture	7.24 ± 1.21^a	7.41 ± 1.11^a	0.56	0.57
Taste	7.58 ± 1.08^a	7.27 ± 1.19^a	–1.04	0.30
Overall acceptability	7.55 ± 1.08^a	7.46 ± 1.07^a	–0.31	0.76

Source: Dalbhagat (2020). Similar letter in the superscript within the row indicate no significant difference between means and was analysed using post-hoc Tukey test at $p < 0.05$.

were given earlier in 2012 by the World Health Organization (WHO) in collaboration with the Global Alliance for improved nutrition (GAIN). In order to curb micronutrient deficiencies such as anaemia, India is also promoting rice and wheat fortification. Thus, the formation of legislation along with standards and monitoring is mandatory in order to specify the rice fortification. Despite this, global implementation of fortification rules on ground performance is still concerning. The data received from 20 national fortification programmes suggested that in 12 countries, less than half fortifications are adequately done as per standards (Mannar and Hurrell 2018). The major barrier regarding this includes unclear government leadership, a frail enforcement

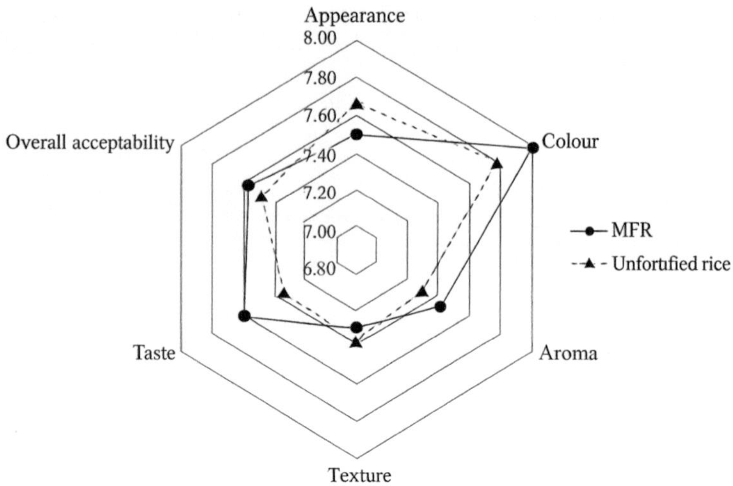

FIGURE 15.3 Radar plot indicating the quality attributes of FR and unfortified rice.

mechanism and insufficient financial aids. Therefore, technical capacity, sustainable leaderships, accountability and funding would aid in improving compliance of these regulatory frameworks. Additionally, quality and sensory attributes such as appearance, aroma, taste and cooking quality impact the consumer acceptability. These are somehow directly associated with the processing conditions resulting in the change in microstructure. Henceforth, there is a requirement of effective monitoring regulation in order to curb down the micronutrient deficiency globally, followed by collaborative working of FR producers and regulatory bodies.

15.10 FUTURE PERSPECTIVES AND CONCLUSIONS

The sensory evaluation of FRK and FR presents several promising future prospects and research gaps. One major area of prospects is conducting longitudinal studies. Most of the current studies evaluate recently developed FR products and the single-time response of the consumer towards these products. On the other hand, longitudinal studies are essential to assess the consumer response and acceptability of fortified foods, mainly FR, with time and with nutritional education. Although the FR is acceptable in terms of its sensorial aspects, the cost is still a hurdle. However, in future, the expanded work within numerous populations, demographics and dietary habits could change into positive acceptance of FR in the market. Additionally, there is a requirement for a standardised sensory lexicon, which positively impacts consumer decisions by constant description of aroma, texture and flavour, making it clear to understand the weak points. This would lead to the production of FR that is comparable in quality to regular rice. It should also be ensured that the cultural relevance of the rice is achieved as well to penetrate the market. Henceforth, modern tool applications such as digital sensory tools assist in automated data collection, digital image creation and analysis, as well as machine learning for enhancing the basis of evaluation of the quality of FR. These tools can also be utilised to understand the product and consumer relationship such as by improving the physicochemical properties of the FR and simultaneously understanding the consumer preferences ultimately driving for a better future.

REFERENCES

Adanse, John, R. S. Kwakudua, K. Bigson, and A. Serwah. 2022. "Quality and Sensory Characteristics of Cookies Fortified with Soybean and Rice Bran Blended Flour." *Asian Journal of Dairy and Food Research* 41 (3): 356–60.

Ayoub, Ali, Yankai Liu, Dennis D Miller, and Syed S. H. Rizvi. 2013. "The Effect of Low Shear on the Development of Fortified Extruded Rice Products." *Starch – Stärke* 65 (5–6): 517–26.

Beinner, Mark A., Gustavo Velasquez-Meléndez, Milene C. Pessoa, and Ted Greiner. 2010a. "Iron-Fortified Rice Is as Efficacious as Supplemental Iron Drops in Infants and Young Children." *The Journal of Nutrition* 140 (1): 49–53.

Beinner, Mark Anthony, Anne Danieli Nascimento Soares, Ana Laura Antunes Barros, and Marlene Azevedo Magalhães Monteiro. 2010b. "Sensory Evaluation of Rice Fortified with Iron." *Food Science and Technology International* 30: 516–19.

Dalbhagat, Chandrakant. 2020. *Studies on Manufacture of Micronutrient Fortified Rice Using Extrusion Technology*. Kharagpur: Indian Institute of Technology Kharagpur.

Dalbhagat, Chandrakant Genu, and Hari Niwas Mishra. 2019. "Effects of Extrusion Process Conditions on System Parameters; Physicochemical Properties and Cooking Characteristics of Extruded Fortified Rice Kernels." *Journal of Cereal Science* 89 (May): 102782.

Dalbhagat, Chandrakant Genu, and Hari Niwas Mishra. 2021a. "Drying Modeling, Cooking Characteristics, Pasting Properties, and Crystallinity of Fortified Rice Kernels." *Journal of Food Processing and Preservation* 45 (6): e15579.

Dalbhagat, Chandrakant Genu, and Hari Niwas Mishra. 2021b. "Effect of the Drying Process on the Color Change, Fissure Development, and Morphology of Fortified Rice Kernels." *Journal of Food Process Engineering* 44 (7): e13719.

Dalbhagat, Chandrakant Genu, A. Nithya, Shubham Mandliya, Siddharth Vishwakarma, and Hari Niwas Mishra. 2024. "Effect of Microwave Drying and Tempering on Color Attributes, Fissure Formation, and Cooking Characteristics of Fortified Rice Kernels." *Journal of Food Science and Technology* 61 (4): 706–16.

FSSAI. 2021. Food Safety and Standards (Fortification of Foods) Regulations, 2018.

Hof, Jessica Rose. 2006. *Vitamin and Mineral Retention and Sensory Evaluation of Extruded Fortified Rice*. "India." 2025. https://www.ffinetwork.org/india-region.

Kaur, Sukhdeep. 2016. "Iron Deficiency Anemia (IDA): A Review." *International Journal of Science and Research* 5 (4): 1999–2003.

Lin, J.-Y., W. Fan, Y.-N. Gao, S.-F. Wu, and S.-X. Wang. 2010. "Study on Volatile Compounds in Rice by HS-SPME and GC-MS" In *Proceedings of the 10th International Working Conference on Stored Product Protection, 27 June to 2 July 2010 Estoril, Portugal* 425:125–134.

Liu, Aihua, Tinghong Ye, Yangyang Zhang, Shipeng Liao, and Xiaokun Li. 2025. "Effects of Potassium Fertilization on Grain Yield, Taste Quality, and Starch Characteristics of Rice (*Oryza Sativa* L.) Grain." *Journal of the Science of Food and Agriculture* 105 (5): 3036–44.

Mamoriska, Sonya, Moch Gelar Hidayat, Cynthia G. Magda, Astri Yuliarti, Eny Cahyaningsih, Eriel Manalu, Sambudi Sambudi, and Rista Yulia Kusuma Putri. 2022. "Characterization of Fortified Rice (Fortivit) and Biofortified Rice (Nutri Zinc)." *JURNAL PANGAN* 31 (2).

Mandliya, Shubham, Jayshree Majumdar, Sourav Misra, Monalisha Pattnaik, and Hari Niwas Mishra. 2023. "Evaluation of Dry Microwave and Hot Water Blanching on Physicochemical, Textural, Functional and Organoleptic Properties of Indian Gooseberry (Phyllanthus Emblica)." *Journal of Food Measurement and Characterization* 17 (3): 2881–91.

Mannar, M. G. Venkatesh, and Richard F. Hurrell. 2018. "Chapter 1- Food Fortification: Past Experience, Current Status, and Potential for Globalization." In *Food Fortification in a Globalized World*, edited by M. G. Venkatesh Mannar and Richard F. Hurrell, 3–11. Academic Press.

Meilgaard, Morten C., B. Thomas Carr, and Gail Vance Civille. 1999. *Sensory Evaluation Techniques*. CRC Press.

Miller, Dennis D., and Ross M. Welch. 2013. "Food System Strategies for Preventing Micronutrient Malnutrition." *Food Policy* 42: 115–28.

Mishra, Abhinav, Hari Niwas Mishra, and Pavuluri Srinivasa Rao. 2012. "Preparation of Rice Analogues Using Extrusion Technology." *International Journal of Food Science and Technology* 47 (9): 1789–97.

Moretti, Diego. 2006. "Iron Fortification of Rice: A Potential Strategy to Counteract Iron Deficiency?" PhD diss., ETH Zurich, 2006.

Moretti, Diego, Michael B. Zimmermann, Sumithra Muthayya, Prashanth Thankachan, Tung-ching Ching Lee, Anura V. Kurpad, and Richard F. Hurrell. 2006. "Extruded Rice Fortified with Micronized Ground Ferric Pyrophosphate Reduces Iron Deficiency in Indian Schoolchildren: A Double-Blind Randomized Controlled Trial 1–3." *The American Journal of Clinical Nutrition* 84 (4): 822–29.

Nithya, A., Chandrakant Genu Dalbhagat, and Hari Niwas Mishra. 2022. "A Comparative Study on the Physicochemical, Cooking and Textural Properties of Fortified Rice Kernels Prepared from Raw and Parboiled Rice." *International Journal of Food Science and Technology* 57 (2): 1325–32.

Nithya, A., Siddharth Vishwakarma, Chandrakant Genu Dalbhagat, and Hari Niwas Mishra. 2024. "Apparent Amylose Content Positively Influences the Quality of Extruded Fortified Rice Kernels." *Carbohydrate Polymers* 338 (March): 122213.

Pande, Ranjana, Srishty Tyagi, Siddharth Vishwakarma, and Shubham Mandliya. 2025. "Dry Microwave and Hot Water Blanching of Underutilized Star Fruit (*Averrhoa carambola*): A Comparative Assessment of Quality Attributes and Sensory Evaluation Using Chemometric Approach." *Food Physics* 2: 100056.

Prom-u-thai, Chanakan, and Benjavan Rerkasem. 2020. "Rice Quality Improvement. A Review." *Agronomy for Sustainable Development* 40 (4): 28.

Prom-u-thai, Chanakan, Benjavan Rerkasem, Shu Fukai, and Longbin Huang. 2009. "Iron Fortification and Parboiled Rice Quality: Appearance, Cooking Quality and Sensory Attributes." *Journal of the Science of Food and Agriculture* 89 (15): 2565–71.

Rai, Anjana, Macha Raja Maharjan, Helen A. Harris Fry, Parbati K. Chhetri, Purna Chandra Wasti, and Naomi M. Saville. 2019. "Consumption of Rice, Acceptability and Sensory Qualities of Fortified Rice amongst Consumers of Social Safety Net Rice in Nepal." *PLoS ONE* 14 (10): 1–17.

Scheuchzer, Pornpimol, Victoria N Syryamina, Michael Bruce Zimmermann, Christophe Zeder, Laura Nyström, Maxim Yulikov, and Diego Moretti. 2023. "Ferric Pyrophosphate Forms Soluble Iron Coordination Complexes with Zinc Compounds and Solubilizing Agents in Extruded Rice and Predicts Increased Iron Solubility and Bioavailability in Young Women." *The Journal of Nutrition* 153 (3): 636–44.

Sereewat, Pornpimol, Chankavee Suthipinittham, Sirirat Sumathaluk, Chureerat Puttanlek, Dudsadee Uttapap, and Vilai Rungsardthong. 2015. "Cooking Properties and Sensory Acceptability of Spaghetti Made from Rice Flour and Defatted Soy Flour." *LWT – Food Science and Technology* 60 (2): 1061–67.

Sinkinson, Cécile. 2017. "Triangle Test." In *Discrimination Testing in Sensory Science*, edited by Lauren Rogers, 153–70. Elsevier.

Standard, Indian. 2021. "भारतीय मानक प्रबलित चावि दाना-लवलिलटि." www.standardsbis.in.

Varela, Paula, and Gastón Ares. 2012. "Sensory Profiling, the Blurred Line between Sensory and Consumer Science. A Review of Novel Methods for Product Characterization." *Food Research International* 48 (2): 893–908.

Wahengbam, Elizabeth Devi, Brian D. Green, and Manuj Kumar Hazarika. 2019. "Fortification of Zinc in a Parboiled Low-Amylose Rice: Effects of Milling and Cooking." *Journal of the Science of Food and Agriculture* 99 (7: 3434–42.

Woods, Bo-Jane, Sonia Gallego-Castillo, Elise F. Talsma, and Daniel Álvarez. 2020. "The Acceptance of Zinc Biofortified Rice in Latin America: A Consumer Sensory Study and Grain Quality Characterization." *PLoS One* 15 (11): 1–21.

16 GMP/GHP and HACCP Protocols in Manufacturing Fortified Rice Kernels and Fortified Rice

Nithya A., Malla Vandana, Chandrakant Genu Dalbhagat, and Hari Niwas Mishra

16.1 INTRODUCTION

Micronutrient malnutrition (MNM) or "hidden hunger" is a major public health concern, especially in low- and middle-income countries, impacting millions of people globally. It occurs because of the inability to have enough of the necessary vitamins and minerals. It results in compromised immune function, impaired thinking, reduced cognitive function, and reduced general well-being. Rice fortification has emerged as a cost-effective and scalable intervention to address MNM. It is known to be effective in regions where rice is a staple food consumed daily by large segments of the population (Piccoli et al. 2012; Junaid-Ur-Rahman et al. 2022). Fortification of rice is not as simple as adding required vitamins and minerals, but involves several processing steps. Each poses distinct food safety hazards, be it biological, physical, and chemical hazards. Hence, to ensure that the final product is safe, stable, and nutritionally beneficial, it involves tight control at all phases of production, without which fortified rice will not only fail to provide its intended health benefits, but it may even pose possible health risks. To overcome this, it is necessary to implement a structured food safety system. This system includes Good Manufacturing Practices (GMP), which establish hygienic, controlled manufacturing processes. Good Hygiene Practices (GHP), which ensure sanitation and minimize possibilities of contamination, and Hazard Analysis and Critical Control Points (HACCP), which analyse and assess systematically food safety hazards and manage them. By harmonizing these principles, manufacturers can uphold safety standards and deliver the high-quality product that is safe and wholesome.

DOI: 10.1201/9781003661245-16

16.1.1 Understanding GMP, GHP, and HACCP

Food safety is a critical and essential factor in fortified rice production. Robust food safety system with stringent protocols is needed to achieve it. GMP, GHP, and HACCP systems work together in unison to make sure that all steps of production that are from receiving raw materials to final packaging are conducted under safe controlled environments.

- **Good manufacturing practices (GMP)**: They are the standard rules or guidelines of sanitation in food manufacturing, processing, packing, or holding food in a food facility. GMP is the pillar of safety in any food processing system, with standard procedures guaranteeing that products are processed and inspected consistently to meet established quality standards. The effective implementation of GMP in food processing serves to reduce errors, chances of contamination, and quality issues in the goods manufactured.
- **Good hygiene practices (GHP)**: They supplement GMP by emphasizing cleanliness surrounding the facility and among employees. It calls for frequent sanitizing of equipment and working areas, disposal of waste, personal cleanliness of workers, and pest control. These practices create a hygienic working condition and space that guarantees the safe manufacture of fortified rice.
- **Hazard analysis and critical control points (HACCP)**: It is a preventive approach that addresses identifying and controlling the risk prior to when it becomes a threat. Every step of the process is examined in detail to determine the potential biological, chemical, or physical hazards. Critical control points (CCPs) are determined and controlled to keep them at safe levels. The system also includes ongoing monitoring and corrective actions to maintain product safety and quality.

By incorporating GMP, GHP, and HACCP systems in the production line, fortified rice manufacturing facilities can produce good quality and safe products. On a longer run, they contribute to the nutrition security by making the safe and healthy fortified products available to the targeted population (De Oliveira et al. 2016; Varghese et al. 2022). This chapter details the implementation of food safety protocols in producing fortified rice and FRK. It also provides insights for food manufacturers, researchers, and policymakers engaged in large-scale fortification.

16.2 GOOD MANUFACTURING PRACTICES IN FORTIFIED RICE PRODUCTION

GMP guarantee the safety, quality, and uniformity of fortified rice by controlling each step, from raw material procurement to final product analysis. They prevent contamination, enhance shelf life, and decrease food recalls and rejections. Important aspects are facility layout, sanitation, equipment care, and documentation. Every

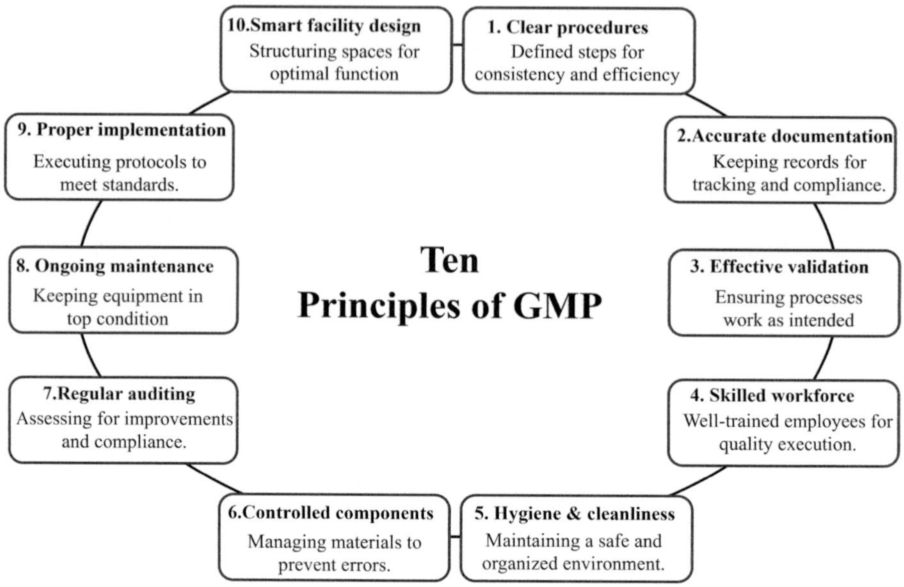

FIGURE 16.1 Key principles of GMP in the food processing industry.

fortified rice plant must create a site-specific GMP plan that addresses its specific operations and hygiene requirements. The key principles of GMP are summarized in Figure 16.1.

16.2.1 Facility and Infrastructure Requirements

GMP-compliant plant infrastructure reduces contamination risk and guarantees hygienic, organized, and efficient processing. The facility design should have a unidirectional material flow and visitors to prevent cross-contact between the raw materials and the finished product. Physical divisions or designated areas must definitively segregate every phase of production, i.e., all unit operations, to avoid contamination and cross-contamination. The buildings should be of proper size and constructed to allow easy cleaning, maintenance, and sanitary operation. All surfaces, including floors, walls, and ceilings, must be smooth, non-absorbent, and conducive to disinfection. Environmental controls like temperature, humidity, and air filtration systems ensure product stability. HEPA filters and positive pressure air supply systems must be present in cases where aseptic handling is necessary. Air-handling systems should have dust and contaminant control capabilities. Proper lighting should be maintained throughout the facility, including emergency lighting systems. Equipment should be made of food-grade, corrosion-free materials like stainless steel to ensure easy access and maintenance. Pest control devices, like mesh windows, air curtains, and purposefully located traps, should be installed to prevent insects and rodents from entering. The plant must also be provided with basic facilities like potable water, adequate drainage, good ventilation, and temperature control to facilitate hygiene and product quality. All these, when properly in place, help achieve a hygienic plant infrastructure for guaranteed safety, nutritional quality, and consumer acceptability of fortified rice products (Awasti, Sunkesula, and Hammam 2022; Sameen et al. 2023; Walia et al. 2023; Meghwal, Heddurshetti, and Biradar 2017).

16.2.2 Equipment and Process Controls

In rice fortification, equipment design and process control are vital for ensuring product safety, quality, and compliance with GMP. Equipment used at each stage, whether for mixing, extrusion, or drying, must be suitable for food processing. Machines that come into direct contact with food should be made of food-grade, non-toxic, and corrosion-resistant materials. Surfaces must be smooth, non-reactive, and easy to clean to avoid chemical interactions or contamination. Proper layout is equally as important as the construction of equipment. Equipment should be arranged to allow a smooth workflow, minimize cross-traffic, and provide easy access for cleaning and maintenance. Design features that allow thorough cleaning are essential, especially in areas prone to residue buildup. Cleaning-in-place (CIP) systems are ideal for pipelines and tanks, whereas other equipment, such as extruders and blenders, must be cleaned manually. Regular cleaning schedules, complemented by microbial tests (e.g., surface swabs), ensure hygiene and prevent contamination hazards.

Process control is important in ensuring that product quality is consistent. Parameters such as ingredient dosing, temperature, moisture content, and screw speed need to be monitored and controlled at all times. Automation systems such as SCADA (Supervisory Control and Data Acquisition) or PLC (Programmable Logic Controllers) allow real-time monitoring and alarms, enabling operators to take corrective measures in a timely manner when deviations happen (Harbud 2022). This guarantees the fortified rice within quality and safety specifications. Preventive maintenance of all the equipment and periodic calibration of weighing and measuring instruments ensure process precision and prevent delays in production. Every maintenance and calibration procedure needs to be documented and retained for GMP and audit purposes. By choosing the equipment carefully and planning automation of the layout and preventive maintenance, the operation of the fortification plant can attain maximum product quality and food safety levels, lending towards the general success and scalability of rice (BIS 2021a; BIS 2021b).

16.2.3 Raw Material Selection and Handling

Raw material safety and quality form the basis of successful rice fortification. They directly influence the nutritional value, shelf life, and safety of the final product. All raw materials, including rice grains and premixes of micronutrients, need to be provided by certified suppliers who follow strong quality control and traceability systems. All incoming consignments of raw materials must be adequately tested and inspected. This means they must undergo testing for moisture content, microbial contamination, pesticide residue, and heavy metals to meet pre-established safety and quality standards. Standard specifications need to be well-defined for every material, with major characteristics such as identity, purity, active ingredient content, and test methods. Good storage practices are essential. The materials are stored in neat, dry, and pest-free areas with sufficient ventilation. Storage conditions such as temperature and humidity levels should be monitored from time to time. The first-in-first-out (FIFO) rule should always be followed to avoid using expired or spoiled materials.

Hygiene plays a significant role in handling, especially in weighing and mixing. Proper handling techniques should be trained among all the staff, and equipment should be sterilized and cleaned regularly to prevent contamination and maintain uniformity of the nutrients. To guarantee consistent quality, every batch of raw material must be inspected by a quality control system. For imported ingredients, the first batch may be analysed thoroughly, and later batches may be accepted on a Certificate of Analysis (CoA) basis, assuming that quality is guaranteed and transportation did not reduce the material's quality. At least one batch per year for each supplier should be thoroughly tested to verify CoA validity. Retention samples need to be stored for a minimum of one year beyond the expiration date of the product. Lastly, all documents, including test reports, CoAs, and storage records, have to be retained for one year beyond the shelf life of the product. There should also be a recall plan with periodic mock recalls to ensure preparedness (Walia et al. 2023; Harris 2010; Holah 2023). With these steps, rice fortification facilities can provide safe, high-quality products while satisfying regulatory requirements (DFPD 2022).

16.2.4 Personnel Hygiene and Training

In rice fortification plants, the role of personnel in maintaining hygiene and safety is fundamental. The employees working in production must be trained periodically on hygiene procedures, safe handling of raw materials and premixes, and the fundamental aspects of GMP and HACCP. Training is not a one-time thing; it has to be continuous, with records appropriately maintained and audited periodically to ensure everyone remains updated. Only experienced and qualified people should be working or overseeing critical phases in the production line. For those handling hazardous materials, additional training is required to ensure that the product and individuals are safe. Production area cleanliness is not to be compromised. Everyone should be clean, wearing light-coloured uniforms, gloves, and hairnets. No jewellery or personal accessories are allowed to remain within processing areas. Handwashing, which is fundamental, is probably one of the most significant habits, especially after breaks or bathroom trips. Regular health examinations ensure that nobody employed in the facility carries any contagious diseases. If sick, one should be reassigned temporarily until cleared to report back.

Consuming food, fluids, and smoking should be prohibited in processing areas to prevent contamination. Maintenance personnel and visitors must also abide by hygiene practices and use protective equipment when entering sensitive areas. It is advisable to conduct periodic inspections or audits to determine whether hygiene measures are practised rather than assumed. Most importantly, creating a workplace culture prioritizing safety and cleanliness has a concrete impact. When employees know the "why" of the procedures and see the management leading by example, it is less challenging to uphold high standards on a daily basis (Owusu-Apenten and Vieira 2023; Meghwal, Heddurshetti, and Biradar 2017; Holah 2023).

16.3 GOOD HYGIENE PRACTICES IN FORTIFIED RICE MANUFACTURING

The stringent adherence to GHP is necessary while producing the FRK and fortified rice to safeguard nutritional quality and safety. GHP involves protocols aiming at maintaining cleanliness, preventing contamination, and ensuring that the equipment, personnel, and the operating environment comply with the rigorous sanitary standards in all the stages of manufacturing. As the FRK and fortified rice are intended to provide to the anemic population, including children and pregnant women, lapses in hygiene may compromise the product quality and ultimately affect consumer health. The general hygienic and sanitary practices to be followed by food business operators are well discussed in Schedule 4 of the Food Safety and Standards (Licensing and Registration of Food Businesses) Regulation 2011 (FSSAI 2011). This regulation mandates implementing hygienic practices in the production unit along with GMP. The necessary facilities in the production unit must be established to maintain the premise in a hygienic state, including equipment sanitation, waste management, personnel hygiene, and environmental control to avoid microbial contamination. Regular internal audits and compliance checks help establish a strong GHP environment for continuous improvement and accountability.

16.3.1 Personal Hygiene Standards

Maintaining personal hygiene among the working personnel is the foundation of food safety in FRK and fortified rice production, as it is the first critical barrier to avoid contamination. Product contamination due to human factors can be neglected if rigorous personal hygiene standards are implemented in the production unit. The FSSAI Regulation (FSSAI 2011) mandates all the FBOs to undergo regular medical screening of the personnel at least annually and report the illness and contagious conditions. Personnel with symptoms such as open wounds, skin disease, or diseases likely to be transmitted through food must be prohibited in the production and packaging areas. The manufacturing unit should have a system of periodic medical examinations or health checkups to ensure that they are free from any contagious, communicable, or infectious disease. The factory staff should be vaccinated against infectious diseases and epidemics.

Providing dedicated facilities such as handwashing areas, clean uniforms, hand gloves, face masks, footwear, and hair restraints like nets and caps is essential to prevent inadvertent contamination. Personnel handling the raw material, intermediate product, and finished products or the operators must routinely wash and sanitize their hands before starting the work and following any activity that may cause contamination of microorganisms, such as eating, smoking, or using the toilet/washroom. While operating the equipment/machinery, the working staff should refrain from smoking, chewing, spitting, sneezing, or coughing over the FRK and fortified rice. Certain activities can be avoided, such as scratching the nose and rubbing eyes, ears, mouth, and body parts, to prevent contamination of microorganisms. The visitors should be discouraged from visiting the manufacturing unit; however, they may visit the food handling areas with proper care, i.e., wearing protective clothes, hair caps, shoe covers, and facemasks. Beyond basic cleanliness, employees must be trained to follow the personal hygiene practices in the manufacturing unit. Daily checks or inspections with written records should be done to strengthen and reinforce compliance. Moreover, the structured training and periodic audits ensure employees understand and follow the personal hygiene protocols.

16.3.2 Cleaning and Sanitation Protocols

The unit manager must prepare the cleaning and sanitation protocols for each equipment/machinery used in the production unit. The cleaning, maintenance, and sanitation records should be maintained, which indicate the details of the specific areas to be cleaned, and sanitation protocol to be followed for specific equipment and cleaning materials to be used. In some instances, the equipment and parts in contact with the food material must be cleaned and sterilized at a set frequency. The cleaning chemicals should be food-grade and stored separately from the raw ingredients as per the instructions and guidelines provided by the manufacturer. Clear tagging of the containers or areas is given to avoid contamination risk. The sanitation programme specifically targets high-risk zones where the potential growth of microorganisms is expected. Employing colour codes for the tools and equipment reduces the risk of contamination by ensuring the segregation of utensils used from raw materials to end products. Floor and drainage areas are specifically targeted to sanitize as they are the harbour for the growth of pathogenic microorganisms. Training the employees enhances cross-contamination prevention and improves product quality.

16.3.3 Cross-Contamination Preventions

Cross-contamination prevention is imperative in FRK and fortified rice manufacturing units to avert the transfer of harmful substances, viz., physical, chemical, and microbiological, between the raw materials, equipment, and final products. The manufacturing facility must have separate zones for storing raw materials and finished products, processing and packaging section, quality testing facility, etc. It should follow the FIFO principle for the raw materials and finished products. Dedicated, hygienically designed equipment should be used to produce FRK and fortified rice. All equipment and tools must be cleaned and appropriately sanitized before switching the batches to prevent cross-contamination, especially when the ingredients are changed. Magnetic filters, metal detectors, and screens must be employed to catch foreign objects (e.g., metal or glass). Operators should follow the personnel hygienic practices (preventing wearing watches, jewellery, bangles, pens, etc.) to maintain production integrity. Airflow management and proper ventilation in the processing area should also be controlled. The air filters must be installed to allow the filtered air to enter the processing zone. The pressure differential (positive pressure in the processing zone) and hygienic barriers such as air curtains prevent particulate migration.

16.3.4 Storage and Transportation Hygiene

The dedicated storage facility should be in the manufacturing unit for raw materials and finished products (FRK and fortified rice). The storage area must be clean, pest-free, and environment-controlled (temperature and RH). A designated storage facility with a clear FIFO system enables timely utilization of the stored materials and maintains quality. Packaging materials must be stored on pallets or shelves in a clean, dry area. The SOPs for storing FRK and fortified rice should be developed and followed stringently. Record keeping must include batch codes, inbound/outbound dates, and details of environmental conditions (DFPD 2019). Pest control measures should be done periodically. Holes, drains, and other places where the pest may get access should be sealed. The birds, animals, and pets should not be allowed in the manufacturing units.

16.4 HACCP IMPLEMENTATION IN FORTIFIED RICE PRODUCTION

HACCP is a systematic preventive approach to food safety that addresses physical, chemical, and biological hazards as a means of prevention rather than finished product inspection. In fortified rice production, where essential micronutrients such as iron, vitamin B_{12}, folic acid, vitamin A, and zinc are incorporated into rice kernels, HACCP becomes vital to ensure food safety, nutritional efficacy, and regulatory compliance.

16.4.1 Principles of HACCP

At the outset, foundational prerequisites like GMP, personal hygiene, standard operating procedures, and personnel training are established to create a robust base for HACCP development. Five preliminary tasks are completed before implementing the seven HACCP principles (Figure 16.2). The seven HACCP principles encompass hazard analysis, identifying critical control points (CCPs), setting critical limits, monitoring procedures, implementing corrective actions, verification procedures, and maintaining records and documentation.

The HACCP system is structured around seven core principles, each providing a foundation for managing food safety risks in a scientific and structured manner:

- **Conduct a hazard analysis**: Identify the hazards that can occur at any stage of production. In the production of fortified rice, potential biological hazards (such as microbial contamination), chemical hazards (such as over-fortification or contamination by cleaning agents), and physical hazards (such as metal fragments from equipment) must be analysed at every stage.
- **Determine the critical control points (CCPs)**: Identify points in the production process where controls are essential to prevent or eliminate hazards or reduce them to an acceptable level. In fortified rice, typical CCPs include fortificants premix addition, blending, and packaging.
- **Establish critical limits**: Establish the maximum or minimum values to which biological, chemical, or physical parameters must be controlled at a particular CCP. For example, iron content must not exceed the recommended dietary intake.

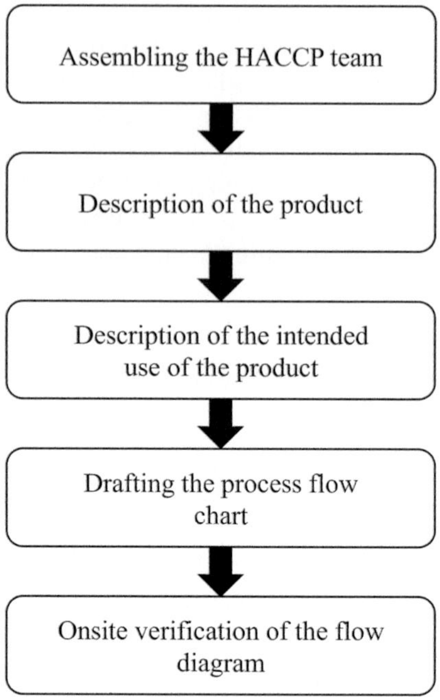

FIGURE 16.2 Five preliminary tasks for HACCP implementation.

- **Establish monitoring procedures**: Develop activities and procedures for the monitoring of CCPs continuously. This includes real-time monitoring of processing parameters such as temperature, nutrient dosage, and physical integrity of rice kernels.
- **Establish corrective actions**: Define specific actions to take when a CCP deviates from its critical limit. This ensures that unsafe products are not released and root causes are addressed.
- **Establish verification procedures**: Apply methods, procedures, and tests in addition to monitoring to confirm that the HACCP system is working effectively. Some examples include internal audits, calibration of equipment, and microbiological sampling.
- **Establish record-keeping and documentation**: Keep a record of hazard analysis, CCPs, critical limits, monitoring systems, corrective actions, and verification results.

This provides traceability, transparency, and compliance with national and international food safety regulations.

16.4.2 Step-by-Step HACCP Application

HACCP system implementation in the production of fortified rice is a stepwise series of sequential steps each designed to fit the unique processes, equipment, and fortification methods used in the plant. In this section, a systematic approach towards the implementation of HACCP in a fortified rice production facility is presented:

Step 1: Assemble the HACCP Team

The first step in HACCP implementation is to form a multidisciplinary team comprising food technologists, quality assurance personnel, production managers, microbiologists, and fortification experts. The team should collectively know about the product, production processes, potential hazards, and control measures.

Step 2: Describe the Product

Clearly define the nature of the fortified rice, including types of fortificants used, physical and chemical characteristics, packaging formats, shelf life, and target consumers. This information forms the basis for hazard identification and risk assessment.

Step 3: Identify Intended Use

Know how the product would be consumed by the public, such as how it would be prepared (e.g., boiling or steaming), and determine vulnerable groups like children, pregnant women, or those with micronutrient deficiencies.

Step 4: Construct a Flow Diagram

Develop a detailed flow diagram outlining all steps from raw material receiving, cleaning, fortificants mixing, extrusion, drying, blending, packing, storage, and distribution. Supporting processes such as sanitation and waste management should also be mapped in the flow diagram.

Step 5: On-Site Verification of the Flow Diagram

Carry out a walk-through of the processing plant to ensure that the flow diagram is correct. Involve operating staff to validate each step of the process.

Step 6: Conduct Hazard Analysis (Principle 1)

Examine each step of the process for potential hazards. For example, the extrusion process may involve chemical hazards due to thermal degradation of nutrients, whereas the packaging stage may introduce physical contaminants like shreds of plastic.

Step 7: Determine Critical Control Points (CCPs) (Principle 2)

Identify CCPs where control is essential to eliminate or minimize hazards. In the case of fortified rice, these points might include where nutrients are mixed, metal detection systems, and the final inspection before packaging.

Step 8: Establish Critical Limits (Principle 3)

Define clear, specific measurable limits for each CCP. For instance, micronutrient content range (e.g., 2.8–4.25 mg/kg iron) and minimum extrusion temperature (e.g., 85°C) can be set to guarantee the safety and effectiveness of the product.

Step 9: Establish Monitoring Procedures (Principle 4)

Create standardized protocols for monitoring each CCP. Methods such as visual inspection, use of checklists, real-time sensor monitoring, and periodic sampling for laboratory testing could be used.

Step 10: Establish Corrective Actions (Principle 5)

Plan clear corrective actions such as isolating non-compliant batches, stopping the production line, and root cause investigation. Reprocessing or disposal of non-conforming products must be documented.

Step 11: Establish Verification Procedures (Principle 6)

Validate the performance of the HACCP plan through regular audits, lab testing, and third-party audits. Verification confirms that the system operates as intended and critical limits are always met.

Step 12: Establish Documentation and Records (Principle 7)

Comprehensive records of all HACCP activities must be kept. These include addition logs for nutrient premix, monitoring checklists, calibration records, training logs, audit reports, and records of corrective action. These records are critical for traceability, accountability, and continuous improvement.

16.5 CHALLENGES AND BEST PRACTICES IN ENSURING FOOD SAFETY

16.5.1 Common Challenges in GMP, GHP, and HACCP Compliance

Ensuring food safety by complying with GMP, GHP, and HACCP requirements in the manufacture of fortified rice often poses significant practical challenges. One of the major issues is the inconsistent or inappropriate implementation of safety protocols. This is often due to a lack of awareness, limited training of personnel, or limitations in the infrastructure of the plant. Most of the workers do not understand the hygiene standards or CCPs completely, which may increase the risk of contamination when handling raw materials and micronutrient premixes. Sometimes, implementation of food safety practices in food processing facilities is often met with resistance from workers because of unfamiliarity or fear of increased workload. Limitations in budget also hinder the adoption of food safety protocols especially wherever upgradation of machinery and training of personnel is required. Another problem is ensuring proper documentation and interdepartmental collaboration. Scaling up operations makes it tougher, and educating external stakeholders and having a company-wide cultural overhaul to prioritize safety and quality requires ongoing efforts. Poorly maintained documentation results in gaps like missing batch record or sanitation logs, which weaken the traceability and accountability. Another major hindrance is the lack of reliable testing methods to check for microbial loads, heavy metals, or the extent of micronutrient degradation.

16.5.2 Best Practices for Food Safety in Fortified Rice Manufacturing

To address the challenges in ensuring food safety protocols, fortified rice manufacturers must adopt a comprehensive set of best practices that integrate quality, hygiene, and safety at every level of production. Firstly, training activities should be performed regularly for all concerned personnel such as operators, supervisors, and members of quality assurance. The training must emphasize food safety fundamentals such as personal hygiene and general GMP/HACCP-specific requirements of fortified rice manufacturing. Training records shall be kept and reviewed from time to time. Strong documentation and record-keeping systems are necessary for traceability and compliance with laws. Standard operating procedures must be made for each process from raw material intake till final packaging. All the process must be carried out adhering to the SOPs. Regular inspections and audits help in detection of possible hazards and enhance adherence to food safety requirements. Proper, detailed documentation enhances traceability of foods, and computerized systems can enhance accessibility and dependability. Employee empowerment support provides a sense of responsibility in employees, and they will feel free to report safety concerns. Process improvement on an ongoing basis such as feedback loops and ongoing process checks enables organizations to refine their food safety processes with time. Finally, a well-defined and frequently simulated recall procedure must be in place. Mock recalls help reveal the vulnerabilities of the recall process and put the facility in a position to respond immediately and responsibly if an actual crisis were to occur. By overcoming challenges and implementing such best practices, fortified rice manufacturers can ensure food safety, remain compliant with domestic and international regulations, and build consumer confidence in the fortified foods they supply.

16.6 PROFORMA FOR GMP, GHP, AND HACCP IMPLEMENTATION

The HACCP protocols for manufacturing FRK (Table 16.1) and fortified rice (Table 16.2) address potential hazards across each processing step, including control of physical, chemical, and microbiological risks through preventive measures like sieving, fumigation, micronutrient analysis, and metal detection. The specific control limits should be monitored under GMP and GHP.

16.7 CONCLUSIONS

The inclusion of GMP, GHP, and HACCP protocols is central to the safe and efficient production of fortified rice and fortified rice kernels. The food safety systems, when used appropriately, minimize the risk of contamination and preserve the quality of fortified rice during processing and distribution. Each of the three systems has a specific role to play like GMP, which focuses on controlling the manufacturing process, GHP, which guarantees cleanliness and hygiene, and HACCP, which involves a systematic process of hazard identification and control. Though practical application can be challenging, especially in settings where resources are limited, strict compliance with these standards is critically important. It builds consumer confidence as

TABLE 16.1
HACCP Protocols for Fortified Rice Kernels (FRK) Manufacturing

Process Steps	Potential Hazard	Critical Limits or Preventive Measures	Monitoring of GMP/CCP
Raw material intake	Presence of foreign bodies and infestation	Free from foreign bodies	Sieving and visual inspection
Rice broken	High moisture content	13% (Maximum)	Lab test (ISO 712-2009)
Vitamin and mineral premix	Micronutrients content	Folic acid – 12.5 g/25kg Vit B_{12} – 0.13 g/25 kg Micronized iron – 4,250 g/25 kg	Laboratory test
	Appearance	Free flowing powder	
	Colour	Off-white	
	Odour	Faint	
Raw Material storage	Infestation	Pest traps Fumigation	Collecting and measuring the number of insects
	Moisture content	As above	
Sieving of rice broken	Metal, stones, plastic, paper, insects, etc.	Metals (<7mm) – magnets Other foreign bodies – sieving and visual inspection	Collecting and measuring the number of contaminants
Weighing of ingredients	Incorrect weighing	Check weighing	Periodical validation with standard weights
Grinding of rice broken	Metal to metal contact Broken sieve	Passing through screen – limit 7mm	Routine check to confirm no screen damage and no metal filings on the screen
	Improper size of rice flour	Passing through sieve size (250 μm)	Checking the amount of overruns
Mixing and conditioning of rice flour with VMP and moisture	Improper mixing	Predetermined mixing time, mixing speed, and amount of water	Moisture content and micronutrient analysis
	Metal-to-metal contact	Visual inspection	Collection of any metal pieces
Extrusion	Incorrect heating Incorrect feeding Incorrect cutting	As per optimum conditions	Regular monitoring of temperature, screw speed, feed rate, and cutter speed
Drying	Improper drying (drying air temperature, velocity, and grain layer thickness)	As per optimum drying conditions	Monitoring of drying air temperature and velocity
Polishing	Foreign body (emery particles)	Sieving and preventive maintenance	Collection and measurement of the contaminants

(Continued)

TABLE 16.1 (*Continued*)
HACCP Protocols for Fortified Rice Kernels (FRK) Manufacturing

Process Steps	Potential Hazard	Critical Limits or Preventive Measures	Monitoring of GMP/ CCP
FRK and normal rice blending	Non-homogeneous mixing	Predetermined normal rice, FRK, and mixing screw speed	Homogeneity testing
Packaging	Metal foreign bodies	Passing the pouches through metal detectors (MD)	Monitoring the number of MD rejections
	Incorrect net weight	Predetermined flow rate	Periodic checking of weight
Product storage	Pest infestation Cross-contamination	Pest traps Insectocutor Fumigation	Collecting and measuring the amount of insects

TABLE 16.2
HACCP Protocols for Fortified Rice (FR) Manufacturing

Process Steps	Potential Hazard	Critical Limits or Preventive Measures	Monitoring of GMP/CCP
Raw materials intake	Presence of foreign bodies and infestation	Free from foreign bodies	Sieving and visual inspection
Rice	High moisture content	13% (Maximum)	Lab test (ISO 712-2009)
	Microbial hazard	SPC 100,000 cfu/g Y&M 1,000 cfu/g	ICC no. 125, AACC 42-11 ICC no. 146, AACC 42-50
FRK	High moisture content	12% w/w (Max) at the time of receipt	Lab test (ISO 712-2009)
	Degradation of micronutrients	Iron – 425 mg/100 g FA – 1250 ug/100 g Vit B_{12} – 12.5 ug/100 g	Lab test
	Microbial Hazard	TPC – 10,000 cfu/g Y&M – 100 cfu/g	ICC no. 125, AACC 42-11 ICC no. 146, AACC 42-50
Raw material storage	Infestation	Pest traps Insectocutor Fumigation As above	Collecting and measuring the number of insects
	Moisture content	Normal rice – 13% FRK – 12%	Lab test (ISO 712-2009)
	Mixing up of materials	Dedicated storage area for NR and FRK Materials covered Proper labeling of materials First in/First out	Periodical auditing and visual inspection

(*Continued*)

TABLE 16.2 (*Continued*)
HACCP Protocols for Fortified Rice (FR) Manufacturing

Process Steps	Potential Hazard	Critical Limits or Preventive Measures	Monitoring of GMP/CCP
Movement of materials	Foreign body Mixing up of materials	Keep the materials covered Proper labeling of materials for traceability	Periodical auditing and visual inspection
Sieving of raw materials	Metal, stones, plastic, paper, insects, etc.	Metals (<7mm) – magnets Other foreign bodies – sieving and visual inspection	Collecting and measuring the number of contaminants
Blending	Non-homogeneous mixing	Predetermined normal rice, FRK, and mixing screw speed	Homogeneity testing Micronutrient analysis
Packaging	Metal foreign bodies	Passing through MD	Monitoring the number of MD rejections
	Incorrect net weight	Predetermined flow rate	Periodical checking of weight
	Segregation of FRK	Preventing vibration	Homogeneity testing of packaged product
Product storage	Pest infestation Cross-contamination	Pest traps Insectocutor Fumigation	Collecting and measuring the number of insects
	Segregation of FRK	Proper stacking height Preventing vibration	Periodical homogeneity testing
Distribution to houses	Segregation of FRK	Proper stacking height and preventing vibration during transport	Periodical homogeneity testing
	Mixing up of batches	Proper labeling of materials for traceability	Periodical auditing and visual inspection
Storage in-house	Segregation of FRK	Dedicated storage area away from vibration and disturbances	Visual inspection/periodical homogeneity testing
	Increase in moisture content	Storage in a closed container away from moisture	Visual inspection/periodical moisture content analysis
Cooking of FR	Leaching of micronutrients	Rinse only once Cook with the exact amount of water (don't pour out the excess water)	Periodical micronutrient analysis of cooked rice

well as protects public health by ensuring that the fortified rice achieves both safety and nutritional levels. Essentially, the effective implementation of these protocols converts the process of fortified rice production from a technology to an accessible public health intervention. Over time, the benefits of well-executed food safety systems go beyond just meeting regulations; they contribute directly to measurable improvements in population nutrition and food security.

REFERENCES

Awasti, Nancy, Venkateswarlu Sunkesula, and Ahmed R. A. Hammam. 2022. "Food Safety Management Systems in the Food Processing Industry." In *Biological and Chemical Hazards in Food and Food Products*, edited by Santosh K. Mishra, Megh R. Goyal, and Manju Gaare, 241–262. Apple Academic Press, New York. https://doi.org/10.1201/9781003189183-13.

Bureau of Indian Standards (BIS). 2021a. "Fortified Rice-Specification (IS 17780 : 2021)." https://standardsbis.bsbedge.com/BIS_SearchStandard.aspx?Standard_Number=17780&id=0

Bureau of Indian Standards (BIS). 2021b. "IS 17782: 2021: Fortified Rice Kernels Specifications." https://standardsbis.bsbedge.com/BIS_searchstandard.aspx?Standard_Number=IS+17782&id=34819

De Oliveira, C. A. F., A. G. Da Cruz, P. Tavolaro, and C. H. Corassin. 2016. "Food Safety: Good Manufacturing Practices (GMP), Sanitation Standard Operating Procedures (SSOP), Hazard Analysis and Critical Control Point (HACCP)." *Antimicrobial Food Packaging*, 129–39. https://doi.org/10.1016/B978-0-12-800723-5.00010-3.

DFPD. 2019. *Pilot Scheme: Fortification of Rice and its Distribution under Public Distribution System*. New Delhi: Department of Food and Public Distribution, Government of India.

FSSAI. 2011. General Hygienic and Sanitary practices to be followed by Food Business Operators, Food Safety and Standards (Licensing and Registration of Food Businesses), Regulations 2011. https://fssai.gov.in/cms/hygiene-requirements.php.

Harbud, Nick. 2022. "Process Automation Systems." In *Integration and Optimization of Unit Operations Review of Unit Operations from R&D to Production: Impacts of Upstream and Downstream Process Decisions*, edited by Barry A. Perlmutter — Integration and Optimization of Unit Operations Perlmutter, 191–218. Elsevier. https://doi.org/10.1016/B978-0-12-823502-7.00021-9.

Harris, James R. 2010. "Good Manufacturing Practices (GMP) and Related FDA Guidelines." In *Pharmaceutical Manufacturing: Regulations and Quality,* edited by Shayne Cox Gad, 1–42. John Wiley & Sons, Inc., Hoboken, New Jersey. https://doi.org/10.1002/9780470571224.PSE374.

Holah, John. 2023. "Principles of Hygienic Practice in Food Processing and Manufacturing." In *Food Safety Management: A Practical Guide for the Food Industry, Second Edition*, edited by Veslemøy Andersen, Huub Lelieveld and Yasmine Motarjemi January, 587–613. Academic Press, Elsevier. https://doi.org/10.1016/B978-0-12-820013-1.00029-2.

Junaid-Ur-Rahman, Syed, Muhammad Farhan, Jahangir Chughtai, Adnan Khaliq, Atif Liaqat, Imran Pasha, Samreen Ahsan, Saira Tanweer, Kanza Saeed, Ayesha Siddiqa, Tariq Mehmood, Ayesha Ali, Shoaib Aziz & Nimra Sameed. 2022. "Rice: A Potential Vehicle for Micronutrient Fortification." *Clinical Phytoscience* 8 (1): 1–14. https://doi.org/10.1186/S40816-022-00342-3.

Meghwal, Murlidhar, Uday Heddurshetti, and Ravikumar Biradar. 2017. "Good Manufacturing Practices for Food Processing Industries: Principles and Practical Applications." In *Food Technology: Applied Research and Production Techniques*, edited by Murlidhar Meghwal, Megh R. Goyal, Mital J. Kaneria, 4–28. Apple Academic Press, New York. https://doi.org/10.1201/9781315365657-1.

Ministry of Consumer Affairs: Department of Food & Public Distribution (DFPD). 2022. "Standard Operating Procedure (S.O.P) for Quality Management Protocols for Fortified Rice Kernels (FRK) and Fortified Rice (FR)." https://anaajkharid.in/Images/FRK/SoP%20for%20FRK%20and%20FR%2015.03.2022.pdf.

Owusu-Apenten, Richard, and Ernest Vieira. 2023. "Food Safety Management, GMP & HACCP." In *Elementary Food Science*, edited by Richard Owusu-Apenten and Ernest R. Vieira, 217–36. Cham: Springer International Publishing. https://doi.org/10.1007/978-3-030-65433-7_10.

Piccoli, Nina Beretta, Nils Grede, Saskia de Pee, Anusara Singhkumarwong, Eveline Roks, Regina Moench-Pfanner, and Martin W Bloem. 2012. "Rice Fortification: Its Potential for Improving Micronutrient Intake and Steps Required for Implementation at Scale." *Food and Nutrition Bulletin* 33 (4): S360–S372.

Sameen, Aysha, Amna Sahar, Farwa Tariq, Usman Mir Khan, Tayyaba Tariq, and Bushra Ishfaq. 2023. "Food Microbial Hazards, Safety, and Quality Control: A Strategic Approach." In *Food Microbial and Molecular Biology: From Fundamentals to Applications*, edited by Saher Islam, Devarajan Thangadurai, Jeyabalan Sangeetha, and Zaira Zaman Chowdhury, 345–380. Apple Academic Press, New York. https://doi.org/10.4018/978-1-7998-7415-7.ch004.

Varghese, Suni Mary, Salvatore Parisi, Rajeev K. Singla, and A. S. Anitha Begum. 2022. "Food Safety and Quality Control in Food Industry." In *Trends in Food Chemistry, Nutrition and Technology in Indian Sub-Continent*, edited by Suni Mary Varghese, Salvatore Parisi, Rajeev K. Singla, and A. S. Anitha Begum, 31–44. Cham: Springer International Publishing. https://doi.org/10.1007/978-3-031-06304-6_5.

Walia, Ankita, Rahul Mehra, Naveen Kumar, Tejinder Pal Singh, and Harish Kumar. 2023. "Good Manufacturing Practices and Safety Issues in Functional Foods and Nutraceuticals." In *Bioactive Components: A Sustainable System for Good Health and Well-Being*, edited by Monika Thakur and Tarun Belwal, 167–82. Singapore: Springer Nature Singapore. https://doi.org/10.1007/978-981-19-2366-1_11.

17 Bioavailability, Efficacy, and Effectiveness of Fortified Rice

Considerations on Dose of Fortification, Iron Regulation and Safety

Prashanth Thankachan

17.1 INTRODUCTION

Micronutrient malnutrition, also known as "hidden hunger," affects over two billion people worldwide, with low- and middle-income countries bearing the brunt of its effects (Bailey et al. 2015). Food fortification, or adding back particular vitamins and minerals that are inadequate in the diet to staple foods, is one of the best ways to deal with the problem (Allen et al. 2006). Rice is a good vehicle for micronutrient fortification because it is a staple food for more than half of the world's population, particularly in Asia and parts of Africa (WHO and FAO 2006).

Determining the bioavailability, efficacy, and effectiveness of such fortified staples before their use in public health programs is a crucial consideration when implementing large-scale public health programs or targeted interventions to reduce micronutrient malnutrition in the population. To effectively translate into an impact that lowers the number of malnourished in the population, a careful analysis of the programmatic, cost, and actual benefits must be conducted.

Bioavailability is the proportion of a nutrient that is absorbed, digested, absorbed, and metabolized via regular pathways to become available for biological functions. Understanding the bioavailability of micronutrients from fortified rice or other food staples in the context of the food's cultural, behavioral, and societal interactions is crucial because micronutrient absorption can be significantly influenced by the food matrix, fortifying form, and preparation methods (Zimmermann and Hurrell 2002). High bioavailability ensures that the targeted populations will benefit the most from fortified foods and that the least amount of additional nutrients will be required to make up for dietary deficiencies.

Efficacy is defined as the extent to which a specific intervention produces a favorable result under ideal, controlled conditions. To demonstrate a direct causal relationship

DOI: 10.1201/9781003661245-17

between consumption and biological effects, such as elevated iron status or hemoglobin concentrations, or changes in iron stores or iron status over time, this approach is primarily borrowed from the pharmaceutical sciences and applied to the field of food science. In order to demonstrate causal effects that affect a biomarker of interest or a decrease in the prevalence of a deficiency at the conclusion of the recommended intervention period, these studies adopt the design of a randomized control trial. By using controlled settings and blinding participants, these studies are also intended to prevent or control for a number of biases. Since there was no difference between the control and intervention groups at baseline, these studies look for differences (positive of negative) between groups at the end of the study period. Therefore, any differences observed between groups are the result of the intervention alone as all other influencing factors are controlled. Several fortified rice efficacy studies have been carried our globally and for the evidence base for scaleup to effectiveness studies.

Effectiveness, as defined by Habicht et al. (1999), is the degree to which an intervention yields the intended outcomes in real-world situations or operational contexts. Even if a fortified product is bioavailable and effective under controlled conditions (Pelletier et al. 1993) within an efficacy framework, behavioral, logistical, financial, or policy issues may hinder it from producing improvements at scale. Therefore, effectiveness studies assess the practical impact and direct any necessary program changes before scaling up as a national program, which necessitates significant financial outlays and stakeholder alignment to produce a successful program.

17.2 COMPONENTS OF THE STUDY DESIGN

Because the objectives of bioavailability, efficacy, and effectiveness evaluations differ significantly, each requires a unique design approach.

17.2.1 Assessing Bioavailability with Stable Isotopic Methods

Design

The use of stable isotopic tracers in metabolic studies has revolutionized bioavailability evaluation. For example, in order to understand the bioavailability of iron-fortified rice, stable isotopic tracers can be used and is either added to rice kernels while being extruded or added to the whole meals prior to consumption. Participants eat the meals that are labeled with the isotope of iron, 14 days post dosing, and blood samples are taken to measure how much of the micronutrient was absorbed. The calculation uses the shift in isotopic enrichment in iron from baseline values in the hemoglobin pool to calculate bioavailability. These studies are expensive, small, often with 20–50 people, and are carefully controlled, helping us understand which forms of micronutrients (chemical forms) are best, and how preparation methods, infections, or other dietary factors can help or hinder absorption.

Measurements

The absorbed iron isotope enters the bone marrow and is used in the synthesis of hemoglobin, and new red blood cells produced post administration of the iron isotope carry the isotopic labels (Thankachan et al. 2007) and are seen in the RBC in circulation. The blood samples are collected, iron is extracted and purified from the blood matrix, and the change in isotopic enrichment is measured using a Thermal Ionization Mass Spectrometer.

Population

The study is usually conducted on healthy adults or specific target groups with low baseline nutritional status (e.g., women of reproductive age). Because isotope techniques are intensive and costly, the sample size is small ($n = 20–50$), and confounding control measures include standardizing meals, following fasting guidelines, managing absorption enhancers or inhibitors (e.g., vitamin C, phytates), and having a good history to check for inflammation or infection from a few days prior to the day of test meal administration.

17.2.2 Efficacy Trials

Design

The most reliable research method for establishing causality is randomized controlled trials or RCTs. Participants are randomly assigned to receive either fortified rice or non-fortified rice (placebo). In this method, there are no discernible differences between the control and intervention groups, and their baseline characteristics will be identical. Since all other factors are considered in this study design, it is hoped that the study subjects will exhibit a positive or negative, or no shift at the end of the intervention period. A between-group analysis will determine whether the observed difference is statistically significant and whether the interventional product is effective.

Metrics

Depending on how the intervention affects physiological parameters, one looks for changes in specific biomarkers, like hemoglobin, serum ferritin, retinol-binding protein, or zinc concentrations, over the course of the intervention period (typically 3–9 months, or longer) or a decrease or rise in a biomarker, such as serum ferritin or soluble transferrin receptors, respectively. It is also possible to look for a decrease in the prevalence of deficiencies in the study population in both the control and intervention groups. It is arguable that the intervention group should experience this change. The control group may occasionally also show changes, either positively or negatively, which could have an impact on the study's final between-group statistical analysis.

Population

Women, adolescents, and children are among the at-risk groups with the highest rates of micronutrient deficiencies. These groups would be the target group for fortified rice interventions because they bear the greatest burden of deficiency.

Blinding

This is a crucial step in removing selection bias, and it blinds all study participants, researchers, and staff. Assessors should, at the very least, be blinded in such cases, even though this may be practically challenging to implement. The control and interventional products must have the same appearance, be packaged in identical bags with the same markings, and be blinded by color into either a blue or red group or any other colors based on the number of study groups, as determined by a non-study personal who will be in possession of the blinding color code key. Color codes need to be stored in a sealed envelope that are only be opened once the study is completed and all available data has been examined for differences between groups. When the study is unblinded, based on change observed, it can be decided if the intervention is efficacious or not. Ideally, a statistical difference is seen in the outcome variable if the intervention has worked (positive or negative) as at the start of the study the group was identical with no differences in any of the study variables.

Compliance

To make sure that the interventional fortified food is being consumed by the family or population under study, one can measure the weekly disappearance of the fortified staple from the kitchen shelf, conduct direct observation, or record compliance over the phone. This data is crucial to evaluate the consumption of intervention to be as per the study design and the lack of effect if seen, which may be attributed to low compliance rather than a lack of efficacy or vice versa.

17.2.3 Evaluations of Effectiveness

Design

Effectiveness trials move from the lab/controlled environments to real-world conditions. Here, entire communities or regions are randomized to receive the intervention. These studies ask the question, does fortified rice improve health outcomes when scaled up to the entire district? These studies have broad indicators that are beyond biomarkers – how many people are eating the rice, if their health improves, and if the program is cost-effective. Importantly, they also talk to families, community leaders, and health workers to understand what's working and what isn't along with all other stakeholders in the pipeline. These studies are crucial before national rollout, ensuring that the program is both impactful and practical. In this instance, the design methodology typically consists of quasi-experimental trials or cluster randomized trials, with a pre-post design (Habicht et al. 1999). This is better suited to represent a wide region and be applied both at the individual and community or population level.

Measures

Much more general metrics, including cost-effectiveness, program coverage, utilization rates, nutrient status, and morbidity outcomes (like the prevalence of anemia), are documented in this study approach.

Population

In order to participate in this type of research, participants must be big, diverse groups that often cover entire communities or regions and are representative of the target populations in the real world. Rice consumption variability, socioeconomic factors, and political and programmatic variances are some of the data collection challenges that must be considered before the study begins.

Mixed Methods

Program barriers and user perceptions are gathered through qualitative evaluations (e.g., focus groups), which are supplemented by quantitative research that can be incorporated into the design to gain insights into how best to implement the program for maximum benefit. These research methods can also help identify the obstacles and facilitators to the intervention from the viewpoints of the family, community, and society as well as from all stakeholders who might have direct or indirect interactions with the interventional product.

17.3 PROGRAM COMPLEXITIES

Moving from efficacy to effectiveness and ultimately to a long-lasting public health impact presents a number of challenges:

17.3.1 Control of Quality

Since the entire ecosystem – which includes rice millers, technology companies that manufacture extruders and blenders, premix manufacturers and suppliers, and regulators who provide the standards – needs to mature to the point where they can cooperate to achieve good quality control in this process, ensuring the highest possible quality of the fortified rice kernels is extremely difficult. It is challenging to ensure the consistency, stability, and sensory acceptability of the fortified rice. According to the requirement, rice fortification requires quality control (QC) to guarantee that the premix used is within the tolerances for particle size and nutrient composition, and that the fortified rice consistently and safely provides the desired nutritional benefits. Extrusion technology has been shown to produce rice kernels that taste and look a lot like raw/parboiled rice, but it is still difficult to maintain quality control across production batches (Moretti et al. 2005), across differing geographies, and also across differing grain shapes and sizes (diverse rice grains in India). In the Indian ecosystem for rice fortification, positive change has been made possible quickly through the ongoing process of engagement with all stakeholders and supportive interaction.

17.3.2 Customer Acceptance

Cultural food preferences have a major impact on the adoption of fortified foods. Even when the nutritional benefits are well known, consumers may reject fortification-induced changes in color, texture, taste, and aroma (Neufeld et al. 2017). Unsubstantiated claims like "artificial plastic rice" have caused consumers to lose faith in the product, which has hindered uptake. Addressing acceptability issues requires a strong communication strategy that is coordinated and disseminated uniformly throughout the value chain.

17.3.3 Policy and Regulation Frameworks

For fortification programs to be successful, coordination between food producers, regulatory agencies, donors, and health ministries is required. Disjointed policies or a lack of quality monitoring or enforcement resources could hinder fortification efforts (WHO 2009).

17.3.4 Economic Considerations

The affordability of fortified rice for vulnerable populations is very essential. Cost analyses must consider supply chain logistics and consumer willingness to pay in addition to production costs (Das et al. 2013). The ultimate cost burden should be minimal or non-existent for the recipient within a focused public health approach and needs to be provided by the Government.

17.3.5 Observation and Evaluation

Longitudinal year-on-year monitoring systems are required to track program impact, reach, and compliance over time. However, in many places, surveillance capacity is limited, which hinders these efforts (Fiedler et al. 2012). At the moment, India's national rice fortification program lacks a strong evaluation and monitoring pipeline to handle problems at the field level, any negative effects that may exist, and how to respond to them. It is necessary to implement data collection methods on how the population's nutritional status has changed over time in parallel with the expansion of the rice fortification program.

17.4 OTHER FACTORS AND CONSIDERATIONS

17.4.1 Setting the Correct Fortification Dose

Adding or enriching our everyday foods with micronutrients like iron to rice or vitamin A to cooking oil, or vitamin d to milk-might seem like a simple idea to prevent deficiencies in the population. But deciding exactly how much of a micronutrient to add to a food to fortify is anything but simple. Food fortification can only work as intended if the dose or level of the nutrient to be added is set to the appropriate levels; if not, it can either fall short of solving the problem of deficiencies in the population or, worse, an excess can cause harm to the population consuming it and can affect a large number of people adversely.

Many people, especially in low-income countries, don't get enough essential micronutrients from their daily diet due to a lack of availability of nutrient dense foods or due to economic and social constraints. Such populations live off cereal based, low micronutrient dense diets that fail to meet their daily dietary requirements.

This nutrient inadequacy in the diet of the population is bridged by food fortification approaches, which adds the missing nutrients to foods people eat daily and is considered one of the most practical ways to address this problem at a population level. Food fortification been done successfully for years, with the best example being the micronutrient iodine added in salt to prevent goiter with the use of iodized salt, or folic acid in flour to protect babies from neural tube defects. Fortification is a public health measure that reaches the population, without asking the population to change what they eat, or with their existing sociocultural behavior and practices. But to make it work effectively, the amount of nutrient that needs to be added has to be right (Allen et al. 2006) for the population being considered for this preventive approach. Careful consideration and planning are needed to ensure food fortification as a preventive strategy has the desired impact over time in the population.

Every vitamin or mineral has a recommended amount that the population for a given age and physiological state should consume each day to meet their dietary requirements to remain in homeostatic balance (losses from the body replenished from the diet). One of the crucial information needed to design a fortification strategy is to have data on the dietary consumption of food in the population and from this data compute nutrient inadequacy in the population. The nutrient inadequacy data provides an estimate of how much a particular micronutrient is being consumed or not being consumed in comparison to the daily dietary recommendations, and this provides information on the level of the dose needed to be set in the fortified food for a given micronutrient.

Also needed is the data on the amount of food vehicle consumed by the population, and the bioavailability of micronutrients when added into that food needs to be taken into consideration. One important criterion while choosing the food vehicle for fortification is that the population needs to consume the vehicle in adequate amounts on a daily basis, and thus, staples like rice, wheat flour, salt, and oil are common fortification vehicles. For example, in India, the average person eats 250–300 g of rice per day; to calculate how much iron should be added to that amount of rice to help meet daily needs, WHO suggests using the 95th percentile of consumption; in other words, the amount of rice that most people eat is at the high end, and the iron levels are computed from inadequacy data at this intake of rice daily, to ensure avoiding too much of micronutrients in those who eat more of the fortification vehicle (WHO and FAO 2006).

The nutrient premix for food fortification comes in various biochemical forms and have varied physical, chemical, and solubility characteristics. Based on these chemical and physical characteristics, some are absorbed better by the body than others. For example, ferrous sulfate is a water-soluble iron compound that is well absorbed but causes organoleptic and sensory changes in some foods. On the other hand, ferric pyrophosphate is insoluble in water and poorly acid soluble, and hence, it doesn't interact with the food components leading to very good sensory and organoleptic characteristics; however, it is poorly absorbed (WHO 2016). To account for the poor bioavailability, twice the amount of ferric pyrophosphate needs to be considered in the calculation of the dose to overcome the low bioavailability. Also, other physical characteristics such as particle size play an important role in the bioavailability profile of the nutrient premix; the lower the particle size, the higher the bioavailability; however, the cost is also higher and may impact the fortification program implementation.

Additionally, during the process of manufacturing the fortified food, the nutrient premix can interact with heat, light, moisture, other food components, and time to final consumption, leading to varying levels of degradation of the nutrient. To offset these losses, sufficient overages need to be added to account for these processing losses. Thus, the final dose of the micronutrient to be added to a fortified food needs to consider all these factors to arrive at the correct dose of fortification to ensure optimal fortification.

Although it appears as a simple straightforward approach, several challenges exist in arriving at the right dose for fortification and includes

Limited population data on actual nutrient intake and nutrient inadequacy

- Unknown bioavailability data.
- Unknown data on losses during storage or cooking.
- Unknown if the fortified food will be acceptable in the population.
- Variation in food habits between regions or seasons.

Choosing the right dose of micronutrients is the most important step in food fortification and involves good scientific evidence, knowledge of local diets, culture, and food practices.

17.4.2 Iron Regulation and Safety in the Context of Fortified Food Consumption

Iron is a micronutrient that can be beneficial and, at the same time, harmful if present in excess within biological systems; hence, the levels of iron in the human body have to be tightly regulated such that all iron in the body is in the bound form to a protein (transport, storage, or a functional protein) to prevent it from reacting with other chemical constituents in the body, leading to adverse interactions within biological systems.

In the context of food fortification, the dose that is added into the food being fortified is set at a level that is primarily aimed at mitigating the nutrient inadequacy in the diet. The dose chosen is such that it is well within the 25%–40% of the daily dietary requirements, and therefore, when considered along with all other food consumed in the daily diet, the net iron consumed per day from the total diet (fortified food and all other foods) will be well within what a person is required to eat on a daily basis if he/she had access to a diverse nutrient dense balanced diet.

However, in a given population, there may be some individuals who may be consuming a diverse diet and meeting their dietary requirements but may not be iron deficient or suffering from iron-deficiency anemia, and hence may not need the additional iron from the fortified food. In the event, such individuals also do consume iron-fortified foods, given that the iron absorption at the duodenum is tightly regulated wherein the absorption of iron is upregulated in deficient states and downregulated in iron-replete states, the excess dietary iron will not be absorbed and will be excreted out in the stool. This regulation is carried out at the cellular interface of the duodenal cells of the micro villi, where as a first step, iron from food is reduced to Fe^{2+} by duodenal cytochrome B prior to transport into duodenal enterocytes by divalent metal transporter 1 (DMT1). Once inside enterocytes, iron is either stored as ferritin or exported into the bloodstream by ferroportin, the cellular iron exporter. Ferroportin's activity is supported by hephaestin and ceruloplasmin, which oxidize Fe^{2+} to Fe^{3+} for loading onto transferrin, and the plasma iron transport protein to be shuttled across all the cells in the body through the bloodstream, delivering the iron to the cells for functional use.

In iron-replete states, there is a downregulation of the expression of the DMT1, reducing iron flux into enterocytes. Also, hepcidin, a liver-derived peptide hormone, is the central regulator of systemic iron homeostasis. It binds to ferroportin, triggering the internalization and degradation, decreasing iron loading onto transferrin from the enterocyte and thereby reducing the absorption of iron from the gut into the bloodstream. This tight regulation of iron absorption at the duodenal interface plays a crucial role in preventing excess iron entering the human body, and iron status plays a crucial role in regulation. Due to this innate regulatory process in preventing iron entry into the body in normal iron-replete individuals, it is unlikely to lead to a state of excess even if dietary iron consumed exceeds the recommended dietary levels, and therefore, the risk of adverse events arising from iron excess from fortified food is non-existent. However, in individuals with disorders of iron regulation such as Hereditary Hemochromatosis and β-Thalassemia, there is lower hepcidin secretion, leading to excessive intestinal iron absorption and systemic iron overload. In those suffering from such disorders, as the dose of iron in the fortified food is similar or lower than what one would expect from a balanced and diverse diet, fortified foods will not contribute to additional harm than the ill effects of the disorder itself or significantly contribute to the iron overload burden in them from fortified foods.

This is an area that is being actively researched to generate causal data on adverse effects in those with such disorders, while from a view of caution, as iron-containing foods are restricted in individuals with such disorders, iron-fortified foods are also generally advised to be restricted or avoided.

17.5 CONCLUSIONS

Assessing the bioavailability, efficacy, and effectiveness of micronutrient-fortified rice is an essential multi-step process needed to transform scientific evidence into successful public health nutrition strategies. By closely examining the design elements and programmatic challenges at every stage, it is possible to increase the likelihood that fortified rice programs will significantly reduce micronutrient deficiencies in vulnerable Indian populations.

REFERENCES

Allen, Lindsay, Bruno de Benoist, Omar Dary, and Richard Hurrell. 2006. *Guidelines on Food Fortification with Micronutrients*. WHO and FAO.

Bailey, Regan L., Keith P. West Jr, and Robert E. Black. 2015. "The epidemiology of global micronutrient deficiencies." *Annals of Nutrition and Metabolism* 66(2): 22–33.

Das, Jai K., Rehana A. Salam, Rohail Kumar, and Zulfiqar A. Bhutta. 2013. "Micronutrient fortification of food and its impact on woman and child health: A systematic review." *Systematic Reviews* 2: 1–24.

Fiedler, Klaus, Florian Kutzner, and Joachim I. Krueger. 2012. "The long way from α-error control to validity proper: Problems with a short-sighted false-positive debate." *Perspectives on Psychological Science* 7(6): 661–669.

Habicht, Jean-Pierre, Cesar Victora, and J. Patrick Vaughan. 1999. "Evaluation designs for adequacy, plausibility and probability of public health programme performance and impact." *International Journal of Epidemiology* 28(1): 10–18.

Moretti, Diego, Tung-Ching Lee, Michael B. Zimmermann, Jeannette Nuessli, and Richard F. Hurrell. 2005. "Development and evaluation of iron-fortified extruded rice grains." *Journal of Food Science* 70(5): S330–S336.

Neufeld, Lynnette M., Shawn Baker, Greg S. Garrett, and Lawrence Haddad. 2017. "Coverage and utilization in food fortification programs: Critical and neglected areas of evaluation." *The Journal of Nutrition* 147(5): 1015S–1019S.

Pelletier, David L., Edward A. Frongillo Jr, and Jean-Pierre Habicht. 1993. "Epidemiologic evidence for a potentiating effect of malnutrition on child mortality." *American Journal of Public Health* 83(8): 1130–1133.

Thankachan, Prashanth, Sumithra Muthayya, Thomas Walczyk, Anura V. Kurpad, and Richard F. Hurrell. 2007. "An analysis of the etiology of anemia and iron deficiency in young women of low socioeconomic status in Bangalore, India." *Food and Nutrition Bulletin* 28(3): 328–336.

WHO. 2016. *Guidelines on Food Fortification with Micronutrients*, edited by Lindsay Allen, Bruno de Benoist, Omar Dary and Richard Hurrell, WHO, Geneva, Switzerland.

WHO. 2009. *Recommendations on Wheat and Maize Flour Fortification Meeting Report: Interim Consensus Statement*, 1–3. Geneva: Department of Nutrition for Health and Development (NHD), WHO.

WHO and FAO. 2006. *Recommended Nutrient Intake and Safe Levels of Intake*. WHO, Geneva, Switzerland.

Zimmermann, Michael B., and Richard F. Hurrell. 2002. "Improving iron, zinc and vitamin A nutrition through plant biotechnology." *Current Opinion in Biotechnology* 13(2): 142–145.

18 National Standards on Vitamin-Mineral Premix, Fortified Rice Kernel, Fortified Rice, and Plant Equipment and Machinery

Lavika Singh, Balasubramanian K., Pradeep Sharma, and Suneeti Toteja

18.1 INTRODUCTION

This chapter outlines the national standards governing Vitamin and mineral premixes (VMP), Fortified rice kernel (FRK), fortified rice (FR), and plant equipment and machinery. It elaborated on regulations, quality benchmarks, and Indian Standards set by the Food Safety and Standards Authority of India (FSSAI) and the Bureau of Indian Standards (BIS) to ensure food safety, nutritional adequacy, and industry standardization.

Originally established as the Indian Standards Institution (ISI) on 6 January 1947, BIS was formed with the core objective of promoting standardization and quality control in industry and commerce. Recognizing the growing need for a stronger legal framework to support standardization, the BIS Act, 1986 was enacted by the Parliament, replacing ISI with BIS as a statutory body. The Bureau of Indian Standards Act 2016 (effective 12 October 2017) superseded the 1986 Act, establishing BIS as the national standards body for standardization, conformity assessment, and quality assurance of goods, articles, processes, systems, and services and for matters connected therewith or incidental thereto (The Bureau of Indian Standards Act 2016, 2016). Together, these legislative milestones reflect India's evolving quality infrastructure, positioning BIS as a central to promoting trust, safety, and competitiveness. BIS Standards are formulated by subject specific technical committees, comprising all stakeholders including manufacturers, consumers, technologists, regulators, government bodies, research organizations, and testing laboratories through the principles of consensus.

DOI: 10.1201/9781003661245-18

FSSAI regulates food fortification and specifies mandatory and voluntary fortification of staple food items such as rice, wheat flour, milk, and oil. The Food Safety and Standards (Fortification of Foods) Regulations, 2018, lays down the permissible levels of vitamins and minerals in fortified foods, labeling requirements, and other compliance mandates (FSSAI 2018).

BIS complements FSSAI by setting quality specifications for food products and related equipment. BIS certification ensures compliance with quality requirements, safety measures, and standardized manufacturing processes. Together, FSSAI regulations and BIS standards establish consumer confidence, facilitate industrial compliance, and improve the overall quality and safety of food products.

18.2 REGULATORY AND STANDARDS FRAMEWORK IN INDIA

In India, rice fortification is governed by FSSAI guidelines and BIS's quality benchmarks. Through Food Safety and Standards Act, 2006, FSSAI provides legally binding norms and BIS issues voluntary but critical quality and performance standards. Together, they ensure that FR and its related commodities meet the necessary nutritional and safety requirements. This integrated framework establishes consistency across production, testing, labeling, and distribution, especially for social security programs like the Public Distribution System (PDS), Integrated Child Development Services (ICDS), and Mid-Day Meal Scheme (MDMS).

18.2.1 Food Safety and Standards Authority of India (FSSAI)

FSSAI, established under the Food Safety and Standards Act, 2006, regulates the manufacture, storage, distribution, sale, and import of food products to ensure safe and wholesome food for human consumption through its science-based standards (Food Safety and Standards Act 2006, 2006).

With a view to ensure the effective rice fortification in entire chain starting from preparation of VMP to manufacturing of FR, FSSAI has developed standards for 'Rice flour for preparation of Fortified Rice Kernel', 'Vitamin Mineral Premix for Preparation of Fortified Rice Kernel', and 'Fortified Rice Kernel' under Food Safety and Standards (Food Products Standards and Food Additives) Regulation, 2011, and specifications for FR is laid down in Food Safety and Standards (Fortification of Food) Regulation, 2018. These regulations specify both mandatory and voluntary fortificants. Mandatory fortificants include iron (as micronized ferric pyrophosphate or sodium feredetate), folic acid, and vitamin B_{12}. Optional micronutrients include zinc, vitamin A, and B-complex vitamins like thiamine, riboflavin, niacin, and pyridoxine (FSSAI 2011b, 2018).

FSSAI also issues guidelines on labeling (such as the use of the '+F' logo), packaging, shelf life, and storage practices. Through its scientific panel on Methods of Sampling and Analysis, it standardizes testing protocols for nutrient analysis and food safety to ensure compliance and quality assurance across the FR supply chain.

Additionally, FSSAI has approved methods for the testing and validation of fortified food products to ensure compliance with regulatory standards. The scientific panel on Methods of Sampling and Analysis prescribes methods for determining the presence and stability of essential nutrients in fortified products. These methods include:

a. Determination of iron, folic acid, and vitamin B_{12} in FR by ICP-MS and LC-MS/MS.
b. Determination of iron, folic acid, and vitamin B_{12} in FRK by ICP-OES and LC-MS/MS.
c. Determination of iron in FRK by Atomic Absorption Spectroscopy (AAS).
d. Determination of iron, folic acid, and vitamin B_{12} in VMP for FRK by AAS and HPLC.

FSSAI also notifies food-testing laboratories, issues licenses and registration to food businesses, conducts inspections and audits of food businesses, and promotes consumer awareness about food safety and hygiene.

18.2.2 Bureau of Indian Standards (BIS)

BIS is India's national standards body under the Bureau of Indian Standards Act, 2016, which develops and publishes standards and runs certification schemes for a wide range of products and processes (The Bureau of Indian Standards Act 2016, 2016). Its Food and Agriculture Department (FAD) formulates national standards in the field of food and agriculture through the Food and Agriculture Division Council (FADC) to ensure the quality and safety of FR distributed through various social security programs, and in the open market, BIS through its food grains, allied products, and other agriculture produce sectional committee, FAD 16, and Agriculture and Food Processing Equipment Sectional Committee, FAD 20, has formulated following Indian Standards on rice fortification ecosystem:

- **IS 17781:2021**: Vitamin and Mineral Premix for manufacturing FRK – Specification
- **IS 17782:2021**: Fortified Rice Kernels – Specification
- **IS 17780:2021**: Fortified Rice – Specification
- **IS 17853:2022**: Equipment for Manufacture of Fortified Rice Kernel – Specification
- **IS 17854:2022:** Equipment for Manufacture of Fortified Rice – Specification

These standards prescribe physicochemical properties, permissible contaminant levels, microbial thresholds, nutrient retention requirements, methods of test and sampling, along with the requirements for equipment used in manufacture of FRK and FR, which also defines hygienic practices, packaging standards, and labeling protocols that manufacturers must follow to obtain BIS certification. Though voluntary, compliance with these standards reflects the commitment of manufacturers in serving high-quality products. Standards may be made mandatory by government for procurement and distribution programs. BIS thus acts as a crucial enabler of quality assurance and public health protection in India's rice fortification mission. These Standards are reviewed at least every 5 years or earlier if needed (BIS 2018b).

18.2.3 Regulation vs Standard

Regulations (e.g., those issued by FSSAI) are legally enforceable, and all food businesses manufacturing or distributing FR must abide by them otherwise, face penalties, license suspensions, or legal action. Regulations cover labeling, consumer protection, nutrient levels, and health and safety issues.

On the other hand, Standards (e.g., those issued by BIS) are technical guidelines that are voluntary unless referenced in legislation or any government policy. They provide detailed quality and safety requirements, manufacturing process, testing procedures, and performance metrics. Adherence to them shows a manufacturer's dedication to quality and safety. BIS standards become legally binding when referenced in regulations or incorporated into government schemes. For high-risk commodities or government procurement, adherence to relevant BIS standards may be mandated.

In essence, standards guarantee technical and quality compliance, while regulations guarantee legal accountability. Together, they create a strong framework that guarantees FR is safe and nutritionally beneficial.

18.3 VITAMIN-MINERAL PREMIX (VMP)

VMP is a combination of micronutrients in desired proportions, used as fortificant in FRK manufacturing. BIS published IS 17781:2021 'Vitamin and Mineral Premix for manufacturing FRK – Specification (BIS 2021c), and FSSAI prescribed 'Vitamin-Mineral Premix for Preparation of Fortified Rice Kernel' under regulation 2.4.6.24(B) of the Food Safety and Standards (Food Product Standards and Food Additives) Regulations, 2011 (FSSAI 2011b).

18.3.1 Composition

As per FSSAI and IS 17781, the VMP must contain micronized ferric pyrophosphate or sodium iron (III) EDTA (NaFeEDTA) as the iron source, folic acid, and vitamin B_{12} as either Cyanocobalamin or Hydroxocobalamin (BIS 2021c). These three micronutrients address deficiencies such as anemia and neonatal defects. Additional nutrients permitted are zinc oxide, retinyl palmitate (vitamin A), thiamine (vitamin B_1), riboflavin (vitamin B_2), nicotinamide or nicotinic acid (vitamin B_3), and pyridoxine hydrochloride (vitamin B_6). Specifications ensure nutritional value and stability throughout processing, distribution, and storage.

18.3.2 Physicochemical Requirements

Moisture content must not exceed 11% by mass, as determined by the method specified in IS 16072 'Determination of moisture content in milk powder and similar products (Routine method)'. To ensure the bioavailability of iron, the particle size of micronized ferric pyrophosphate must be 1–3 µm (D90 particles) (BIS 2021c). These specifications ensure the quality, stability, and uniform blending of the VMP while making FRK. As per FSSAI regulations, the VMP must also conform to microbiological limits of Aerobic Plate Count (i.e., 10^4 cfu/g) and limit of yeast and mold count (i.e., 10^2 cfu/g) to prevent spoilage and degradation of micronutrients.

18.3.3 Ingredients

Micronutrients in the VMP must be in food-grade chemical forms. Micronutrient compounds must be precisely weighed and blended to ensure homogeneity while formulating VMP. As per IS 17781, a white base or carrier compatible with the food matrix and non-reactive with micronutrients must be used in VMP. VMP may contain approved food-grade emulsifiers, hydrocolloids, binding agents, acid regulators, and antioxidants conforming to Food Safety and Standards (Food Products Standards and Food Additives) Regulations, 2011 (BIS 2021c). These additives would enhance the physical stability and shelf life of the VMP. All ingredients must comply with the standards and be free from contaminants, undesirable odors, and foreign matter. Production must follow. Furthermore, the VMP must be produced in premises built and maintained under hygienic practices under IS 2491 'Food hygiene — General principles — Code of practice' (BIS 2021c).

18.3.4 Packaging and Storage

Appropriate packaging and storage of the VMP are vital for preserving its quality and effectiveness. As per IS 17781, VMP must be packed in non-reactive, opaque, and food-grade containers preventing exposure to light, oxygen, moisture, and contaminants. The packaging material should be impervious to water vapor and gas (BIS 2021c).

Storage conditions for VMP must ensure its stability and effectiveness throughout its shelf life. It should be stored in a clean, dry, and cool area with temperature not exceeding 25°C, with relative humidity below 60% to avoid clumping or microbial growth (BIS 2021c).

18.3.5 Labeling

As per IS 17781:2021 and the Food Safety and Standards (Labeling and Display) Regulations, 2020, each package must be legibly and indelibly marked with product name, manufacturer details, date of packing, batch or lot number, net quantity, best-before date, micronutrients with forms and concentrations, and storage instructions. Manufacturers must provide a Certificate of Analysis (CoA) with each batch, detailing the chemical composition (BIS 2021c).

18.3.6 Test Methods for Analysis of Fortificants in VMP

For quantification of iron, IS 17781 recommends methods such as AOAC 944.02 or AACC 40–70.01, which require AAS (BIS 2021c). Similarly, FSSAI prescribes the method FSSAI.VMP-FRK.16.008.2023, also using AAS. Additionally, AOAC 984.27 is recognized by BIS, which employs Inductively Coupled Plasma (ICP) Emission Spectroscopy for iron estimation. For vitamin B_9 (folic acid), FSSAI recommends the test method FSSAI.VMP-FRK.16.009.2023, which uses HPLC for accurate measurement. In the case of vitamin B_{12} (Cyanocobalamin), BIS refers to IS 16640 and FSSAI prescribes FSSAI.VMP-FRK.16.010.2023, both of which utilize HPLC as the preferred analytical technique (FSSAI 2023b). BIS has recently published indigenous method, IS 7529: 2025 for determination of vitamin B_{12} in foodstuffs by HPLC- UV (BIS 2025a) and IS 7234: 2025 for determination of vitamin B_9 (folates) in foodstuffs by ultra-high performance liquid chromatography and tandem mass spectrometry (BIS 2025b).

18.4 FORTIFIED RICE KERNELS (FRK)

FRK are extruded rice-shaped granules containing added micronutrients to address nutritional deficiencies. They are generally produced using extrusion technology and are intended to be blended with non-FR in fixed ratios (such as 1:100 or 1:50) to prepare FR.

18.4.1 Composition

As per IS 17782:2021 and 2.4.6.24(C) of the Food Safety and Standards (Food Products Standards and Food Additives) Regulations, 2011, the mandatory micronutrients in FRK include iron, folic acid, and vitamin B_{12}. Iron may be added either as micronized ferric pyrophosphate (at 280–425 mg/kg for 1:100 blending ratio) or as sodium iron (III) EDTA (NaFeEDTA) at lower levels (140–212.5 ppm) due to higher bioavailability. Folic acid is added at 750–1,250 ppb, and vitamin B_{12}, either as Cyanocobalamin or Hydroxocobalamin, at 7.5–12.5 ppb (BIS 2021b; FSSAI 2011b). Optional fortificants such as zinc, vitamin A, thiamine, riboflavin, niacin, and pyridoxine are also permitted within defined limits.

18.4.2 Ingredients

As per IS 17782, rice flour at an initial moisture content of 11%–12%, prepared from clean, broken rice, is used for producing FRK. Rice flour must be white to off-white in color, free-flowing, and devoid of foreign matter or undesirable odor. As per IS 17782, rice flour must have an average particle size of less than 250 μm (preferably less than 150 μm). Clause 24A of the Food Products Standards and Food Additives Regulations, 2011 specifies that a minimum 90% of the rice flour must pass through a 60-mesh sieve (250 μm). The moisture content must not exceed 14%, the acid-insoluble ash must be $\leq$0.15%, and the alcoholic acidity should be $\leq$0.18%. Rice flour must also conform to the limits prescribed under the Food Safety and Standards (Contaminants, Toxins and Residues) Regulations, 2011 (FSSAI 2011a). As per IS 17782, VMP used must comply with specifications laid down in IS 17781:2021. To ensure stability, binding and enhanced extrusion quality approved food-grade emulsifiers, hydrocolloids, binding agents, antioxidants, and acid regulators conforming to Food Safety and Standards (Food Products Standards and Food Additives) Regulations, 2011 may be added (BIS 2021b). Water used for dough preparation during extrusion should conform to IS 10500.

18.4.3 Requirements

Physicochemical Requirements

FRK must resemble raw/parboiled rice in shape, size, and color and should not have off-odors or visible physical defects. The strength and integrity of FRK are essential during blending and cooking, so the physical properties must be aligned with the behavior of raw/parboiled rice. Texture and resilience during washing and cooking are critical attributes that ensure seamless substitution. For ensuring stability and storability, IS 17782 specifies a moisture limit of 12% by mass and limit of broken kernels and foreign matter. Complete absence of damaged, chalky, and discolored grains is also enforced under regulations. FRK must comply with the Food Safety and Standards (Contaminants, Toxins and Residues) Regulations, 2011 (FSSAI 2011a), including maximum permissible limits for heavy metals, pesticide residues, naturally occurring toxic substances, and aflatoxins. The uric acid content must not exceed 100 mg/kg.

Cooking Characteristics

The cooking characteristics of FRK are a crucial determinant of its performance in real-world settings. The solid loss during cooking must not exceed 10%. Furthermore, after cooking, the FR must retain at least 90% of iron, 60%–80% of folic acid, and not less than 60% of vitamin B_{12} (BIS 2021b).

Microbiological Requirement

To ensure safety and quality, FRK must conform to strict microbiological standards. The aerobic plate count must be $\leq 10^4$ colony forming units (cfu)/g, and the yeast and mold count must not exceed 10^2 cfu/g (BIS 2021b; FSSAI 2011b).

18.4.4 Packaging and Storage

FRK must be packed in clean, dry, food-grade water-resistant containers, providing a barrier against light, oxygen, and contaminants. Typically, coextruded multilayer laminated pouches or HDPE-lined woven sacks are used. FRK should be stored in a cool and dry place, with ambient temperatures not more than 25°C and relative humidity below 60% (BIS 2021b).

18.4.5 Labeling

Each package of FRK must display the product name, manufacturer's name and address, batch number, date of manufacture, best-before date, and net weight, micronutrient composition (chemical forms and levels), and the blending ratio (e.g., 1:100 or 2:100). As per FSSAI regulations and IS 17782, FRK is not meant for direct consumer sale and is intended only for industrial blending purposes, and the same shall be marked on the product package (BIS 2021b).

18.4.6 GMP/GHP Guidelines for Manufacturing of FRK

The Good Manufacturing Practices (GMP) and Good Hygienic Practices (GHP) guidelines detailed in IS 17782 assist manufacturers in production of high-quality, safe, and effective FRK, thus efficiently contributing to national nutrition programs and public health initiatives. On receipt, all raw materials (broken rice and VMP) must be inspected for quality, correct documentation, and absence of contamination. Rice should be stored under controlled ambient conditions for rice and VMP in cool, pest-free areas (BIS 2021b).

During grinding, broken rice is pulverized into fine flour with a particle size preferably below 150 microns. Equipment used for grinding must be thoroughly cleaned and securely assembled to prevent foreign material inclusion. Proper monitoring of feed rates and sieve integrity is crucial to achieve consistent flour quality. In the mixing stage, rice flour, VMP, and approved additives are blended under controlled conditions ensuring even micronutrient distribution. Water is uniformly sprayed to condition the mix, followed by adequate resting time before extrusion.

Extrusion must be done at carefully set parameters such as temperature, screw, and cutter speeds to produce kernels of uniform size and density. After extrusion, the kernels are immediately cooled, dried, and stored properly to prevent moisture absorption or contamination. These measures collectively protect the nutritional and physical quality of FRK, intended for social security programs (BIS 2021b).

18.4.7 Test Methods for Analysis of Fortificants in FRK

IS 17782 refers to AOAC 944.02 or AACC 40-70.01, which require AAS for quantification of iron. FSSAI prescribes the method FSSAI.FRK.16.004.2023 using Inductively Coupled Plasma – Optical Emission Spectrometer (ICP-OES) and FSSAI.FRK.16.007.2023 using AAS. BIS also recommends AOAC 984.27, which employs ICP Emission Spectroscopy for iron estimation in the form of elemental iron. For vitamin B_9 (folic acid), FSSAI recommends the test method FSSAI.FRK.16.005.2023, which uses LC-MS/MS for accurate measurement. For vitamin B_{12} (cyanocobalamin), BIS refers to IS 16640, which utilizes HPLC, and FSSAI prescribes FSSAI.FRK.16.006.2023, which employs LC-MS/MS (FSSAI 2023a). BIS recently published indigenous method, IS 7529: 2025 for determination of vitamin B_{12} in foodstuffs by HPLC-UV (BIS 2025a) and IS 7234: 2025 for determination of vitamin B_9 (folates) in foodstuffs by ultra-high performance liquid chromatography and tandem mass spectrometry (BIS 2025b).

18.5 FORTIFIED RICE

18.5.1 Composition

FR is produced by blending FRK with unfortified, milled raw or parboiled rice, typically done in a ratio of either 1:100 or 2:100 by mass. A tolerance of ±10% is permitted on the stated blending ratio to account for operational variability. The aim of blending FRK with conventional rice is to deliver essential micronutrients to large population segments through a widely consumed staple. The end product should be physically indistinguishable from ordinary rice and maintain similar taste, texture, and cooking behavior (BIS 2021a).

18.5.2 Fortification Requirements

Mandatory Micronutrients

IS 17780 and the Food Safety and Standards (Fortification of Foods) Regulations, 2018 prescribe three mandatory micronutrients for FR: iron, folic acid, and vitamin B_{12}. As per IS 17780 and FSSAI regulations, iron must be added as either micronized ferric pyrophosphate (28–42.5 ppm) or sodium iron (III) EDTA (14–21.25 ppm), the latter required at lower levels, due to higher bioavailability. Folic acid must be added at 75–125 ppb, preventing neural tube defects and supporting red blood cell formation. Vitamin B_{12}, such as cyanocobalamin or hydroxocobalamin, is added at level of 0.75–1.25 ppb, supporting nerve function and DNA synthesis (BIS 2021a).

Mandatory micronutrient levels are based on Recommended Dietary Allowances (RDA) and estimated daily rice consumption patterns in India. A uniform distribution of FRK in the FR is essential for achieving consistent fortification.

Optional Micronutrients

Optional micronutrients permitted under IS 17780 and FSSAI regulations include zinc (as zinc oxide), 10–15 ppm, vitamin A (as retinyl palmitate), 500–750 ppb RE, thiamine (vitamin B_1), 1.0–1.5 ppm, riboflavin (vitamin B_2), 1.25–1.75 ppm, niacin (vitamin B_3), 12.5–20 ppm, and pyridoxine (vitamin B_6), 1.5–2.5 ppm (BIS 2021a).

18.5.3 Physicochemical Requirements

In order to ensure quality and safety of FR, IS 17780 lays down limits for moisture content, foreign matter, uric acid, and aflatoxin B1. Uric acid in FR should not exceed 100 ppm, and the limit of aflatoxin B1 is set at 10 ppb (BIS 2021a; FSSAI 2011a). IS 17782 also specifies limits for grain refractions, such as weevilled, damaged, and chalky kernels, specified separately for milled raw and parboiled rice (BIS 2021a).

18.5.4 Packaging and Storage

Packaging and storage guidelines ensure that FR remains nutritionally potent and physically sound until consumption. FR must be packed in a durable, clean, food-grade material. As per IS 17780, HDPE or PP woven sacks may be used, conforming to IS 16208. These bags must have water vapor transmission rate (WVTR) of 20–50 g/m²/day and oxygen transmission rate (OTR) of 1.9–8 cc/m²/day to prevent micronutrient degradation. Test methods for estimation of WVTR and OTR have also been specified. Storage areas must be clean, cool, dry, and well-ventilated, with temperatures not exceeding 25°C and relative humidity below 60% (BIS 2021a).

18.5.5 Labeling

As per IS 17780 and the Food Safety and Standards (Fortification of Food) Regulation (2018), the label must include the product name (Fortified Rice), fortification logo (the '+F' symbol), blending ratio (e.g., 1:100 or 1:50), micronutrient list and levels with chemical forms, name and address of the manufacturer, packer, or blender, date of manufacture, best-before date, etc., along with storage and cooking instructions. An advisory statement 'People with Thalassemia may take under medical supervision' must be specified on the package. Labels must be printed in food-grade ink and in a format that is legible, indelible, and resistant to smudging during transportation and storage. These requirements ensure product traceability, support informed use, and enhance public trust in the fortified product (BIS 2021a; FSSAI 2018).

18.5.6 GMP/GHP Guidelines for Manufacturing of Fortified Rice

As per IS 17780, the blending of FRK with raw or parboiled rice must ensure uniform mixing by checking the cleanliness of equipment and setting accurate feed rates. Both batch and continuous blending require monitoring to maintain homogeneity. Hazards such as non-uniform mixing and metal contamination are controlled through process checks and magnetic detectors. During packaging, seal integrity must be ensured by temperature and machine cleanliness. Manual packaging must be conducted hygienically. Potential hazards like inaccurate weight or metal contamination are managed through calibrated balances and metal detectors (BIS 2021a).

18.5.7 Test Methods for Analysis of Fortificants in Fortified Rice

IS 17780 refers to AOAC 944.02 or AACC 40-70.01, which employs AAS for quantification of iron. Similarly, FSSAI prescribes the method FSSAI.FR.16.001.2022 using ICP-MS. Additionally, AOAC 984.27 is referred to by BIS, which employs ICP Emission Spectroscopy for iron estimation. For vitamin B_9 (folic acid), FSSAI recommends the test method FSSAI.FR.16.002.2022, which uses LC-MS/MS for accurate measurement. For vitamin B_{12} (cyanocobalamin), BIS refers to IS 16640 utilizing HPLC, and FSSAI prescribes FSSAI.FR.16.003.2022, which employs LC-MS/MS (FSSAI 2022).

18.6 PLANT EQUIPMENT AND MACHINERY

18.6.1 Equipment for the Manufacture of FRK

IS 17853 specifies requirements of the equipment used in FRK manufacture, covering micropulverizers, mixing unit, extruder, and dryer. The detailed specifications include material, constructional, and performance requirements for each unit (BIS 2022b). Material requirements for these components are given in Table 18.1.

Micropulverizer

The constructional requirements of the micropulverizer ensure the proper design and functionality. Components such as hopper, grinding chamber, cyclone, control panel, blower, and motor should be designed properly to facilitate efficient and safe operation (BIS 2022b). The hopper regulates feed intake and incorporates a magnet for metal impurity removal. The grinding chamber should have arrangements for the installation of wear-resistant screen and hammers with proper clearance. Dust collecting bags/filters are required to collect the fine dust released out of the cyclone (BIS 2022b).

TABLE 18.1
Material Requirements for Equipment Used in FRK Manufacturing

Micropulverizer	Mixer	Extruder	Dryer
Hopper, grinding chamber, Screen, Hammer, Cyclone, Pipes and fittings, and Supporting frame shall be made of stainless steel grade X04Cr19Ni9	Mixing chamber, cover/lid, shaft, paddle, helical ribbon, and supporting frame and structure shall be made of stainless steel grade X04Cr19Ni9 Gasket shall be made of food-grade rubber	Barrel, extruder screw, and die-head shall be made of wearing, abrasion and corrosion resistant, food-grade nitriding steel of grade EN41B or 38CrMoAl or any other similar material Die cutter, body cover, supporting frame, and structure shall be made of stainless steel grade X04Cr19Ni9	Belt of dryer shall be made of food-grade material

Mixing Unit

The constructional requirements of the mixer ensure efficient and standardized mixing of vitamins, minerals, and flour. The mixer's capacity and type should uphold consistent and high-quality mixing operations. Elements such as the mixing chamber design, water spray arrangement, and lump breaking mechanism should work continuously to prevent lumps formation and ensure uniform blending. Components like paddles and lids with food-grade gaskets maintain hygiene and prevent spillage and contamination. The water dosing unit should be provided with suitable arrangement to spray water through the water dosing pump and automatic flow control system. The integration of water dosing and automatic feeding units with a comprehensive control panel facilitates precise control over mixing parameters, enhancing operational efficiency and product quality. Compliance with material standards and motor size specifications ensures the mixer's reliability and safety, further bolstering its constructional integrity. Performance requirements ensure that the mixer can achieve specific outcomes must uniformly blend key nutrients like iron, folic acid, and vitamin B_{12}, ensuring nutritional consistency. The mixer should distribute moisture uniformly without forming lumps to maintain product texture and quality (BIS 2022b).

Extruder

The constructional requirements for the extruder play a vital role in ensuring its efficiency and precision in manufacturing of FRK. The barrel design (square or circular) with heating elements maintains the optimal temperature for extrusion. A wear-resistant screw design supporting higher pressure, with precise feeder control, ensures smooth material conveyance and uniform product output. The feeder shall be provided with suitably designed hopper to deliver wet material to the screws. Temperature and pressure sensors in the die head guarantee consistent extrusion quality. Specific die and cutter designs based on rice variety prevent kernel sticking and enable adjustable cutting speeds. A control panel with Supervisory Control and Data Acquisition (SCADA) system enables real-time monitoring and control of critical parameters like torque, die-pressure, barrel zone, die head temperature, etc. Compliance with material standards and proper motor sizing ensures reliable and safe operation, underpinning the extruder's constructional integrity (BIS 2022b).

Dryer

The constructional requirements for the dryer are pivotal for maintaining the quality of final FRK. The standard stipulates the use of a continuous electrical type dryer with a multilayer vibratory arrangement to ensure efficient drying. Temperature and humidity control with proper heating and cooling arrangement, and airflow management through installation of blower of suitable size, prevent undesirable outcomes like color alteration and fissure formation. Sensors (for temperature and RH), air flow meter, and a control panel allow precise monitoring and adjustment of crucial parameters, enhancing operational precision. Compliance with belt material requirement and appropriate motor sizing guarantees dependable performance (BIS 2022b).

18.6.2 Equipment for the Manufacture of FR

IS 17854 stipulates the requirements for both batch and continuous type blending units depending upon the capacity and requirement. Blender types include:

a. Continuous screw/ribbon blender;
b. Drum blender;
c. Continuous drum blender;
d. Paddle blender; and
e. Octagonal blender.

All the blending components such as hopper, lid screw/ribbon paddle, containers, and agitator must be made up of stainless steel of grade X04Cr19Ni9 to ensure the durability (BIS 2022a).

The agitator must provide uniform mixing without damaging either FRK or raw rice. The blender must be equipped with a magnet to trap the metal impurities. The blending unit should have a control system to feed FRK and rice at a predefined rate, with LCD/LED/HMI screen for process parameters. A SCADA system is recommended to record FRK and rice feed rate, output capacity, etc. Performance requirements specify the desired outcomes and quality standards for the blended product. The output ratio of FRK to normal rice should be maintained at either 1:100 or 2:100, as per the ratio selected (BIS 2021a).

18.7 IMPLEMENTATION OF REGULATIONS AND STANDARDS

18.7.1 Mode of Implementation

Food business operators are responsible for integrating fortification into their production lines with strict compliance with FSSAI regulations. Industries are also encouraged to comply with quality assurance and control protocols set by BIS.

Monitoring and enforcement are carried out by FSSAI to conduct inspections, test products, and perform market surveillance. Penalties are imposed on entities that do not meet the mandatory standards. To support enforcement, capacity building and training are vital. Training programs are provided to manufacturers and laboratory analyst on fortification technologies, sampling techniques, and analytical instruments. BIS conducts dissemination programs for the all five standards on VMP, FRK and FR, and its equipment and also participates in training programs conducted by FCI and professional training programs conducted by 'Innovation Hub on Rice Fortification' of IIT Kharagpur.

18.7.2 Challenges

Despite a robust regulatory framework, the implementation of rice fortification in India faces several challenges. Technical issues include variability in FRK quality and compatibility with different rice varieties, which may cause consumer rejection due to changes in texture, color, or cooking behavior. Maintaining precise blending ratio requires specialized equipment, and errors in blending can lead to nutrient overdosing or underdosing.

A major constraint is the infrastructure and equipment gap across the FR ecosystem. While BIS has certified numerous FRK manufacturing and blending units, coverage remains limited compared to national targets. Many rice mills, especially smaller units, lack standardized blending and drying equipment, essential for maintaining uniform micronutrient distribution and product stability. Although BIS offers robust certification systems, the voluntary nature of these standards limits participation. BIS Certification is mostly pursued by larger or more quality-conscious manufacturers.

Inadequate segregation infrastructure complicates operations, making it difficult to separate fortified and non-FR during storage and distribution. Cold and dry storage facilities critical for preserving the nutrient integrity are often insufficient, particularly in rural and decentralized supply chains.

Quality control and compliance are also problematic. FSSAI faces challenges in monitoring decentralized production due to limited inspection capacity. Non-compliance and unclear understanding of regulations among small manufacturers further hinder compliance. Consumer acceptance is also an issue, as taste, appearance, and perceived artificiality of FR discourage some consumers. Low awareness of the health benefits of FR and cultural resistance to altering traditional rice consumption practices further reduce acceptance.

18.8 COMPLIANCE AND ENFORCEMENT

Compliance with Indian standards and enforcement of regulations are vital for ensuring the quality, safety, and effectiveness of FR and related commodities across the supply chain. FRK manufacturers are encouraged to use VMP as per IS 17781, produce FRK using machinery complying to IS 17853, and obtain a BIS license for IS 17782. Likewise, rice mills that blend FRK with conventional rice should comply with IS 17780 and utilize BIS certified blending equipment as per IS 17854.

These certifications ensure standardization, traceability, and accountability in production and blending operations.

Despite the voluntary nature of standards, BIS has granted 76 licenses for IS 17782 and 12 licenses for IS 17781. BIS has also issued 3 licenses for FRK manufacturing equipment (IS 17853) and 50 licenses for blending equipment (IS 17854).

For certification, the BIS has established a robust and fully digitized system managed through five Regional Offices, 36 Branch Offices, and 10 state-of-the-art laboratories. Numerous third-party labs are also recognized through its Laboratory Recognition Scheme (LRMS). For efficient application processing, test data management, and license issuance, BIS has digitized its entire certification and laboratory process through the Manak Online Portal and Laboratory Information Management System (LIMS).

The FSSAI strengthens compliance through regular laboratory testing tests to verify the blending ratio, levels of mandatory micronutrients, and physical quality of VMP, FRK, and FR. FSSAI maintains and regularly publishes an approved list of food testing laboratories that are competent to analyze mandatory fortificants in VMP, FRK, and FR. FSSAI has released sampling guidelines of FR and related products. Surprise audits are conducted regularly in collaboration with the National Accreditation Board for Testing and Calibration Laboratories (NABL), and appropriate punitive actions are enforced against non-compliant laboratories. *FoRTrace* (Fortified Rice Traceability), a digital platform developed by FSSAI, enables end-to-end tracking of FR from VMP preparation to final blending.

Inspections and surveillance are conducted by FSSAI at manufacturing units, warehouses, and retail outlets. Food Safety Officers (FSO) conduct surveillance audit of Public Distribution System (PDS) operations. Violations under the Food Safety and Standards Act, 2006, penalties are imposed based on the severity of the offense. Misbranding or nutritional deficiencies in FR may cause penalties from ₹1 to 3 lakhs, while severe offenses such as adulteration or fraud can lead to fines up to ₹10 lakh and even imprisonment. Non-compliant batches may be seized, and suppliers that repeatedly violate fortification norms may be blacklisted. FSSAI's Fortification Monitoring System (FMS) is a real-time digital platform that collects compliance data from state nodal agencies and suppliers.

In spite of robust frameworks, several enforcement challenges exist such as disintegrated supply chains across states, a shortage of trained inspectors, struggle of maintaining nutrient levels during supply chain, and a general lack of awareness among millers and distributors.

18.9 BIS CERTIFICATION

Under the BIS Act, 2016 and the Rules and Regulations framed there under, BIS has the authority to undertake Conformity Assessment of products, services, systems, processes, and articles under the relevant schemes specified in the BIS (Conformity Assessment) Regulations, 2018. Under Scheme I, manufacturers can apply for a license to use the BIS Standard Mark, i.e., ISI mark (BIS 2018a). Eligibility is assessed based on manufacturers' ability to have the necessary production infrastructure, effective process controls, and adequate quality control and testing facilities in line with relevant Indian Standard. Further, the product must comply with the requirements laid down in Indian Standard, and this is verified by BIS before grant of license. Compliance with the relevant Indian standard is demonstrated through testing, either in third-party laboratories, at the manufacturing facility, or through a combination of both. After successful verification, BIS grants the license to the manufacturer.

The first step in BIS certification process is the identification of relevant BIS standard applicable to their product category. Then, the manufacturer will submit an application to the BIS including plant layout, manufacturing equipment, process flowchart, and production capacity details of product specifications. Following the application, BIS official undertakes a thorough inspection of the manufacturing premises focusing on the infrastructure, quality control systems, and overall compliance with the relevant Indian Standards. Additionally, product samples are collected and tested in BIS/BIS recognized laboratories to evaluate whether the product meets the parameters laid down in the Indian Standard.

Upon satisfactory preliminary inspection and test results, the BIS grants the certification license, authorizing the use of the BIS Standard (ISI) Mark. After certification, the BIS continues to monitor the manufacturing process through regular market and factory surveillance to ensure that quality standards are consistently maintained. Samples of ISI-marked products are drawn from both the market and licensed manufacturing units and tested in BIS/BIS recognized laboratories. Furthermore, BIS officials carry out factory inspections to ensure ongoing compliance with the terms of the license and the approved Scheme of Inspection and Testing.

Corrective actions are required when any non-conformity is identified or when the licensee's performance is not satisfactory. Failure to take appropriate corrective measures, repeated instances of non-compliance, or continued substandard performance may lead to suspension of the lice by BIS.

18.10 GLOBAL STANDARDS FOR FORTIFICATION

Global Standards for FR and related products have been defined to ensure that FR provides essential micronutrients in an efficient and safe manner, especially in rice dependent and micronutrient deficient countries.

18.10.1 Codex Alimentarius (FAO/WHO)

The Codex Alimentarius, jointly established by the FAO and WHO, is the principle reference for food standards. At present, there is no specific Codex standard for VMP, FRK, or FR. However, the Codex Guidelines on Food Fortification (CAC/GL 9-1987) provide general principles for fortification of staple food items. These standards permit rice fortification for national programs, provided the fortified product meets safety and efficacy criteria.

The WHO recommends fortification of rice with iron and possibly with other vital micronutrients, such as folic acid, zinc, vitamin A, and B complex vitamins. These recommendations are especially targeted at rice dependent and micronutrient deficient populations. In the 2018 guidelines for fortification of rice with iron, WHO recommended micronized ferric pyrophosphate as the best form of iron to use in rice fortification because of its stability and sensory characteristics. As per this guideline, the proposed iron level of FR is 28–42.5 mg/kg, based on regional dietary intakes (WHO 2018).

The World Food Program (WFP) and UNICEF also have an influential role on the implementation of rice fortification standards, especially for school feeding programs and emergency food assistance. Moreover, UNICEF in partnership with the Global Alliance for Improved Nutrition (GAIN) offers technical support to countries for the implementation of rice fortification programs. They also recommend the harmonization of standards, enhanced quality assurance systems, and stronger public–private partnerships to promote the scalability and sustainability of FR programs globally.

18.11 CONCLUSIONS

Bringing out comprehensive Indian Standards on VMP, FRK, FR, and their machinery and establishing FSSAI regulations on VMP, FRK, and FR is a vital step toward ensuring the quality, safety, and effectiveness of rice fortification programs. These standards serve as a foundational framework and support manufacturers for ensuring consistency in formulation, manufacturing practices, blending processes, and final product quality across the FRK value chain.

Specifications for the VMP specify the types, forms, and concentrations of nutrients that must be included to address micronutritional needs. The standards for FRK emphasize nutrient stability, uniformity, and compatibility with local rice varieties and cooking practices. Requirements for FR ensure that the final blended product delivers the intended nutritional benefits without compromising taste, appearance, and safety. Indian Standards for FRK manufacturing machinery and blending equipment guide the extrusion of FRK, blending ratios, and packaging, thus rendering them equally significant. These standards ensure the reliability, scalability, and hygiene of the entire production process. Together, these standards facilitate the effective implementation of mandatory and voluntary rice fortification programs, enhance consumer trust, and contribute to national nutrition goals.

Thus, national standards provide a robust regulatory and operational framework, which supports the scalability of rice fortification as a proven public health intervention. Their implementation, together with monitoring and stakeholder coordination, is essential for achieving long-term goal, combating hidden hunger and improving public health.

REFERENCES

BIS. 2018b. *Bureau of Indian Standards (Conformity Assessment) Regulations-2018*. New Delhi: Bureau of Indian Standards.

BIS. 2018a. *Bureau of Indian Standards Rules-2018*. New Delhi: Bureau of Indian Standards.

BIS. 2021a. *Fortified Rice — Specification (IS 17780 :2021)*. New Delhi: Bureau of Indian Standards, Government of India. https://standardsbis.bsbedge.com/BIS_SearchStandard.aspx?Standard_Number=17780&id=0

BIS. 2021b. *Fortified Rice Kernels — Specification (IS 17782:2021)*. New Delhi: Bureau of Indian Standards, Government of India. https://standardsbis.bsbedge.com/BIS_searchstandard.aspx?Standard_Number=IS+17782&id=34819

BIS. 2021c. *Vitamin and Mineral Premix for Manufacturing Fortified Rice Kernels — Specification (IS 17781:2021)*. New Delhi: Bureau of Indian Standards, Government of India. https://standardsbis.bsbedge.com/BIS_searchstandard.aspx?Standard_Number=IS+17781&id=34818

BIS. 2022a. *Equipment for Manufacture of Fortified Rice — Specification (IS 17854:2022)*, New Delhi: Bureau of Indian Standards, Government of India. https://standardsbis.bsbedge.com/BIS_searchstandard.aspx?Standard_Number=IS+17854&id=35288

BIS. 2022b. *Equipment for Manufacture of Fortified Rice Kernel — Specification (IS 17853:2022)*. New Delhi: Bureau of Indian Standards, Government of India. https://standardsbis.bsbedge.com/BIS_searchstandard.aspx?Standard_Number=IS+17853&id=35289

BIS. 2025a. *Determination of Vitamin B9 (Folates) in Foodstuffs by Ultra-High Performance Liquid Chromatography and Tandem Mass Spectrometry–Method of Test (IS 7234:2025)*. New Delhi: Bureau of Indian Standards, Government of India. https://standardsbis.bsbedge.com/BIS_searchstandard.aspx?Standard_Number=IS+7234&id=53327

BIS. 2025b. *Determination of Vitamin B12 in Foodstuffs by HPLC- UV – Method of Test (IS 7529:2025)*. New Delhi: Bureau of Indian Standards, Government of India. https://standardsbis.bsbedge.com/BIS_searchstandard.aspx?Standard_Number=IS+7529&id=53444

Food Safety and Standards Act. 2006. Ministry of Law and Justice (Legislative Department). New Delhi: Government of India.

FSSAI. 2011a. *Food Safety and Standard (Contaminants, Toxins and Residues) Regulation*. Vol. 0. New Delhi: Food Safety and Standards Authority of India.

FSSAI. 2011b. *Food Safety and Standards (Food Products Standards and Food Additives) Regulations, 2011*. New Delhi: Food Safety and Standards Authority of India.

FSSAI. 2018. *Food Safety and Standards (Fortification of Food) Regulation (2018)*. New Delhi: Food Safety and Standards Authority of India.

FSSAI. 2022. *Methods for Testing of Fortificants (Iron, Folic Acid and Vitamin B12) in Fortified Rice*. New Delhi: Food Safety and Standards Authority of India.

FSSAI. 2023a. *Methods for Testing of Fortificants (Iron, Folic Acid and Vitamin B12) in Fortified Rice Kernel (FRK)*. New Delhi: Food Safety and Standards Authority of India.

FSSAI. 2023b. *Methods for Testing of Fortificants (Iron, Folic Acid and Vitamin B12) in Vitamin Mineral Premix for Fortified Rice Kernel (FRK) and Iron in FRK by AAS*. New Delhi: Food Safety and Standards Authority of India.

The Bureau of Indian Standards Act. 2016. Ministry of Law and Justice (Legislative Department). New Delhi: Government of India.

WHO. 2018. *Guideline: Fortification of Rice with Vitamins and Minerals as a Public Health Strategy*. Geneva: World Health Organization. https://apps.who.int/iris.

Index

G

H

I

J

K

L

M